MULTIPLE EMULSIONS

THE WILEY BICENTENNIAL–KNOWLEDGE FOR GENERATIONS

Each generation has its unique needs and aspirations. When Charles Wiley first opened his small printing shop in lower Manhattan in 1807, it was a generation of boundless potential searching for an identity. And we were there, helping to define a new American literary tradition. Over half a century later, in the midst of the Second Industrial Revolution, it was a generation focused on building the future. Once again, we were there, supplying the critical scientific, technical, and engineering knowledge that helped frame the world. Throughout the 20th Century, and into the new millennium, nations began to reach out beyond their own borders and a new international community was born. Wiley was there, expanding its operations around the world to enable a global exchange of ideas, opinions, and know-how.

For 200 years, Wiley has been an integral part of each generation's journey, enabling the flow of information and understanding necessary to meet their needs and fulfill their aspirations. Today, bold new technologies are changing the way we live and learn. Wiley will be there, providing you the must-have knowledge you need to imagine new worlds, new possibilities, and new opportunities.

Generations come and go, but you can always count on Wiley to provide you the knowledge you need, when and where you need it!

WILLIAM J. PESCE
PRESIDENT AND CHIEF EXECUTIVE OFFICER

PETER BOOTH WILEY
CHAIRMAN OF THE BOARD

MULTIPLE EMULSIONS

TECHNOLOGY AND APPLICATIONS

Edited by

Abraham Aserin

BICENTENNIAL 1807 WILEY 2007 BICENTENNIAL

WILEY-INTERSCIENCE

A John Wiley & Sons, Inc., Publication

Published by John Wiley & Sons, Inc., Hoboken, New Jersey
Published simultaneously in Canada

For general information on our other products and services or for technical support, please contact our Customer Care Department within the United States at (800) 762-2974, outside the United States at (317) 572-3993 or fax (317) 572-4002.

Wiley also publishes its books in a variety of electronic formats. Some content that appears in print may not be available in electronic formats. For more information about Wiley products, visit our web site at www.wiley.com.

Wiley Bicentennial Logo: Richard J. Pacifico

Library of Congress Cataloging-in-Publication Data:

Multiple emulsions : technology and applications / edited by Abraham Aserin.
p. cm.
Includes index.
ISBN 978-0-470-17093-9 (cloth)
1. Emulsions. 2. Emulsions (Pharmacy) I. Aserin, Abraham.
TP156.E6.M84 2008
660′. 294514—dc22

2007019893

Printed in the United States of America

10 9 8 7 6 5 4 3 2 1

CONTENTS

PREFACE

In Honor of Prof. Nissim Garti's 60th Birthday

In 2006, Prof. Nissim Garti celebrated his sixtieth birthday. By that time he had had a brilliant scientific career spanning more than thirty-five years. No happier way can be found of celebrating a scientist's birthday than by issuing a *Festschrift* in his honor. Such a *Festschrift*—be it a special issue of a journal or a book—generally contains studies that intend to reflect various topics relating to the putative extensive work and all-round interests of the honoree. Usually only one *Festschrift* is issued on a particular anniversary. Nissim has justly been awarded two *Festschrifts*!

The first was a special issue of a scientific journal.[1] The intervening year before the publication of the present book has served only to enhance the renown of this outstanding scientist and to increase the fervor and esteem with which this tribute is offered.

Between the covers of this book are published articles written by colleagues, alumni, and friends in honor of Prof. Garti. They are all one in their affection and esteem for his personality and achievements. Many more would have contributed but for space and schedules. Yet all join us in congratulating Prof. Garti on passing his sixtieth birthday and on amassing a lifetime of work and ground-breaking benchmark accomplishments as one of the most original and innovative scientists in the field of surfactant chemistry.

Whereas the literature cited at the end of this Preface demonstrates ramified and diverse research areas—almost every one of which has long interested Prof. Garti—the present book is focused on one subject: multiple emulsions. This topical choice for the *Festschrift* reflects a challenging area to which Nissim has devoted his talent and efforts over the past two decades, an area in which he pioneered creative ideas and spearheaded and opened up new far-reaching vistas. It is, I trust I may say, the hope of all contributors to this book, that besides being our humble tribute to Prof. Garti, it may stimulate a few readers to pursue the investigation of multiple emulsions.

Before delving into some of the most prominent works of Prof. Garti, it is apposite to say something here about Nissim—the man and scientist. In describing the biography of a distinguished scientist, there is always the challenge of balancing the personal aspects with the professional accomplishments. The reader is referred to the Preface of the first *Festschrift* for a biographical sketch of Prof. Garti, which offers a testimony to his eminence in various facets of surfactant chemistry.

It would not be in accord with Jewish ethics to tell all of a man's praise in his presence—it can be done only in part. This brief Preface is therefore not the place to expand on the personal characteristics of Nissim and the more so as some of them were detailed at the end of this Preface.

However, it is still worth reiterating Professor Garti's deep love for teaching. We know that Nissim, who is not capable of patting himself on the back, will do anything he can to evade premeditated questions intended to highlight his remarkable scientific career. Yet, were we to ask Prof. Garti what would best characterize his life's work, then, modest as he is about his multifarious achievements, he would undoubtedly prefer to emphasize that at heart he is, first and foremost, a teacher.

Presumably two reasons underlie Prof. Garti's devotion to the arduous task of education. First, learning was prized in Nissim's household, his parents striving to imbue their children with a yearning for enlightenment. Nissim still remembers that no sacrifice—including the carefree joys of childhood—was too great in the pursuit of learning. Second, Nissim has a deep and abiding respect for the teachers who directed him toward the experimental sciences. Perhaps these two reasons led Nissim to develop a commitment to the teaching profession.

In the Preface of the first *Festschrift*, we have elaborated on the inferior status of the teacher, on the way experimental work has to be done, and on the assistance tendered by Nissim to colleagues and young scientists. In the present Preface we would like to deal with two other points. First, we discuss the seemingly contradictory method of Nissim's teaching: Prof. Garti is considered a preeminent teacher, beloved by all his students without exception. It seems prima facie rather paradoxical because, after all, Prof. Garti is a very demanding and tough teacher, somewhat reluctant to give compliments or high grades. So what is the magic of Nissim?

It seems that Prof. Garti is the consummate pedagogue who teaches by personal example. Few teachers have given more of themselves to their students than Nissim. He is never too busy to help young scientists, to give them the gift of knowledge, how to proceed on the paths they had embarked upon, always with kindness and a sincere interest in them and their work. Nissim tries to help even low-performing students to cope with the myriad problems they face in their scientific work on an almost daily basis. He provides his assistance, sparing no effort and expecting no recompense.

He will juggle his hectic schedule in order to respond to others in need of advice. He is always ready to put his personal library at his students' disposal and they say that his books are frequently of more value than those of the university library. Any attempt to thank Nissim is usually a source of embarrassment to him. A teacher devoted to his students, Prof. Garti will not cancel a single lesson even if it involves great inconvenience for him. Nissim's students are impressed with his fairness and integrity, and with his scientific rigor and intellectual grasp. Those who have passed through Prof. Garti's hands can appreciate his refusal to yield to perfunctory or superficial performance as part of his instruction and research work. Above all, he invigorates his students to launch and conduct their research projects, to raise the proper questions, and to try to gain otherwise unavailable insights by employing innovative measuring techniques. Prof. Garti teaches them repeatedly that investigation yoked too tightly to a specific scientific model is almost always liable to fail. They should therefore seek less trite approaches for their apparent insoluble research problems. In a word, he encourages them to aspire to the highest standards of scientific excellence.

The legacy of a scientist is usually measured in the number of his publications. Prof. Garti thinks that the contribution of a true scientist should be valued via the students who came into his circle of influence, who learned from him, and who would be ready to emulate his way even if it is difficult to follow because of its high standards and exacting nature. We know, however, that Nissim, even after meticulously checking and testing his experimental data and their analysis again and again, always thinks that his work still lacks that perfection that would have been possible given more ample leisure. Indeed many a potential scientific paper that could readily be published was postponed or even abandoned by Nissim, since in his opinion just a certain minor point was not sufficiently well-founded. This is the reason why Prof. Garti disdains what might be called tawdry scientific methodology—hastily publishing premature articles that rely on a maze of rather implausible hypotheses with only scant experimental data behind them. Nissim asserts that it would have been salutary if the authors of such papers had noticed that in this way science does not advance in the least.

Most educators have failed to study diligently how students use their knowledge and training in out-of-classroom (or lab) settings when these may be the uses that matter most. Nissim, however, initiates his students not only into careful research but also into the practical life of the professional. For example, he instructs them how to acquire the competence and confidence to prepare and submit articles for publication in scientific journals and how to see their works properly through the publishing process, a task that without assistance can be frustrating and discouraging even to a highly motivated student. Even when Nissim himself writes the manuscript of a scientific paper, he feels that every student who has participated in the research work is just as involved and helpful in its completion and getting it published. Moreover Prof. Garti would generously help students and colleagues with their own writing in order to achieve clarity of exposition.

The second point relates to science education. The most widely claimed goals of science education are: to produce citizens who are scientifically literate and to support citizens in understanding reports and discussions of science (especially about issues that may be controversial) that appear in the popular media. At the core of science education activity we find also two other tasks:

1. To attack pseudoscience beliefs and ideas that have been gaining ground over the last two decades, mainly because of the entertainment industry.
2. To improve the ability of the general public to distinguish between fact and fiction, since the visual media have blurred the distinction between these two concepts.

Nissim's sense of concern and commitment to the furtherance of science has always extended beyond his own students to include the "educated layman." So he does his best to achieve these goals.

As a part of his endeavor to bring academic interests and scientific achievements to a large audience and to widen the circle of informed readers and listeners, Prof. Garti is always ready to find time, whether to lecture on science to interested groups or to write scientific articles in popular journals. Nissim is endowed with an extraordinary blend of oratorical power and intellectual acuity that, together with his infectious enthusiasm and eloquent writing, almost inevitably enables him to imbue his listeners and readers with a keen interest in science. Moreover Nissim also tries to direct their attention to research findings that are rarely communicated to the general public directly by scientists.

Such lectures or papers have to be prepared carefully and thoroughly. Nissim knows how to elucidate points of cardinal significance and how to translate from scientific language into something palpable and meaningful to the "common person." Indeed Nissim's limpidity and economy of expression, combined with the restriction of technical material to the minimum, enable him to simplify intricate scientific topics.

Prof. Garti also knows how to whet the appetite of his audience for more information. In lectures he succeeds in igniting the interest of his listeners by utilizing highly entertaining and informative presentations communicated in a voluble, exuberant style. Frequently he tells vivid stories about how science works and about the strengths and frailties of prominent scientists. Far from least, these stories reveal the significance of scientific achievements more than the usual dull reports. It should, however, be noted that Prof. Garti gets far and away the best response from his listeners and readers when he teaches them to be skeptical and to accept as truth only statements that are irrefutable through scientific scrutiny.

Nissim's solution to the prevailing problem of disgruntled listeners who pester the lecturer with questions is very typical and instructive. Most lecturers regard them as too clever by half and needing to be taken down a peg or two. Prof. Garti always tries the opposite attitude toward these smart alecs. Generally, he follows a two-pronged strategy. First, he would treat them with considerable respect and tolerance, never being invidious or talking down to them. Second, he would use his dry and witty sense of humor to relieve tension during his lectures.

We now turn to the seminal contributions of Prof. Garti to the study of multiple emulsions. Nissim's research interests could never be limited to any single scientific problem or theme, and this only becomes fully apparent upon examining the scope of his over 300 refereed publications, invited symposia, book chapters, and reviews. However, for Nissim, multiple emulsions have been a favorite object of investigation ever since the early 1980s, and he has become one of the most active scientists in this burgeoning area.

Moreover there is an interesting relation between Prof. Garti's skills as a scientist and teacher regarding multiple emulsions. It is known that persuading students to be involved in multiple emulsions research is a formidable educational challenge. Far be it for me to argue that such an investigation is too difficult for a graduate student, since the underlying concepts in multiple emulsions science are quite easy to grasp and yet experimental work on these emulsions is undoubtedly one of the most demanding and enervating research tasks students may encounter within surfactant chemistry. This is simply because the continuous, long-term (in the order of weeks and even months!) monitoring need for following up any changes in the multiple emulsion stability, and furthermore in the rate of addendum release, virtually confines the student to the lab as long as such an experiment is going on. Nevertheless, some students have been lured by Nissim's enthralling descriptions of multiple emulsions, and the main results of their M.Sc. and Ph.D. works are detailed in the following.

These multi-compartment liquid dispersions are considered adequate vehicles for controlled and sustained delivery of entrapped addenda, such as drugs. Thus a drug that is dissolved or suspended in the internal aqueous phase of a W/O/W emulsion is forced to diffuse across the oil phase prior to being released into the body fluids.

Multiple emulsions have been the subject of numerous investigations due to their importance as an emerging and promising technology for slow and controlled release of active ingredients and as a major scientific challenge regarding the preparation and improvement of the kinetic stability of these inherently thermodynamically unstable entities. As a testimony to the evolution of the field of multiple emulsions, one may merely peruse the subject matter of the articles in this book.

Prof. Garti maintains a preeminent position in the research of multiple emulsions as is manifested by his nearly 40 articles concerning them. These scientific papers exhibit the gripping combination—typical of Nissim's research work—of brilliant and ingenious insights together with rigorous treatment of even the minutest details, which are acquired by hard and systematic experimental work.

Several highlights achieved by Prof. Garti are herewith outlined more or less in chronological order.

In 1983 Nissim had already tackled the problem of how to overcome the unpleasant taste of drugs such as chlorpromazine-HCl. Oral administration of a bitter-tasting medication to a child is obviously a cumbersome task. The solution suggested by Prof. Garti was to dissolve the drug in the inner phase and to release it throughout the oil phase in the presence of synthetic gastric juice. A child will swallow medication willfully when the outer water phase contains synthetic flavors that simulate attractive tastes such as that of strawberry or raspberry. It should be noted that the drug release in this case is neither controlled nor targeted.[2]

The release of electrolytes and drugs from multiple emulsions can, in principle, proceed via two possible mechanisms:

1. The oil layer separating the inner and outer aqueous phases behaves as a semipermeable membrane: The release of solutes from a W/O/W multiple emulsion, for instance, only occurs through breakdown of the multiple droplets as a result of osmotic flow of water to the inner phase and consequent coalescence of the droplets.
2. The oil layer behaves as a permeable membrane and the solute migrates by diffusion from the inner emulsion.

Prof. Garti has shown[3] that the diffusion mechanism is a predominant factor in the migration of electrolytes from the inner to the outer phase in multiple emulsions. The release of the electrolyte is affected by its hydrophobicity and concentration but not by the viscosity of the internal phase.

In another paper, written in collaboration with D. Whitehill,[4] it was demonstrated that the addition of NaCl to multiple emulsions causes droplet shrinkage due to loss of internal water. Micelle transport seemed to be the primary mechanism.

A kinetic model, adapted from that of Higuchi for release of dispersed drugs from polymeric matrices, was found to be suitable for the release of electrolytes from multiple emulsions. The existence of a diffusion-controlled mechanism was experimentally confirmed. This mechanism is facilitated as the concentration of reverse micelles formed in the oil phase increases.[5]

In another investigation it was shown that whereas the presence of electrolytes in the outer aqueous phase has no bearing on the control of drug leaching from the inner aqueous phase even when there is no osmotic pressure gradient between the two aqueous phases—implying that the system is not controlled solely by this factor—the presence of electrolytes in the internal water phase can retard the drug migration. Electrolytes that can cause salting-in of the emulsifier to the oil phase strengthen the oil–water interface and thereby decrease drug transport.[6]

Additional milestones in Prof. Garti's work regarding the stabilization of multiple emulsions should be mentioned:

Solid oil (paraffin wax) is superior to liquid oil in formulations of multiple emulsions as a stable rigid oil membrane is formed that hinders the migration of additives from the inner water phase.[7]

Prof. Garti was among the first scientists to understand the significance of the replacement of monomeric surfactants (usually blends of hydrophobic and hydrophilic amphiphiles) by polymeric emulsifiers. Such multi-anchoring macromolecules provide strong steric stabilization capabilities via the formation of thick and flexible interfacial films. An early highlight of Nissim's work in this area comprises the utilization of polysiloxane-graft-poly(oxyethylene) to stabilize W/O/W multiple emulsions. Stable, small droplet size W/O emulsions were formed with hydrophobic comb-grafted copolymers adsorbed at the inner interface. The outer interface of the W/O/W multiple emulsions was stabilized by hydrophilic comb-grafted copolymers with similar structures but with high-density grafting and long poly(oxyethylene) chains. The release rates of additives from such multiple emulsions were very slow.[8] Silicone-based surfactants impart unusual mechanical stability to W/O/W multiple emulsions that makes them possible candidates for slow release systems for agricultural applications.[9]

In addition to synthetically tailor-made polymeric amphiphiles, naturally occurring biopolymers are considered for stabilization of multiple emulsions, especially where these emulsions are needed for food and cosmetic applications. Thus the protein bovine serum albumin (BSA) significantly improves the mechanical and steric stability of W/O/W emulsions when used in a blend with a nonionic surfactant, such as Span 80. It is assumed that the two amphiphiles act synergistically by forming an interfacial complex—presumably a thick, strong gelled film that confers resistance to rupture and elasticity on the inner droplets. Based on Garti's modification of Higuchi's model, it was concluded that BSA has a double role:[10]

1. At the inner phase, BSA provides a mechanical barrier to the release of small molecules from the internal interface. The release proceeds mainly via reverse micellar transport. The presence of BSA reduces the chance of reverse micelle formation and thus decreases the release rate of entrapped addenda within the emulsion droplets.
2. At the outer phase, BSA impedes coalescence via steric stabilization.

Stable water in vegetable oil emulsions can serve as basic preparations for food-grade W/O/W multiple emulsions. A novel way of stabilization of such systems is to use submicronal α-form, crystallized hydrogenated fat (tristearin) homogeneously dispersed in the oil phase.[11] The crystals should be submicronal in size to effectively adsorb and accommodate at the interface. Large crystals flocculate in the continuous oil phase.[12] Yet these solid fat particles cannot sufficiently stabilize the W/O emulsion, and they have to be blended with a lipophilic surfactant (polyglycerol polyricinoleate, PGPR). The combination of fat microcrystals and PGPR has the following advantages:

1. Aggregation and flocculation processes are inhibited.
2. PGPR serves as a cross-linker or a bridge between the fat particles and water.
3. PGPR facilitates the anchoring of the fat particles in the oil phase while dangling itself in the water phase.
4. PGPR functions as an α-tending crystal structure modifier.

Improved stability can be obtained by utilizing the emulsifier PGPR in the internal water phase and a protein-polysaccharide hybrid (as a substitute for the common nonionic hydrophilic monomeric emulsifiers) in the external interface. For instance, whey protein isolate (WPI) forms soluble complexes with hydrocolloids such as xanthan gums or galactomannans at selected pH values and weight ratios. These hybrids with specific interface recognition capabilities stabilize multiple emulsions mainly via steric interactions at the oil-water external interface. However, at high gum levels, the emulsions become more elasticized, uncomplexed gum migrates to the bulk, and the depletion mechanism will dominate. The stabilizing effect of such hybrids is synergistic. Thus droplets of WPI/xanthan-based multiple emulsions are, respectively, one-fourth or one-eighth smaller than those of multiple emulsions based solely on either WPI or xanthan.[13]

Excellent stability to coalescence was obtained via interaction of WPI with modified pectin.[14]

The effects of WPI/polysaccharide conjugates on the stabilization of multiple emulsions are the subject of recent studies. It was observed, for example, that the release of vitamin B_1, entrapped in the core of W/O/W multiple globules, is hampered due to the biopolymer adducts formed at the interface.[15] Similar adducts stabilize O/W/O multiple emulsions and serve as efficient barriers against release of addenda contained in the inner phase.[16]

In Prof. Garti's opinion, employing multiple emulsions should not be restricted to pharmaceuticals, nutraceuticals, and cosmetics. He hopes that other promising and exciting applications will be available in near future. The microencapsulation of fine boron particles in W/O/W multiple emulsions may serve as such a feasible example.[17]

Last, the results of a very recent investigation concerning a novel type of multiple emulsion, dubbed "emulsified microemulsion" (EME), are herewith shown. Since the kinetic stability of the internal phase is inversely proportional to its droplet size, it is conceivable that nanosized droplets (microemulsions) would improve the system stability. One potential application of such emulsified microemulsions is the formation of submicronal injectable liquid W/O/W preparations for controlled drug release.

This concept has barely been implemented before now due to the tedious experimental work necessary to ascertain that the nano-droplets remain intact after the secondary emulsification stage as well as after prolonged storage. Moreover the very fast exchange of the monomeric surfactants between the internal and external interfaces induces concurrent emptying of the core phase and leaching of the inner water phase.

This second problem has been solved skillfully in Prof. Garti's lab by using glycerol monooleate as surfactant, R(+)-limonene/ethanol 1:1 (by weight) as the oil phase, and water/glycerol (9.15:0.85) (wt/wt) as the water phase. Samples of this microemulsion were further emulsified, utilizing the hydrophobic graft copolymer Pluronic F127 (PEO_{99}-PPO_{67}-PEO_{99}, where PEO stands for polyethylene oxide and PPO for polypropylene oxide). Pulsed-gradient-spin-echo NMR has shown that glycerol monooleate and R(+)-limonene are present in the inner phase of the emulsified microemulsion even after the second emulsification process. Electrical conductivity measurements have demonstrated that at least 60 wt% of the internal water phase remained confined after the second emulsification stage. SAXS (small angle X-ray scattering) measurements have shown that the inner microemulsion in the emulsified microemulsion is more ordered than the original L_2-phase. This surprising finding may be attributed to the effect of the polymeric surfactant. Cryo-TEM (transmission electron microscopy) images of the emulsified microemulsion provide direct evidence for the existence of spherical globules in the inner phase, having a mean diameter of 250 ± 50 nm in rather good agreement with corresponding results by DLS (dynamic light scattering): approximately 200 ± 50 nm, which is about 10 to 20 times smaller than the size of conventional multiple emulsion globules. The investigated system has remained stable during up to 12 months of storage at ambient temperature. This promising system is now being studied to explore its ability to retain bioactive addenda and release them in a controlled manner.[18]

It seems appropriate to conclude with a personal note. Prof. Garti is to me a teacher and a mentor. The scope and magnitude of his influence on me are without equal. As a senior member of Nissim's research team, I have a daily contact with him, and I would like to share some of my impressions with the readers of this Preface. Prof. Garti is now at the pinnacle of his highly variegated scientific achievements. As Nissim enters his seventh decade of life, I can assure the readers that he has not yielded to the demands of age. On the contrary, every year Nissim looks younger. None of his stamina has been lost. His energy and willpower may be likened to an ever-flowing fountain and, more wonderful still, they show no signs of diminishing. Prof. Garti's enthusiasm for work, his alert response to new ideas, and his original insights have remained unblemished and keen. He is always the first person to enter the lab in the morning and the last to leave at night, filling his long day with incessant activity.

Nissim never forgets to acknowledge that all his accomplishments and success are due to his wonderful family and, most of all, to his wife, Ricki. She has been standing by Nissim in joy and sorrow, always giving of herself without stint to him and their family.

This *Festschrift* is herewith presented to you, Nissim, as a token of our heartfelt appreciation and admiration. The book is accompanied by our ardent prayer that like our ancient leader, Moses, may "your eyes remain undimmed and your vigor unabated." It is the fervent hope of your friends, colleagues, and students that you will enjoy good health and continuous creative labor for many years to come.

Abraham Aserin, Ph.D.

LITERATURE CITED

1. Aserin A. 2006. Honoring Professor Nissim Garti. *Adv Colloid Interface Sci* (1–3): 128–130.
2. Garti N, Frenkel M, Schwartz R. 1983. Multiple emulsions. Part II: Proposed technique to overcome unpleasant taste of drugs. *J Dispers Sci Technol* 4(3): 237–252.
3. Magdassi S, Garti N. 1984. Release of electrolytes in multiple emulsions: Coalescence and breakdown or diffusion through oil phase? *Colloids Surf* 12: 367–373.
4. Garti N, Magdassi S, Whitehill D. 1985. Transfer phenomena across the oil phase in a water-oil-water multiple emulsion evaluated by Coulter counter. 1. Effect of primary emulsifier on permeability. *J Colloid Interface Sci* 104(2): 587–591.
5. Magdassi S, Garti N. 1986. A kinetic model for release of electrolytes from W/O/W multiple emulsions. *J Controlled Release* 3(4): 273–277.
6. Garti N, Romano-Pariente A, Aserin A. 1987. The effect of additives on release from W/O/W emulsions. *Colloids Surf* 24: 83–94.
7. Magdassi S, Garti N. 1986. Formation of water/oil/water multiple emulsions with solid oil phase. *J Colloid Interface Sci* 120(3): 573–579.
8. Sela Y, Magdassi S, Garti N. 1994. Polymeric surfactants based on polysiloxanes—graft-poly(oxyethylene) for stabilization of multiple emulsions. *Colloids Surf A* 83(2): 143–150.
9. Sela Y, Magdassi S, Garti N. 1995. Release of markers from the inner water phase of W/O/W emulsions stabilized by silicone-based polymeric surfactants. *J Controlled Release* 33(1): 1–12.
10. Garti N, Aserin A, Cohen Y. 1994. Mechanistic considerations on the release of electrolytes from multiple emulsions stabilized by BSA and nonionic surfactants. *J Controlled Release* 29(1–2): 41–51.
11. Garti N, Binyamin H, Aserin A. 1998. Stabilization of water-in-oil emulsion by submicrocrystalline α-form fat particles. *J Am Oil Chem Soc* 75(12): 1825–1831.

12. Garti N, Aserin A, Tiunova I, Binyamin H. 1999. Double emulsions of water/oil/water stabilized by α-form fat microcrystalline particles. *J Am Oil Chem Soc* 76(3): 383–389.
13. Benichou A, Aserin A, Garti N. 2002. Double emulsions stabilized by new molecular recognition hybrids of natural polymers. *Polymers Adv Technol* 13(10–12): 1019–1031.
14. Garti N, Wicker L. 2005. Pectin methylesterase modified pectin interaction with whey protein isolate and stability of double emulsions. Abstracts of Papers, 229th ACS National Meeting, San Diego, CA, March 13–17, 2005, CELL-119.
15. Benichou A, Aserin A, Garti N. 2007. W/O/W double emulsions stabilized with WPI-polysaccharide complexes. *Colloids Surf A* 294(1–3): 20–32.
16. Benichou A, Aserin A, Garti N. 2007. O/W/O double emulsions stabilized with WPI-polysaccharide conjugates. *Colloids Surf A* 297(1–3): 211–220.
17. Berkovich Y, Aserin A, Garti N. 2004. Practical and mechanistic considerations in the use of W/O/W double emulsions for microencapsulation of fine boron particles. *J Dispersion Sci Technol* 25(1): 89–99.
18. Lutz R, Aserin A, Wachtel EJ, Ben Shoshan E, Danino D, Garti N. 2007. A study of emulsified microemulsion by SAXS, cryo-TEM, SD-NMR and electrical conductivity. *J Dispersion Sci Technol* 28(8): forthcoming.

INTRODUCTION TO MULTIPLE EMULSIONS TECHNOLOGY AND APPLICATIONS: AN UPDATE

When it was decided to compile a book in honor of Prof. Nissim Garti on the occasion of his sixtieth birthday, choosing an appropriate topic was rather difficult. The diversity of topics investigated by Prof. Garti and his research group in the surface chemistry field is simply amazing. A most conspicuous theme among these topics is the area of multiple emulsions in which Nissim has become a foremost authority. Moreover this attractive subject provides a good prospect for much demand even if the technology is still waiting for a real breakthrough.

Since William Seifriz described for the first time in 1925 these intricate liquid systems having ternary, quaternary, or more complex structures that he named multiple emulsions, the literature has been "flooded" every year with tens of new examples demonstrating release patterns and control of active ingredients using these systems. Multiple emulsions, at least in theory, have significant potential in many applications because the internal droplets can serve as an entrapping reservoir for active addenda that can be released by a controlled transport mechanism. Many of the potential applications would be realized in the fields of agriculture, pharmaceuticals, cosmetics, and food.

In practice, double emulsions consist of large and polydispersed droplets that are thermodynamically unstable, with a strong tendency for coalescence, flocculation, and creaming.

Efforts have been made to improve emulsion stability and to control the release of active matter. Almost any possible blend of low-molecular weight emulsifiers, oils, cosolvents, and coemulsifiers have been tested. The nonviscous fluid multiple emulsions were always unstable. Only semisolid multiple emulsions, gelled or thickened systems, have long shelf-life and prolonged stability. Biopolymers, synthetic graft and comb copolymers, and polymerizable emulsifiers impart steric or mechanical stabilization to the multiple emulsions and significant controlled release of additives contained in them. Naturally occurring and synthetic macromolecular surfactants that increase the viscosity of each phase of the multiple emulsion and form complexes with the emulsifiers or the oil can lead to formation of systems that will behave much like microcapsules, microspheres, and mesophasic liquid crystals.

This book mostly stresses the recent findings that model the transport phenomena through the different interfaces present in multiple emulsions. It will be useful for each formulator to understand how he can enhance the stability of multiple emulsions. The achievements include (1) new theoretical approaches and modeling to characterize the transport mechanism in multiple emulsions, (2) droplet size reduction and increased shelf-life stability by using polymeric amphiphiles and complex adducts, (3) use of new emulsification techniques to enhance the monodispersibility of the droplets, and (4) potential applications in drug delivery systems where clinical studies have already proved their efficacy.

Thus the chapters, which have been contributed by leading authorities in the field, are arranged logically in sections according to their increasing complexity: I—definitions and properties; II—formation and stability; and III—potential applications with a special emphasis on medical and pharmaceutical applications.

The first chapter by J. Jiao and D. J. Burgess discusses the thermodynamic instability of multiple emulsions as a result of the excess of free energy caused by the formation of the emulsion droplets. In multiple emulsions consisting of three distinct liquid phases, counteracting the effect of the Laplace pressure by electrolyte addition to the inner dispersed aqueous phase will increase the destabilization of the system owing to osmotic pressure. In addition the authors discuss the effects of both osmotic and Laplace pressure as well as the interfacial rheological properties of these complex systems and their stability.

In the second chapter P. Perrin, F. Pringent, and P. Hebraud describe the structure and rheological properties of stable W/O/W multiple emulsions stabilized with polymeric surfactants. By operating confocal microscopy and diffusing wave spectroscopy, they were able to visualize and to study the dynamical properties of the inner dispersed droplets.

The next chapter by L. B. Lawson and K. Papadopoulos is an overview of microscopy techniques, including photomicrography, video micrography, capillary microscopy, and electron microscopy, used to study the stability and transport phenomena in multiple emulsions systems.

The fourth chapter by M. Kanouni and H. Rosano describes the effect of an oil-insoluble solute on the stability of multiple water-in-oil-in-water emulsions. The authors build a theoretical model based on Laplace and osmotic pressure to interpret the transport phenomena in multiple droplets.

Part II focuses on the formation and stabilization of multiple emulsions.

R. Lutz and A. Aserin in chapter five develop a new concept to stabilize W/O/W multiple emulsions by using hybrids (complexes) of protein and polysaccharides that were shown to improve the stability of these complex systems and to better control the transport of the entrapped addendum.

The next chapter by G. T. Vladisavljevic and R. A. Williams is a comprehensive and systematic review of new techniques of preparation of multiple emulsions, emulsions, and microparticles. The authors envisage the ways to form multiple droplets by using membrane emulsification processes and microchannel and microcapillary devices. They also pay special attention to the preparation of solid microparticles via a double emulsion emulsification method using membrane emulsification and microfluidic devices.

A. Benichou and A. Aserin in the next chapter present developments in O/W/O multiple emulsions. The chapter reviews the most recent findings about enhancing stability and reducing droplet sizes for prolonged shelf-life stability. The chapter critically reviews the recent literature and brings some new emerging improvements involving stability and release control issues. Potential applications of O/W/O multiple emulsions in food, cosmetics, and drug delivery are discussed and evaluated.

The third part of the book is dedicated to medical and pharmaceutical applications of multiple emulsions.

First, J. L. Grossiord and M. Stambouli summarize 15 years of intensive research in this field that has been done by the French team. They mainly report the potential applications of these complex systems in drug delivery and detoxification.

A. J. Khopade and N. K. Jain review the anticancer activity of surface-modified fine multiple emulsions. These systems are characterized by nano-sized/micellar internal aqueous phase and highly viscous or solidified oil phase dispersed in an aqueous phase. Surface modification permits new applications in anticancer drug delivery and drug targeting.

Chapter 10 by S. Higashi describes the application of multiple emulsions to transarterial injection chemotherapy. By using membrane emulsification, particularly stable multiple emulsions were formed. From clinical trials it was established that multiple emulsions may be an alternative to conventional chemotherapy with a reduction in significant side effects.

In the next chapter, T. Hino and T. Ohwaki use a two-step pumping emulsification method to prepare lipiodol W/O/W multiple emulsions for transcatheter arterial embolization therapy.

Finally, A. Bozkir and O. M. Saka review the use of multiple emulsions as carriers for delivery systems for antigens and vaccines.

We would like to thank John Wiley & Sons Publishing, Inc. and particularly Mrs. Anita Lekhwani, senior acquisition editor, and her editorial assistant, Mrs. Rebekah Amos, for accepting this book for publication. We would also like to acknowledge with gratitude the contributions of authors and co-authors who have taken part in this enterprise.

It is our hope that the scientific information compiled herein will modestly contribute to a real updated understanding of multiple emulsions and to an improved comprehension of their current and future promising applications.

Abraham Aserin, Ph.D.

CONTRIBUTORS

ABRAHAM ASERIN, *Casali Institute of Applied Chemistry, The Institute of Chemistry, The Hebrew University of Jerusalem, 91904 Jerusalem, Israel*

AXEL BENICHOU, *Casali Institute of Applied Chemistry, The Institute of Chemistry, The Hebrew University of Jerusalem, 91904 Jerusalem, Israel*

ASUMAN BOZKIR, *Ankara University, Faculty of Pharmacy, Department of Pharmaceutical Technology, Tandogan, Ankara, Turkey*

DIANE J. BURGESS, *School of Pharmacy, University of Connecticut, Storrs, CT 06269, USA*

JEAN-LOUIS GROSSIORD, *Laboratoire de Physique Pharmaceutique UMR CNRS 8612, Faculté de Pharmacie, 5 avenue Jean-Baptiste Clément, F 92296 Châtenay-Malabry, France*

P. HÉBRAUD, *P.P.M.D. ESPCI, UMR 7615, 10 rue Vauquelin, 75231 Paris Cedex 05, France*

SHUSHI HIGASHI, *Miyakonojo Regional Medical Center, Miyakonojo, Miyazaki, Japan*

TOMOAKI HINO, *The College of Pharmacy, Kinjo Gakuin University 2-1723 Ohmori, Moriyama, Nagoya 463-8521, Japan*

N. K. JAIN, *Pharmaceutics Research Laboratory, Dr. Harisingh Gour University, Sagar-470003 (M.P.), India*

JIM JÍAO, *Pharmaceutical R&D, PGRD, Pfizer Inc. Groton, CT 06340, USA*

MOUHCINE KANOUNI, *City College of New York, 300 West 135th, New York, NY 10030, USA*

AJAY J. KHOPADE, *Sun Pharma Advanced Research Centre, Tandalja, Baroda 390020 Gujarat, India*

LOUISE BRAUD LAWSON, *Department of Microbiology and Immunology, Tulane University Health Sciences Center, New Orleans, Louisiana, USA*

RACHEL LUTZ, *Casali Institute of Applied Chemistry, The Institute of Chemistry, The Hebrew University of Jerusalem, 91904 Jerusalem, Israel*

TAKAYUKI OHWAKI, *Formulation Research Lab., Eisai Co., Ltd. 1 Kawashimatakehaya-machi, Kakamigahara City Gifu Pref. 501-6195, Japan*

KYRIAKOS PAPADOPOULOS, *Department of Chemical and Biomolecular Engineering, Tulane University, New Orleans, LA 70118, USA*

P. PERRIN, *P.P.M.D. ESPCI, UMR 7615, 10 rue Vauquelin, 75231 Paris Cedex 05, France*

F. PRIGENT, *I.P.C.M.S. GMO, UMR 7504, 23 rue du Loess, 67034 Strasbourg Cedex, France*

HENRI ROSANO, *Hostos Community College, Department of Natural Science, Bronx NY, USA*

ONGUN MEHMET SAKA, *Ankara University, Faculty of Pharmacy, Department of Pharmaceutical Technology, Tandogan, Ankara, Turkey*

MONCEF STAMBOULI, *Laboratoire de Génie des Procédés et Matériaux, Ecole Centrale Paris, France*

GORAN T. VLADISAVLJEVIĆ, *Chemical Engineering Department, Loughborough University, Loughborough, Leicestershire, LE11 3TU, United Kingdom*

RICHARD A. WILLIAMS, *Institute of Particle Science & Engineering, School of Process, Environmental & Materials Engineering, Clarendon Road, University of Leeds, Leeds, LS2 9JT, United Kingdom*

CHAPTER 1

Multiple Emulsion Stability: Pressure Balance and Interfacial Film Strength

JIM JIAO and DIANE J. BURGESS

Contents

1.1 INTRODUCTION

Multiple emulsions, or emulsions having ternary, quaternary, or more complex structures, have been studied since their first description in 1925 (Seifriz, 1925). The simplest multiple emulsions, sometimes called "double emulsions," are in fact ternary systems, having either a water-in-oil–in-water or an oil-in-water–in-oil structure, whereby the dispersed droplets contain smaller droplets of a different phase. Multiple emulsions have a number of potential applications in pharmaceutical, cosmetic, food, and separation sciences. The pharmaceutical applications of multiple emulsions include use as vaccine adjuvants (Gresham et al., 1971), red blood cell substitutes (Zheng et al., 1993), lymphatic drug-targeting vehicles (Yoshioka et al., 1982; Omotosho, 1989), prolonged drug

Multiple Emulsions: Technology and Applications, Edited by Abraham Aserin

delivery systems (Elston et al., 1970; Fukushima et al., 1983; Oza and Frank, 1989; Omotosho, 1990; Vaziri and Warburton, 1994), and sorbent reservoirs in drug overdosage treatment (Frankenfeld et al., 1976; Chiang et al., 1978; Moriomoto et al., 1979, 1982).

As is the case for simple emulsions, multiple emulsions are thermodynamically unstable due to the excess free energy associated with the surface of the emulsion droplets. The excess surface free energy arises as a result of the cohesive forces between the molecules of an individual liquid being greater than the adhesive forces between the liquids (Banker and Rhodes, 1979; Martin et al., 1993). On dispersion, the interfacial area of the dispersed phase liquid increases considerably compared to that of the continuous phase liquid. Consider the interfacial free energy (1.1) associated with the interface between two immiscible liquids:

$$\Delta G = \gamma \Delta A, \tag{1.1}$$

where G is the interfacial free energy, γ is the interfacial tension, and A is the total interfacial area of the dispersed phase. The increase in interfacial area results in a thermodynamically unstable system that tends to revert back to the original two-phase system to minimize interfacial area. The dispersed droplets therefore strive to come together to reduce the surface area, which can result in eventual destruction of the emulsion. In order to minimize this effect, a third component, a surfactant, is added to the system to improve its stability.

Multiple emulsions are complex systems where both water-in-oil (W/O) and oil-in-water (O/W) emulsion types exist simultaneously. In the case of water-in-oil-in-water multiple emulsions, the oil droplets have smaller water droplets within them, and the oil droplets themselves are dispersed in a continuous water phase. Oil-in-water-in-oil multiple emulsions, on the other hand, consist of tiny oil droplets entrapped within larger water droplets, which in turn are dispersed in a continuous oil phase. These systems thus differ from the familiar water-in-oil or oil-in-water simple two-phase emulsions in that they have three distinct phases (Pal, 1996). Multiple emulsions typically require two or more emulsifiers, one that is predominately hydrophobic stabilizing the primary W/O emulsion and one that is predominately hydrophilic stabilizing the secondary O/W emulsion. The hydrophobic and hydrophilic emulsifiers are added to the oil and continuous aqueous phases, respectively. The two emulsifiers may interact at the external water/oil interface and interfere with each other's stabilizing performance (Opawale and Burgess, 1998). In addition the osmotic pressure may affect the stability of W/O/W emulsions that is not observed in simple emulsions. If the osmotic pressure is higher in the internal aqueous phase, water will pass into this phase, with the internal droplets swelling until they rupture and release their contents onto the external phase. Transfer of water from the internal to external aqueous phases can cause shrinkage of the internal droplets to occur if a reverse gradient exists; this can also exert a destabilizing influence (Florence and Whitehill, 1985).

It has been demonstrated that the Laplace pressure works against the stability of simple emulsions (Davis, 1981). For water-in-oil emulsions, the addition of a small quantity of electrolyte to the disperse phase was determined to have a stabilizing effect as a consequence of counteracting the Laplace pressure effect. In W/O/W emulsions, the osmotic pressure generated by the presence of electrolytes in the inner dispersed water phase can cause swelling and ultimately bursting of the inner dispersed droplets, so the impact on multiple emulsion stability is negative. In order to balance these two effects, the concentration of electrolytes has to be high enough to counteract the Laplace pressure but sufficiently low to avoid osmotic effects.

The interfaces are the same in multiple emulsion systems as they are in simple emulsions. For example, one liter of a concentrated emulsion can contain up to 5000 m^2 of interface (the equivalent to a football pitch). So the interfacial area can be enormous because of the large number of droplets in the system. The large interface presents challenges and requires a quick migration of surfactants in the system to stabilize the dispersed phase(s). The structure and properties of the interface can therefore affect many aspects of the physical properties of emulsion systems. This is the main reason why interfacial characteristics are an important area of study in emulsions and especially multiple-emulsions systems. It has been shown that the stabilities of both multiple and simple emulsions are dependent on emulsifier interfacial film strength, ionic strength, and the presence of various additives. It has been experimentally proven that the interfacial film strength can be used as a means to predict emulsion stability (Burgess, 1997).

In this chapter the effects of pressure balance and interfacial rheological properties on the stability of multiple emulsions are discussed.

1.2 MULTIPLE EMULSION PRESSURE PROPERTIES

1.2.1 Osmotic Pressure

For W/O/W multiple emulsions the oil phase can be viewed as a membrane separating the inner and outer aqueous phases at the water/oil interface. The thickness of the oil membrane varies with changes in the multiple emulsion composition. Water can pass through the oily membrane from one aqueous phase to the other depending on the osmotic pressure. A higher osmotic pressure in the internal aqueous phase than in the external continuous aqueous phase causes water to pass into the inner water phase, resulting in swelling of the internal droplets before they eventually burst and release their contents. The reserve also applies: if the osmotic pressure is higher in the external aqueous phase than in the inner aqueous phase, water will transfer from the internal phase to the external aqueous phase, causing shrinkage of the internal droplets. If the osmotic difference across the oil layer is extreme, then the passage of water becomes so rapid that almost immediate rupture of the oil droplets occurs with loss of the internal droplets. When the oil layer ruptures, the inner aqueous phase in the multiple oil droplets disappears instantaneously, mixing with the external aqueous phase and leaving simple emulsions.

The osmotic pressure effect on stability of multiple emulsions has been investigated for almost four decades. W/O/W emulsions, when given in vivo, break down rapidly at the site of injection, with the consequence that no significant delay in response to the entrapped drug is obtained compared to aqueous solutions of the drug (Collings, 1971). It was determined that the premature breakdown of the emulsions in vivo is due to unequal osmotic pressures between the internal and external aqueous phases. The osmotic pressure in the external environment (body fluids) is higher than the internal phase leading to shrinkage of the internal aqueous droplets and/or rupture of the oil layer. Collings (1971) partially solved the problem by incorporating small amounts of sodium chloride in the internal aqueous phase so that this phase was isotonic with the final external phase.

Materials other than electrolytes (e.g., proteins, sugars, and drugs) in the aqueous phase can also exert this effect (Adeyeye and Price, 1990). A variety of materials entrapped in the inner phase of multiple emulsions are found to affect osmotic pressures (Florence and Whitehill, 1982; Cuemen and Zatz, 1988; Garti and Aserin, 1996). The middle phase acts as a semipermeable membrane, and consequently osmotic effects become significant as they control multiple emulsion stability and drug release rates both in vitro and in vivo (Collings, 1971; Davis and Burbage, 1978; Matsumoto and Kohda, 1980; Florence and Whitehill, 1981).

Sodium chloride and other electrolytes added initially in the inner or outer aqueous phase of W/O/W multiple emulsions can migrate across the oil layer and get into the other aqueous phase through molecular migration (Collins, 1971; Chilamkurti and Rhodes, 1980). The migration of the electrolytes induces changes in osmotic pressure over time and consequently alters multiple emulsion stability. It has been observed that multiple emulsions stabilized by Span 83 and Tween 80 are more stable with sodium salicylate incorporated in the inner aqueous phase than with sodium chloride (Jiao et al., 2002). The difference in the stability of the multiple emulsions observed can be attributed to a faster migration of sodium chloride from the inner aqueous phase to the outer aqueous phase and a consequent more significant imbalance in the osmotic pressure compared to that with sodium salicylate.

The transport mechanism of electrolytes through the oily liquid phase has been the subject of many investigations over the past decades. Nevertheless, there remains a lack of a clear understanding as to what and how various formulation parameters of multiple emulsions affect the kinetics and extent of the migration of electrolytes across the middle phase, and thereby influence the osmotic pressure. Partition coefficient, ionization, charge density, molecular weight, and molecular mobility of electrolytes can have some impact on electrolytes' ability to cross the oil phase. The association of electrolytes with the surfactant, which may form inverted micelles in the oil phase, has also been considered (Chilamkurti and Rhodes, 1980).

1.2.2 Laplace Pressure

Laplace pressure arises from the interfacial tension of a mixture of two liquids at a curved interface when one liquid is dispersed as droplets into another liquid. The pressure varies inversely with the radius of curvature and takes the following form:

$$\Delta P = \gamma\left(\frac{1}{r_1} + \frac{1}{r_2}\right), \tag{1.2}$$

where γ is interfacial tension and r is particle radius. For a spherical particle such as a droplet, $r_1 = r_2$, the Laplace equation becomes

$$\Delta P = \frac{2\gamma}{r}. \tag{1.3}$$

Hence a spherical droplet having a radius r in an emulsion will exert greater pressure on the inner concave interface than on the convex side, as expressed in equation (1.3), and the larger surface tension constitutes a larger force pushing inward into the droplet. Because the relationship of ΔP and r is inversed, a smaller radius will result in a larger inward force. Therefore this relationship has important consequences for any curved surface as r becomes very small and γ relatively significant. When this relationship is applied to the context of an emulsion in which two droplets with the same surface tension are connected, the smaller droplet can be expected to experience a greater pressure, driving its collapse and pushing all of its contents into the larger droplet.

When droplet deformation occurs, the Laplace pressure of the deformed droplet will be a function of the radius along the droplet surface. In the extreme cases, as a droplet becomes elongated and cylindrically shaped, the Laplace pressure is reduced to a half that of the original spherical droplet. For multiple emulsions, the Laplace pressure exists in both the inner and multiple droplets. However, because the size of the inner droplets is much smaller, the Laplace pressure on the stability of the inner droplets is much greater than that on the multiple droplets.

The Laplace pressure in the process of emulsification is what causes an emulsion to become thermodynamically inefficient. For an emulsion to form the small, highly curved droplets, extra energy is required to overcome the large pressure that exists in the droplets.

1.2.3 Balance between Laplace Pressure and Osmotic Pressure

Consider a water droplet of radius r containing a certain amount of salt in a solvent (oil phase) in equilibrium at the water/oil flat interface. The film around this water droplet can be assumed (for simplicity) to be impermeable to water and capable of preventing coalescence. The Laplace pressure $2\gamma/r$ of a droplet containing a salt, dispersed in the solvent, will cause shrinkage of the droplet. However, the osmotic pressure will cause swelling of the droplet, leading to a counterbalanced water diffusion. In the ideal case the osmotic pressure is given by

$$\prod_{osm} = mRT = m_o\left(\frac{r_o}{r}\right)^3 RT, \tag{1.4}$$

where m is the molar concentration of salt and m_o refers to the original droplet. The difference between the Laplace pressure and the osmotic pressure can be defined as excess pressure ΔP:

$$\Delta P = \frac{2\gamma}{r} - m_o\left(\frac{r_o}{r}\right)^3 RT. \tag{1.5}$$

The condition that must be fulfilled in order to reach the equilibrium is

$$\frac{d(\Delta P)}{dr} = -\frac{2\gamma}{r^2} + 3m_o\left(\frac{r_o}{r}\right)^3 \frac{1}{r} RT = 0. \tag{1.6}$$

In words, as r decreases, the excess pressure ΔP decreases and equilibrium is reached. Thus

$$2\gamma = 3mRT. \tag{1.7}$$

Equation (1.7) was proposed by Walstra (1996). Walstra's equation shows that an optimal salt concentration in the internal phase exists between the Laplace and osmotic pressures exerted on the inner aqueous droplets.

Stability of W/O/W multiple emulsion containing Span 80 and Tween 80 was evaluated with respect to sodium chloride and sodium salicylate concentrations in the inner water phase (Jiao and Burgess, 2002). In this study we observed that the multiple emulsion droplets deformed and there was coalescence of the inner aqueous droplets as we applied an external force (i.e., a microscopic coverslip) to multiple emulsion samples on a microscope slide. Under certain conditions (e.g., lipophilic surfactant concentration and internal phase osmotic pressure) the destabilized multiple emulsions formed unique metastable structures that had a "dimpled" appearance. The formation of these metastable structures correlated with the real time instability of the W/O/W multiple emulsions investigated. Our study revealed that emulsions with a salt concentrations closer to the optimal value calculated by using (1.7) had maximum stability.

The treatment above is only good for a simplified emulsion system where the osmotic and Laplace pressures are the major forces controlling droplet stability. There are other factors such as viscosity that influence the dynamics of droplet growth. We take these factors into consideration using the general Navier-Stokes equation to mathematically describe droplet expansion:

$$\rho\left(\frac{\partial v}{\partial t}+(v\bullet\nabla)v\right)=-\nabla p(\gamma, t)+\eta\Delta v, \qquad \operatorname{div} v=0, \tag{1.8}$$

where $p(\gamma, t)$ is the total pressure at any given point on the droplet surface. The influence of any other factors can be entered into the equation by way of boundary conditions (Mikhin, Stepanow, and Byakov 2003).

1.3 INTERFACIAL RHEOLOGY AND STABILITY

1.3.1 Interfacial Film and Film Strength

Multiple emulsions require surfactants to stabilize both the internal aqueous droplets and the external multiple droplets. The added surfactants adsorb at the water/oil interfaces, reducing interfacial tension and forming an interfacial film that resists droplet coalescence following droplet contact. It has been shown that the stronger this film is, the more stable are the emulsions, and that the interfacial film plays a more crucial role than the reduction of interfacial tension in maintaining long-term emulsion stability to coalescence (Burgess, 1993). The strength of this film, which can be a monolayer, a multilayer, or a collection of small particles adsorbed at the interface, depends on the structure and conformation of surfactant or emulsifier molecules at the interface (Swarbrick, 1997). The structure and conformation can be affected by formulation variables, including surfactant or emulsifier type and concentration, other additives or levels, storage temperature, ionic strength, and pH. For the film to be an effective barrier, it must remain intact when sandwiched between two droplets. If broken, the film has the capacity to reform rapidly. So the film must possess a certain degree of surface elasticity. It has been shown that interfacial elasticity correlates well with interfacial film strength and can be used to predict the stability of multiple emulsions (Opawale and Burgess, 1997). Knowing the relationship between interfacial properties and emulsion stability enables one to rationally approach the research and development of more stable multiple emulsion systems.

The only way significant amounts of immiscible fluids can be mixed together is if the interfacial layer surrounding the dispersed droplets is occupied by an adsorbed layer of molecules that keep the droplets from coalescing. Figure 1.1 shows the importance of the interfacial layer in emulsion systems for the two main classes of surface-active molecules, surfactants and proteins, that stabilize them. Low molecular weight surfactants, lipids, and emulsifiers self-assemble at interfaces with the appropriate part of the molecule associating with the appropriate hydrophilic or hydrophobic phases. Proteins, on the other hand,

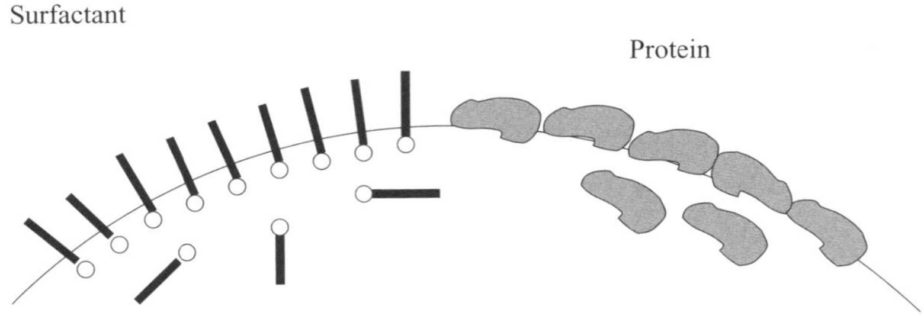

Figure 1.1 Two classes of surface-active materials in stabilizing emulsions: Surfactant and protein.

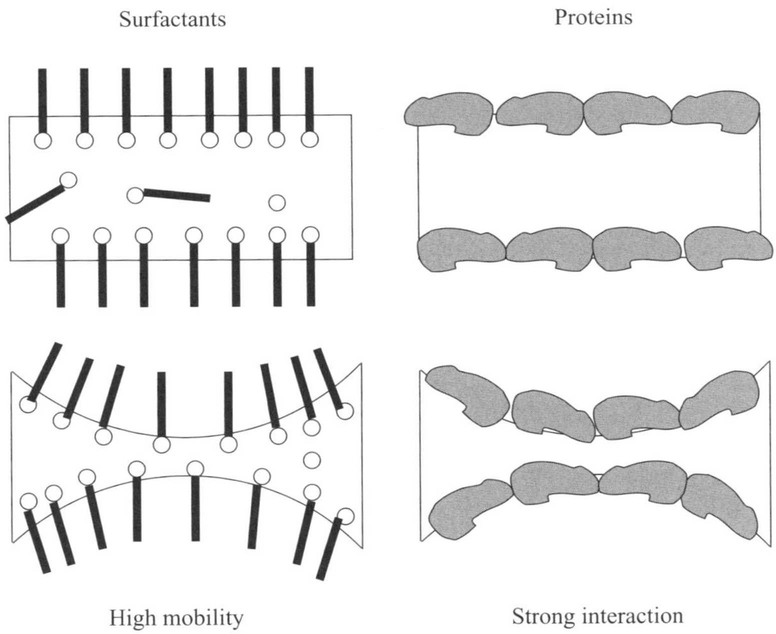

Figure 1.2 Stability mechanism for surfactants and proteins.

are much larger and more complex macromolecules. Proteins will adsorb at an interface but then proceed to unfold, exposing their hydrophobic groups to the hydrophobic phase.

Figure 1.2 shows how these two very different types of molecules stabilize emulsion systems. Surfactants rely on rapid diffusion to dissipate any disturbances to the interface. This rapid motion will drag fluid along into the interlamellar space between droplets, keeping them separated. This activity is known as the Gibbs-Marangoni mechanism. On the other hand, proteins unfold, develop strong interactions with neighboring protein molecules, and effectively form a gel at the interface. The viscoelastic gel can stretch and deform to absorb deformations in the interface, and hence stabilize against coalescence. The main difference between interfaces stabilized by proteins and surfactants is the viscoelasticity of the interface. Therefore interfacial rheology is a useful probe for comparing these two types of interfaces.

1.3.2 Interfacial Tension and Rheology

The interfacial properties exhibited by emulsifier systems are interfacial rheology, tension, and charge (Burgess and Yoon, 1995; Burgess and Sahin, 1997). Interfacial rheology measures the emulsifier film viscosity and/or elasticity and hence the mechanical barrier to droplet coalescence. Interfacial tension is related to emulsion stability through the Gibbs equation (Eq. 1.1). The interfacial charge on emulsion droplets gives a direct measurement of the electrostatic barrier to coalescence. Interfacial rheology, tension, and charge have been used as predictors of emulsion stability (Burgess and Yoon, 1995; Burgess and Sahin, 1997). Cumper and Alexander (1950), Srivastava (1964), and Burgess (1998) have shown that the interfacial rheology of protein films correlates with O/W emulsion stability.

Interfacial Tension Lowering of interfacial tension is one way in which the increased surface free energy associated with the formation of droplets can be reduced. Since surfactant molecules continuously adsorb at the interface, interfacial tension will decrease as a function of time until equilibrium is achieved. Reduction of interfacial tension by the addition of a surfactant can serve to preserve the surface area generated during the dispersion process, thus preventing phase separation. Low interfacial tension enhances the formation of smaller emulsion droplets with narrower size distributions and greater kinetic stability (Burgess and Yoon, 1995). The major requirement of a potential surfactant or emulsifier is that it readily form an interfacial film. A rapid decrease in interfacial tension indicates high interfacial activity and a tendency for fast reformation of the surfactant film after rupture. Rapid reformation of a new interfacial film results in increased resistance to droplet coalescence, and hence emulsion stability is improved (Myers, 1988).

The dynamic process of adsorption of emulsifiers and the equilibrium state of the interfacial film can be measured by the change in interfacial tension as a function of time. Dynamic interfacial tension techniques exist that measure without disturbing the interface. Various such techniques to measure interfacial tension have been reported in the literature (Addison and Hutchinson, 1949; Padday and Russel, 1960). The Wilhelmy plate technique is preferred over other techniques because the values obtained are more accurate than those obtained using other techniques such as the capillary rise or du Nouy ring methods (Padday and Russel, 1960). In the latter two methods, the long equilibration time (3–60 hours) and difficulties in accurately positioning the ring can introduce errors that are not an issue with the Wihelmy plate method.

Interfacial Rheology It has been reported that for long-term emulsion stability to coalescence and phase separation, the strength of the interfacial film is more important than reduction in interfacial tension (Myers, 1988; Swarbrick, 1990; Martin, 1993). Interfacial rheology is the study of the mechanical and flow properties of adsorbed layers at fluid interfaces, and it has been used to quantify film strength (Murray and Dickinson, 1996; Opawale and Burgess, 1998). Interfacial film strength characteristics can be described in terms of viscous (liquid-like) or elastic (solid-like) properties (Warburton, 1993). In an emulsion stabilized with surfactant-type emulsifiers forming monomolecular films, coalescence is opposed by the elasticity and cohesiveness of the films sandwiched between the two droplets. Multilayers confer high resistance of emulsions to coalescence, as arises from the mechanical strength of layering (Myers, 1988). For films to be efficient barriers, they must not thin out and rupture when sandwiched between the two droplets. The film must therefore possess enough elasticity to assist in preserving its integrity (Myers, 1988).

There are two main methods for measuring the interfacial rheological properties of adsorbed layers. They can be either dilational or shear methods. Figure 1.3 shows the principle underlying each method, without going into too much practical detail. Interfacial dilational rheology is determined by measuring the change in interfacial tension due to a specific change in interfacial area. This is a measure of the resistance to compression and expansion of the adsorbed layer. Interfacial shear rheology, on the other hand, can be a direct measure of the mechanical strength of the adsorbed layer. Here the interface is subject to a shear stress, and the measured strain is recorded.

When an element of area covered with soluble material is subject to surface contraction, some of the material escapes into the bulk phase and returns when the interface is expanded in interfacial dilational rheology (Murray and

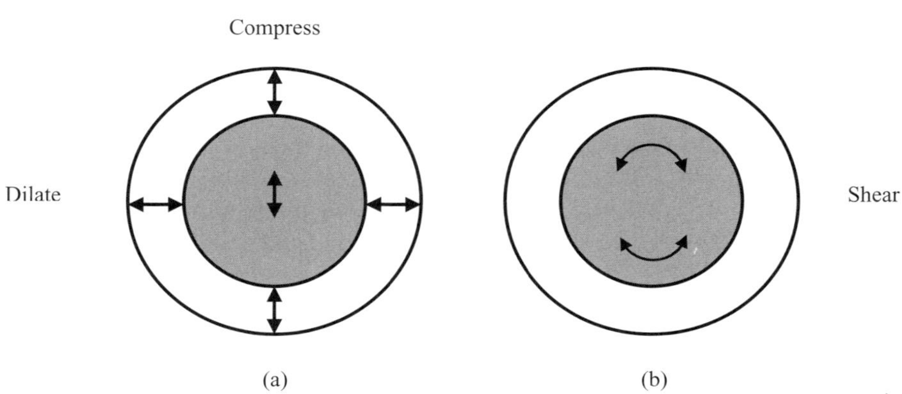

Figure 1.3 Interfacial dilational (*a*) and interfacial shear (*b*) rheology.

Dickinson, 1996). In such experiments the interpretation of interfacial dilational rheology needs to take into account the dilation processes. However, in interfacial shear rheology, a defined interfacial area is sheared to a first-order approximation and is therefore not altered during the shearing process (Sheriff and Warburton, 1975). The interfacial pressure remains constant during the experiment and the material does not diffuse out of or into the interface as a result of rheological measurements. Consequently interfacial shear experiments are less destructive than interfacial dilational experiments, and measurements can provide information on the intramolecular and intermolecular forces acting at the interface (Warburton, 1993). The kinetics of interfacial film formation can also be studied using this technique. Interfacial shear measurements can be performed by several methods: continuous flow, creep compliance, stress relaxation, and oscillation (Warburton, 1993). It is not possible to study film kinetics, and intra- and intermolecular interactions among interfacial molecules, using interfacial dilational techniques, since the interfacial film is continuously destroyed. Using a MK2 surface oscillatory ring rheometer, which operates in the interfacial shear mode, Opawale and Burgess (1998) were able to determine the kinetics of interfacial film formation of Spans (20, 80, 83, and 85) under various conditions (different temperatures, Span concentrations, salts, and macromolecules such as bovine serum albumin and cholesterol).

1.3.3 Multiple Emulsions' Stability and Interfacial Properties

As shown in Figure 1.2, emulsions can be stabilized by surfactants or emulsifiers employing the Gibbs-Marangoni mechanism, which has a very low interfacial viscoelastic modulus, or by protein-like molecules, which employ a viscoelastic mechanism with a naturally high viscoelastic modulus. Both mechanisms result in stable systems individually, but in many commercial emulsions there is often a mixture of these two molecule types.

Figure 1.4 shows an interface stabilized by a mixture of protein and surfactant type of molecules. The surfactants disrupt the strong interactions developed between neighboring protein molecules, effectively weakening the interface. Because the surfactants rely on rapid surface migration, they are constrained by the presence of protein molecules still at the interface. If the protein component is still in the form of a two-dimensional network, effectively caging the surfactant molecules, it can seriously hamper their motion. The net effect is reduced stabilization of each component, and hence the emulsion is reduced in stability.

In principle, it should then be possible to predict the stability of an emulsion system from the interfacial rheology of the continuous phase. Figure 1.5 shows the relative stability to coalescence of an emulsion system stabilized by a protein (beta-lactoglobulin) with increased concentrations of non-ionic surfactant (Tween 20). In this case the presence of surfactants has entirely destabilized the protein emulsion.

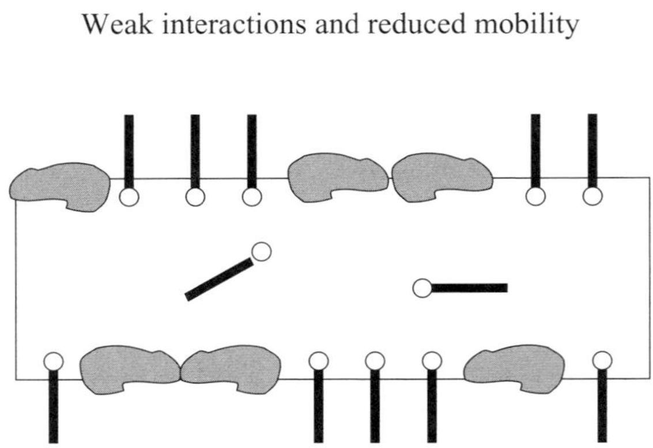

Figure 1.4 Mixed protein and surfactant interfaces: Weak protein interactions and restricted diffusion of surfactants result in reduced stability and probable film rupture.

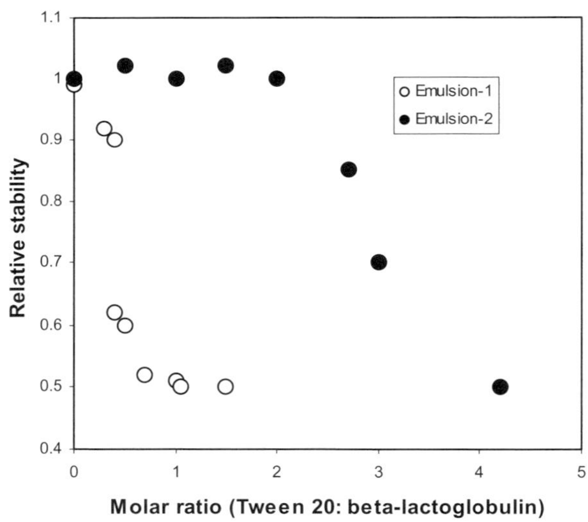

Figure 1.5 Coalescence stability of protein-stabilized emulsions as a function of surfactant (Tween 20) concentration.

1.3.4 Determination of Interfacial Properties

As was mentioned earlier, there are two basic methods of measuring interfacial rheology: dilational and shear (Murray and Dickinson, 1996). The practical and theoretical aspects underlying these measuring methods are briefly discussed here.

Interfacial Dilational Rheology Dilational rheology, as the name suggests, is a method that deals with the expansion and compression of the interface. Simply put, it is a mechanical system that is constructed to allow the interface to be expanded and contracted, usually in a sinusoidal manner, while the interfacial tension is simultaneously monitored. The first such method used a standard Langmuir trough, as shown in Figure 1.6. Barriers normal to the surface are used to gradually compress or expand the interface to control the surface concentration of insoluble monolayers. A small modification to this method allows the barriers to be oscillated sinusoidally, producing small changes in the surface area.

Assume that there is no exchange of surfactant occurs between the surface and the bulk, the compression/expansion cycle will cause a change in the surface tension. As the surface is compressed, the effective surface concentration increases, and the interfacial tension will go down. Conversely, expanding the surface will result in an increase in the surface tension. The relationship between surface area and surface tension is shown in Figure 1.7.

The surface dilational modulus ($|E|$) is given as

$$|E| = A \cdot \frac{d\gamma}{dA}. \tag{1.9}$$

The surface dilational modulus is then split into the elastic (E') and viscous (E'') components. If the surface is purely elastic, then the phase lag (θ) will be zero; if it is viscous, then $\theta = 90$. In practice, the behavior is usually intermediate between the two extremes, and the two components can be calculated as follows:

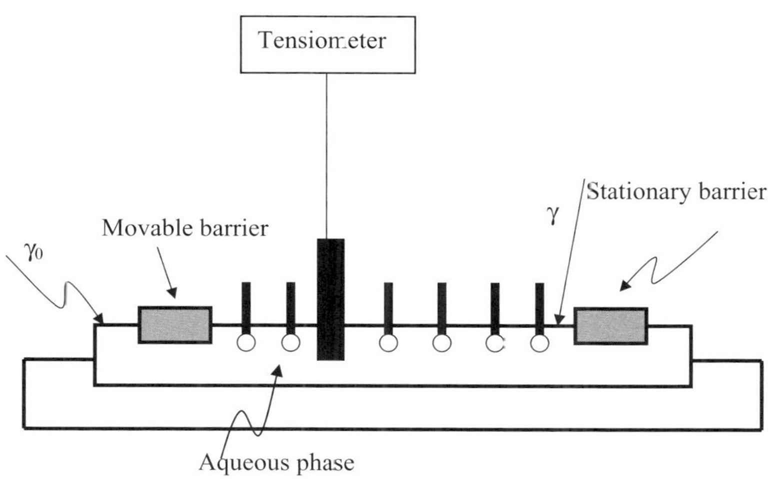

Figure 1.6 Use of a Langmuir trough fitted with oscillating barriers to change the surface area A while simultaneously monitoring surface tension.

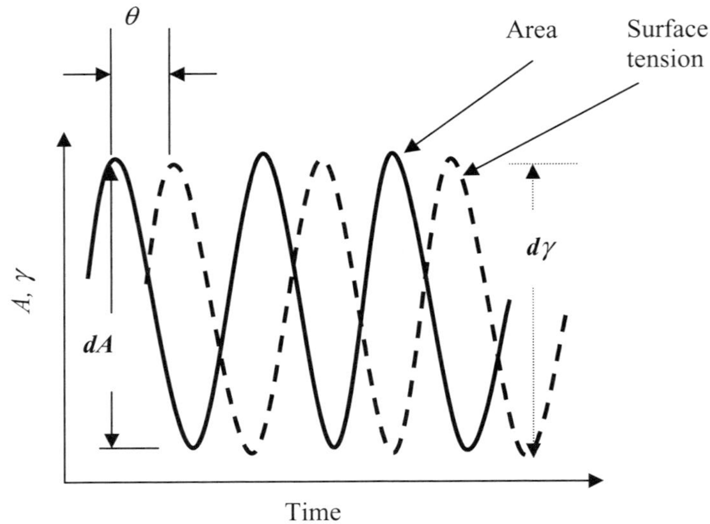

Figure 1.7 Time-dependent relationship between area and surface tension during a typical dilational rheology experiment.

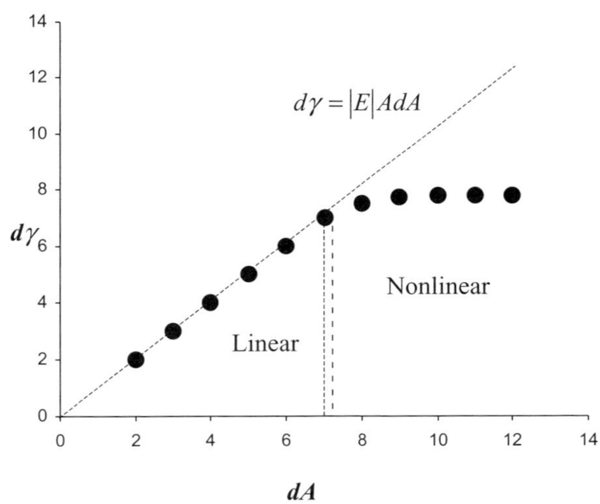

Figure 1.8 Relationship between $d\gamma$ and dA, showing how $d\gamma$ becomes nonlinear as dA is increased and the surface is stretched beyond its elastic limit.

$$E' = |E|\cos\theta, \quad E'' = |E|\sin\theta. \tag{1.10}$$

Care must be taken to ensure that the surface is not overcompressed, or the interfacial layer may collapse. Figure 1.8 shows the relationship between $d\gamma$ and dA in a typical experiment. The experiments should ideally be conducted

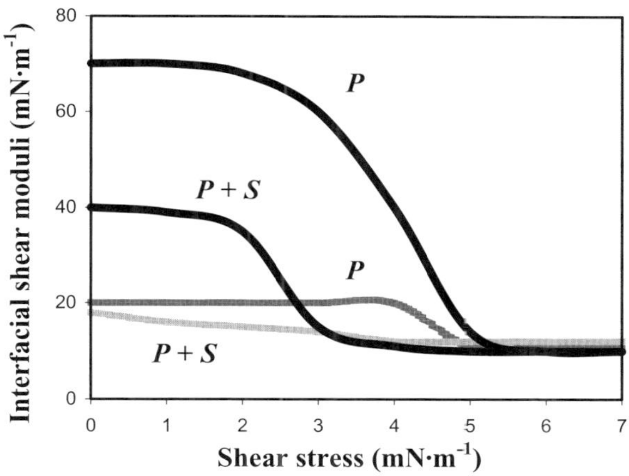

Figure 1.9 Interfacial shear moduli (elastic = solid lines; viscous = dotted lines) for protein alone (*P*) and protein + surfactant (*P* + *S*) at the oil/water interface.

in the linear region, but the length of the linear region does give information about the resistance of the surface to compression and collapse, the same as the stress dependent data shown in Figure 1.9.

A dilational method that has attracted much attention over recent years is the pendant drop method. This method has proved useful for oil/water interfacial rheology. The principle is exactly the same as for standard surface dilational rheology. The interfacial tension is calculated by measuring the size and shape of a liquid drop suspended from a capillary in a less dense fluid. The interfacial area is changed by increasing or decreasing the size of the drop by controlling the liquid flow through the capillary. Changes in the interfacial area and interfacial dilational modules can be calculated. This technique is useful for small sample volumes, for it avoids the hydrodynamic problems encountered when trying to expand/compress the oil/water interface.

A limitation of the dilational method is that it is an indirect method of measurement. The rheological information is inferred from interfacial tension values. The interfacial tension can change through adsorption and desorption effects that are stimulated by the expansion and compression of the interface. For example, for a freely soluble surfactant at high concentrations, there will be a rapid rate of exchange between the surface and the bulk, so at low dilation frequencies, lower than the exchange rate, no changes in surface tension will be detected. At lower concentration, with a much lower exchange rate, changes in surface tension will be measured, but these will be dependent on the exchange rate of the surfactant, not the surface rheological properties: The way to solve this problem is to measure at high frequencies using the surface capillary wave technique. Figure 1.10 shows a schematic of the experimental setup. Capillary waves on the surface of a solution can be induced by thermal or mechanical agitation at high frequencies. The presence of a wave on a surface corresponds to an increase in surface area, the shape of the wave is

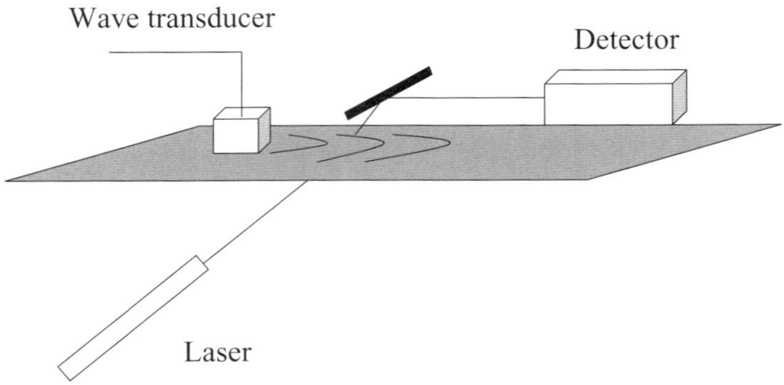

Figure 1.10 Schematic setup for capillary wave-type experiment. The wave transducer excites surface waves. The laser diffraction and detector measure damping of the waves.

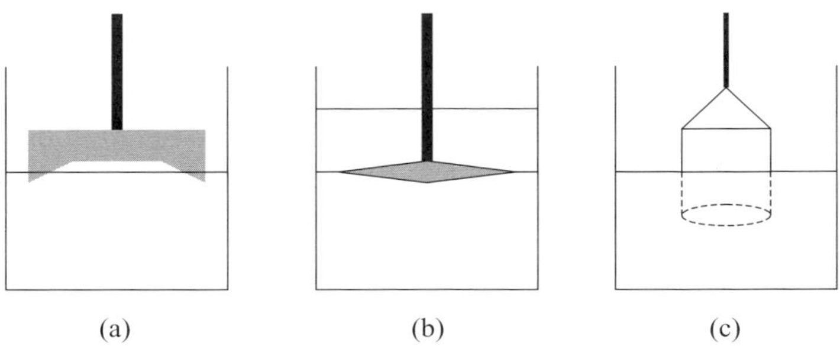

Figure 1.11 Typical geometries for measuring interfacial shear rheology: (*a*) Knife edge for air/water interface; (*b*) bicone for oil/water interface; (*c*) Du Nouy ring for sensitive measurements.

monitored by laser diffraction, and the amplitude/damping of the wave as it propagates along the surface is determined by the elastic and viscous moduli of the surface of the solution. The measurement is difficult to do, and the errors are quite large. Nevertheless, the high-frequency regime required for surfactant systems can in this way be accessed.

Interfacial Shear Rheology In contrast to the dilational technique, the surface shear methods are direct determinations of the mechanical properties of an interface. The simplest approach is a two-dimensional adaptation of standard three-dimensional viscoelastic measurements performed on a standard rheometer. The only difference is the sensitivity and the geometry. Figure 1.11 shows the geometries commonly used for oil-water interfaces.

In interfacial shear rheology, shear stress is applied as a tangential force (F) acting along the interface. Shear stress can be calculated using the equation below,

$$p_{xy} = Ge_{xy}, \tag{1.11}$$

where p_{xy} is the interfacial shear stress. The Du Nouy ring, commonly used for measuring interfacial tension, can be specifically designed to measure the surface shear viscoelasticity through a method developed by Sherriff and Warburton. The light construction of this geometry makes it particulary sensitive to interfaces with very low rheological properties.

The simplest approach is to apply a standard shear rate, that is, to rotate the geometry at constant speed and measure the induced stress. The geometry is normally held by a torsion wire of known torsion strength, and the rotation between the top and the bottom of the wire allows the stress to be calculated. An variation on this approach is to use the canal viscometer approach. An interface is held between two concentrically circular, rigid walls, and the floor of the vessel is rotated to impart motion of the fluid below the interface. The transmission of motion to the interface from the bulk is measured by following the motion of small Teflon particles at the interface. Although this is a time-consuming measurement, the rheological properties of the bulk fluid can easily be accounted for.

However sensitive the technique, continuous rotation will result in disruption of structures created at the interface. A better approach is to use an oscillatory motion, which, if small enough, should not break down any structures formed at the interface. An oscillating stress with known amplitude is applied, and the resultant strain is measured. The stress and strain relationships are shown in Figure 1.12.

The total viscoelastic modulus G^* is given as

$$G^* = \frac{\sigma_0}{\gamma_0}, \tag{1.12}$$

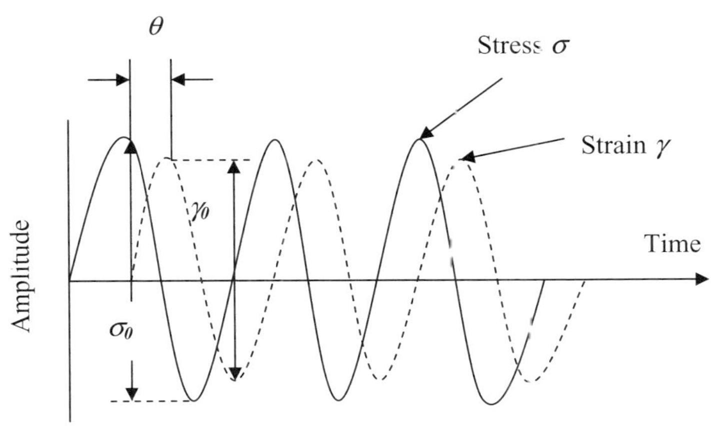

Figure 1.12 Stress–strain relationship for a typical oscillatory surface shear viscoelasticity measurement.

where σ and γ are the amplitudes of the stress and the strain, respectively (γ should not be confused with the surface tension). The stresses and strains are the effective two-dimensional equivalents of the three-dimensional standard viscosity measurements. So the stress here is the applied force per unit distance, and the strain is the distance moved relative to the gap between the geometry and the outer vessel. Similar to the dilational method, if the phase lag is 0 and 90°, the G^* is either totally elastic or viscous, respectively. The elastic (G') and viscous (G'') moduli can be calculated as follows:

$$G' = G^* \cos\theta, \quad G'' = G^* \sin\theta \tag{1.13}$$

The Sherriff and Warburton is a little more complex. A schematic diagram of the apparatus simplifies the complex electronics involved. Basically a galvanometer is fed with a sinusoidal input voltage to oscillate the ring, imparting a known stress to the interface; then a proximity probe is used to measure the strain. The electronics exploit mechanical resonance so that the sensitivity of the interface can be maximized. The mechanical or rheological properties of the interface affect the resonance of the ring as the electronic feedback system keeps the whole system in resonance. The values of the feedback signals are used to calculate the viscous and elastic components of the interfacial viscoelasticity. The light weight nature of this setup allows assessment of very fragile interfaces. The frequency is also not just limited to the resonant frequency; the feedback loops can be used to shift the measurement frequency over a wide range of values.

In summary, measurement of interfacial rheology can take one of two approaches, either dilational or shear. The choice of approach will depend on its suitability to particular applications. The most accurate and reproducible results tend to come from methods that utilize small, reversible applied stresses and strains, thus minimizing any disruption or damage to the interfacial layer.

1.4 CONCLUSIONS

Numerous applications of multiple emulsions in various fields have been reported. More applications need to be realized if multiple emulsions stability is to be fully understood and approaches to stabilize multiple emulsions fully rationalized. The stability of multiple emulsions is influenced by numerous formulation and process variables. As demonstrated in this chapter, long-term multiple emulsion stability is dependent on the osmotic and Laplace pressures of the inner droplets as well as on the pressure balance between them described by the Walstra equation. Stability also equally, in some cases even more, depends on the strength of the interfacial film formed on the interface of droplets of multiple emulsions. This property can be characterized by interfacial rheology.

REFERENCES

Annianson EAG, Wall SN. 1974. Kinetics of step-wise micelle association. *J Phys Chem* 78(10): 1024–1030.

Aronson MP, Petko MF. 1993. Highly concentrated water-in-oil emulsions: Influence of electrolyte on their properties and stability. *J Colloid Interface Sci* 159(1): 134–149.

Abraham SM, Vieth RF, Burgess DJ. 1996. Novel technology for the preparation of sterile alginate-poly-lysine microcapsules in a bioreactor. *Pharm Dev Tech* 1(1): 63–68.

Addison CC, Hutchinson SK. 1949. The properties of freshly forced surface. Part XI. Factors influencing surface activity and adsorption rates in aqueous-alcohol solutions. *J Chem Soc* 61: 3387–3395.

Adeyeye CM, Price JC. 1990. Effect of dextrose in the internal aqueous phase on formation and stability of W/O/W multiple emulsions. *Drug Dev Ind Pharm* 16(6): 1053–1071.

Adeyeye CM, Price JC. 1991. Effect of non-ionic surfactant concentration and type on the formation and stability W/O/W multiple emulsions: microscopic and conductometric evaluations. *Drug Dev Ind Pharm* 17(5): 725–736.

Attwood D, Florence AT. 1983. *Surfactant system: The Chemistry, Pharmacy and Biology*. London: Chapman and Hall, pp 469–650.

Banwan AA, Abumalooh A, Sallam E, Abukalaf A, Jawan O. 1985. A sustained release drug delivery system using calcium alginate beads. *Drug Dev Ind Pharm* 11(2): 239–256.

Banker GS, Rhodes T. 1979. *Modern Pharmaceutics*. New York: Marcel Dekker, pp 329–359.

Becher P. 1965. In: *Emulsions: Theory and Practice*, 2nd ed. New York: Reinhold, p 2.

Benoy CH, Elson LA, Schneider R. 1972. Multiple emulsions, a suitable vehicle to provide sustained release of cancer chemotherapeutic agents. *Br J Pharmacol* 45(1): 135–136.

Binks BP, Clint JH, Fletcher PDI, Rippon S. 1998. Kinetics of swelling of oil-in-water emulsions. *Langmuir* 14: 5402.

Binks BP, Clint JH, Fletcher PDI, Rippon S. 1999. Kinetics of swelling of oil-in-water emulsion stabilized by different surfactants. *Langmuir* 15: 4495.

Biswas B, Haydon DA. 1962. The coalescence of droplets stabilized by viscoelastic adsorbed films. *Kolloid Z* 185: 31.

Blake JR. 1996. Cavitation and bubble dynamics. *J Fluid Mech* 316: 376–377.

Bodmeiser R, Chen H, Paeratakul O. 1989. A novel approach to the oral delivery of micro or nanoparticles. *Pharm Res* 6(5): 413–417.

Boyd J, Parkinson C, Sherman P. 1972. Factors affecting emulsion stability and HLB concept. *J Colloid Interface Sci* 41(2): 359–370.

Brodin AF, Kavaliunas DR, Frank SG. 1978. Prolonged drug release from multiple emulsions. *Acta Pharm Suec* 15: 1–10.

Brodin AF. 1978. Drug release from oil-water-oil (O/W/O) multiple emulsion systems. *Acta Pharm Suec* 15: 111–113.

Bungenberg de Jong MG. 1949. In: *Colloid Science. II. Reversible Systems*, Kruyt GR, eds. New York: Elsevier, pp 335–432.

Burbage AS, Davis SS. 1979. The characterization of multiple (W/O/W) emulsions using a radiotracer technique. *J Pharm Pharmacol* 31: 6.

Burgess DJ, Yoon JK. 1993. Interfacial tension studies on surfactant systems at the aqueous/perfluorocarbon interface. *Colloid Surf B* 1: 283–293.

Burgess DJ. 1993. In: *Biotechnology and Pharmacy*, Pezzuto JM, Johnson ME, Manasse HR, eds. New York: Chapman and Hall, p 116.

Burgess DJ, Yoon JK. 1995. Influence of interfacial properties on perfluorocarbon/aqueous emulsion stability. *Colloids Surf* 4: 297–308.

Burgess DJ, Sahin NO. 1994. Interfacial rheology of ß-casein solution. In: *Structure and Flow in Surfactant Solutions*, Herb CA, Prud'homme RK, eds. ACS Symp Ser, Washington, DC: ACS, pp 380–393.

Burgess DJ, Sahin NO. 1997. Influence of protein emulsifier interfacial properties on oil-in-water emulsion stability. *Pharm Dev Tech* 3: 21–32.

Buscall R, Davis SS, Potts DC. 1979. The effect of long-chain alkanes on the stability of oil-in-water emulsions: the significance of Ostwald ripening. *Colloid Polym Sci* 257: 636.

Chanamai R, Horn G, McClements DJ. 2002. Influence of oil polarity on droplet growth in oil-in-water emulsions stabilized by a weekly adsorbing biopolymer or a nonionic surfactant. *J Colloid Interface Sci* 247: 167.

Chiang C, Fuller GC, Frankenfeld JW, Rhodes CT. 1978. Potential pf liquid membranes for drug overdosage treatment: In vitro studies. *J Pharm Sci* 67: 63–66.

Chilamkurti RN, Rhodes CT. 1990. Transport across liquid membranes: effect of molecular structure. *J Appl Biochem* 2: 17–24.

Collings AJ. 1971. Improvements in or relating to sustained release preparations, British Patent 1,235,667.

Criddle DW, Meager Jr AL. 1955. Viscosity and elasticity of oil surface and oil-water interfaces. *J Appl Phys* 26: 838–842.

Csoka I, Eros I. 1997. Stability of multiple emulsions I. Determination of factors influencing multiple drop breakdown. *Int J Pharm* 156: 119–123.

Cueman GH, Zatz JL. 1988. Multiple emulsions. In: *Controlled Release Systems: Fabrication Technology*, Vol. 2, Hsieh DS, ed. Boca Raton: CRC Press, pp 23–40.

Dalgleish D. 1993. Food emulsions. In: *Emulsions and Emulsion Stability*, Sjoblom J, eds. New York: Marcel Dekker, pp 287–325.

Davies CK, Nash P, Stevens RN. 1980. *Acta Metall* 28: 179.

Davis SS. 1976. The emulsion-obsolete dosage form or novel drug delivery system and therapeutic agent? *J Clin Pharm* 1: 11–27.

Davis SS, Burbage AS. 1977. Electron micrography of water-in-oil-in-water emulsions. *J Colloid Interface Sci* 62(2): 361–363.

Davis SS, Burbage AS. 1978. The particle size analysis of multiple emulsions (water-in-oil-in-water). In: *Particle Size Analysis*, Groves MJ, eds. London: Heyden, pp 395–410.

Davis SS, Round HP, Purewal TS. 1981. Ostwald ripening and the stability of emulsion systems: An explanation for the effect of an added third component. *J Colloid Interface Sci* 80: 508.

Davis SS, Walker I. 1983. Measurement of the yield of multiple droplets by a fluorescent tracer technique. *Int J Pharm* 17: 203–213.

De Luca M, Grossiord JL, Medard JM, Vaution C. 1990. A stable W/O/W multiple emulsion. *Cosmet Toilet* 105(11): 65–69.

Dickinson E, Owusu RK, Tan S, Williams A. 1993. Oil-soluble surfactants have little effect on competitive adsorption of a-lactalbulmin and ß-lactoglobulin in emulsions. *J Food Sci* 58(2): 295–298.

Dickinson E, Evison J, Owusu RK. 1991. Preparation of fine protein-stabilized water-in-oil-water emulsions. *Food Hydrocolloids* 5: 481–485.

Dickinson E, Hong ST. 1994. Surface coverage of ß-lactoglobulin at the oil-water interfaces: Influence of protein heat treatment and various emulsifiers. *J Agric Food Chem* 42: 1602–1606.

Ducoulombier D, Zhou H, Boned C, Peyreleasse J, Saint-Guirons H, Xans P. 1986. Condensed phases and macromolecules—pressure (1–1000 bars) and temperature (20–100°C) dependence of the viscosity of liquid hydrocarbons. *J Phys Chem* 90: 1692.

Economou IG, Heidman JL, Tsonopoulos C. 1997. Mutual solubilities of hydrocarbons and water: III. 1-hexene, 1-octene, C10-C12 hydrocarbons. *AIChE J* 43: 535.

Elbary AA, Nour SA, Ibrahim I. 1990. Physical stability and rheological properties of W/O/W emulsions as a function of electrolytes. *Pharm Indus* 52(3): 357–363.

Elson LA, Mitchley BCV, Collings AJ, Schneider R. 1970. Chemotherapeutic effect of a water-in-oil-in-water of methotrexate on the mouse L1210 leukaemia, *Rev Europ Etudes Clin Biol* 15: 87–90.

Engel RH, Riggi SJ, Fahrenbach MJ. 1968. Insulin: Interstinal absorption as water-in-oil-in-water emulsions. *Nature* 129: 856–857.

Eros I, Balazs J, Peter I, Tasci M. 1990. Investigation of drug containing multiphase emulsions. *Pharmazie* 45: 419–422.

Estrada-Baltazar A, Alvarado JF, Iglesias-Silva G. 1998. Experimental liquid viscosity of decane and octane + decane from 298.15 K to 373.15 K and up to 25 Mpa. *J Chem Eng Data* 43: 441.

Ferreira LAM, Seiller M, Grossiord JL, Marty JP, Wepierre J. 1994. Vehicle influence on in vitro release of metronidazole: Role of W/O/W multiple emulsion. *Int J Pharm* 109: 251–259.

Ferreira LAM, Doucet J, Seiller M, Grossiord JL, Marty JP. 1995. In vitro percutaneous absorption of metronidazole and glucose: comparison of O/W, W/O/W and W/O systems. *Int J Pharm* 121: 169–179.

Ficheux MF, Bonakder L, Leal-Calderon F, Bibette J. 1998. Some stability criteria for double emulsions. *Langmuir* 14: 2702–2706.

Florence AT, Whitehill D. 1985. Stability and stabilization of water-in-oil-in-water multiple emulsions. *ACS Symp Ser* 272 (*Macro- and Microemulsions*): 359–380.

Florence AT, Law TK, Whateley TL. 1985. Nonaqueous foam structures form osmotically swollen W/O/W emulsion droplets. *J Colloid Interface Sci* 107(2): 584–588.

Florence AT, Rogers JA. 1971. Emulsion stabilization by nonionic surfactants: Experiment and theory. *J Pharm Pharmacol* 23: 153–169.

Florence AT, Whitehill D. 1982. Stabilization of water/oil/water multiple emulsions by polymerization of the aqueous phase. *J Pharm Pharmacol* 34: 687–691.

Florence AT, Whitehill D. 1982. The formulation and stability of multiple emulsions. *Int J Pharm* 11: 277–308.

Frankenfeld JW, Fuller GC, Rhodes CT. 1976. Potential use of liquid membranes for emergency treatment of drug overdose. *Drug Dev Commun* 2(4–5): 405–419.

Fukushima S, Juni K, Nakano M. 1983. Preparation of and drug release from W/O/W type double emulsion containing anticancer agents. *Chem Pharm Bull* 31(11): 4048–4056.

Garti N, Aserin A. 1996. Emulsions and microemulsions. In: *Drugs and the Pharmaceutical Sciences*, Vol. 73, Benita S, ed. Dekker, New York, NY.

Garti N, Aserin A, Cohen Y. 1994. Mechanistic considerations on the release of electrolytes form multiple emulsions stabilized by BSA and nonionic surfactants. *J Controlled Release* 29: 41–51.

Geiger S, Jager-Lezer N, Tokgoz S, Seiller M, Grossiord JL. 1999. Characterization of the mechanical properties of a water/oil/water multiple emulsion oily membrane by a micropipette aspiration technique. *Colloids Surf A* 157: 325–332.

Ghaicha L, Leblanc R. 1995. Monolayers of mixed surfactants at the oil-water interface, hydrophobic interactions, and stability of water-in-oil emulsions. *Langmuir* 11: 585–590.

Gresham PA, Barrett M, Smith SV, Schneider R. 1971. Use of a sustained release multiple emulsion to extend the period of radioprotection conferred by cysteamine. *Nature* 234: 149–150.

Griffin WC. 1949. Classification of surface-active agents by HLB. *J Soc Cosmet Chem* 11: 311–326.

Herbert WJ. 1965. Multiple emulsion: A new form of mineral oil antigen adjuvant. *Lancet* 2: 771.

Higuchi WI, Misra J. 1962. Physical degradation of emulsions via the molecular diffusion route and the possible prevention thereof. *J Pharm Sci* 51: 459.

Hou W, Papadopoulos KD. 1997. $W_1/O/W_2$ and $O_1/W/O_2$ globules stabilized with Span 80 and Tween 80. *Colloids Surf A* 125: 181–187.

Hunter RJ. 1986. *Foundations of Colloid Science*, Vol. 2. New York: Oxford University Press.

Jiao J, Rhodes DG, Burgess DJ. 2002. Multiple emulsion stability: Pressure balance and interfacial film strength. *J Colloid Interface Sci* 250: 444–450.

Jiao J, Burgess DJ. 2003. Rheology and stability of W/O/W multiple emulsions containing Span 83 and Tween 80. *AAPS Pharm Sci* 5(1): 62–73.

Jiao J, Burgess DJ. 2003. Ostwald ripening of water-in-hydrocarbon emulsions stabilized by non-ionic surfactant Span 83. *J Colloid Interface Sci* 264: 509–516.

Jiao J, Burgess DJ. 2005. Characterization and analysis of dispersed systems. In: *Drugs and the Pharmaceutical Sciences*, Vol. 149: Injectable Dispersed Systems—Formulation, Processing, and Performance, Burgess DJ, ed. Boca Raton: Taylor and Francis Group, pp 77–124.

Kabalnov AS, Perrzov AV, Shchukin ED. 1987. Ostwald ripening in emulsions 1. Direct observation of Ostwald ripening in emulsions. *J Colloid Interface Sci* 118: 590–597.

Kabalnov AS, Makarov KN, Perrzov AV, Shchukin ED. 1990. Ostwald ripening in emulsions 2. Ostwald ripening in hydrocarbon emulsions: Experimental verification of equation for absolute rates. *J Colloid Interface Sci* 138: 98–104.

Kabalnov AS, Shchukin ED. 1992. Ostwald ripening theory: Applications to fluorocarbon emulsion stability. *Adv Colloid Interface Sci* 38: 69–97.

Kabalnov AS. 1994. Can micelles mediate a mass transfer between oil droplets? *Langmuir* 10: 680.

Kavaliunas DR, Frank SG. 1978. Liquid crystal stabilization of multiple emulsions. *J Colloid Interface Sci* 66: 586–588.

Kawashima Y, Hino T, Takeuchi H, Niwa T. 1992. Stabilization of water/oil/water multiple emulsion with hypertonic inner aqueous phase. *Chem Pharm Bull* 40(5): 1240–1246.

Kita Y, Matsumoto S, Yonezawa D. 1977. Viscometric method for estimating the stability of W/O/W type multiple-phase emulsions. *J Colloid Interface Sci* 62: 87–94.

Kitchener JA, Mussellwhite PR. 1968. *The Theory of Stability of Emulsions*, Sherman P, ed. London: Academic Press, pp 77–125.

Kover T, Csoka I, Eros I. 1997. Formation and stability of multiple phase (W/O/W) emulsions. Part I: Investigation of the primary interface. *Pharmazie* 52: 166–167.

Kover T, Csoka I, Eros I. 1997. Formation and stability of multiple phase (W/O/W) emulsions. Part II: Relationship between primary interface saturation and formation. *Pharmazie* 52(4): 328.

Krog N. 1991. Thermodynamics of interfacial films in food emulsions. In: *Microemulsions and Emulsions in Food*, El-Nokaly M, Cornell D, eds. Washington, DC: ACS, pp 138–145.

Lanier JG, Newman MJ, Lee EM, Sette A, Ahmed R. 1999. Peptide vaccination using nonionic block copolymers induces protective anti-viral CTL responses. *Vaccine* 18(5–6): 549–557.

Laugel C, Chaminade P, Baillet A, Seiller M, Ferrier D. 1996. Moisturizing substances entrapped in W/O/W emulsions: analytical methodology for formulation, stability and release studies. *J Controlled Release* 38: 59–67.

Laugel C, Baillet A, Youenang Piemi MP, Marty JP, Ferrier D. 1998. Oil-water-oil multiple emulsions for prolonged delivery of hydrocortisone after topical application: Comparison with simple emulsions. *Int J Pharm* 160: 106–117.

Law TK, Florence AT, Whateley TL. 1984. Release from multiple W/O/W emulsions stabilized by interfacial complexation. *J Pharm Pharmacol* 36(suppl): 50.

Law TK, Whateley TL, Florence AT. 1986. Stabilization of W/O/W multiple emulsion by interfacial complexation of macromolecules and nonionic surfactants. *J Controlled Release* 3: 279–290.

Li NN, Shrier AL. 1972. Liquid membrane water treating. In: *Recent Development in Separation Science*, Vol.1, Cleveland: Chem. Rubber Co., pp 163–174.

Lifshitz IM, Slyozov VV. 1961. The kinetics of precipitation from supersaturated solid solutions. *J Phys Chem Solids*, 19: 35.

Lin TJ, Kurihara H, Ohta H. 1975. Effects of phase inversion and surfactant location on the formation of O/W emulsions. *J Soc Cosmet Chem* 26: 121–139.

Lin SJ, Wu WH. 1991. Physical parameters and release behaviors of W/O/W multiple emulsions containing cosurfactants and different specific gravity of oils. *Pharm Acta Helv* 66(12): 342–347.

Magdassi S, Frenkel M, Garti N, Kasan R. 1984. Multiple emulsion II: HLB shift caused by emulsifier migration to external interface. *J Colloid Interface Sci* 97(2): 374–379.

Magdassi S, Frenkel M, Garti N. 1984. On the factors affecting the yield of preparation and stability of multiple emulsions. *J Disper Sci Tech* 5(1): 49–59.

Magdassi S, Garti N. 1984. Release of electrolytes in multiple emulsions: Coalescence and breakdown or diffusion through oil phase? *Colloids Surf* 12: 367–373.

Magdassi S, Garti N. 1986. A kinetic model for release of electrolytes from W/O/W multiple emulsions. *J Controlled Release* 3: 273–277.

Martin A, Swarbrick J, Cammarata N. 1993. *Physical Pharmacy, Fourth Edition*, Philadelphia, Pennsylvania: Lee & Febiger, pp 371–373 and 487–490.

Matsumoto S, Kita Y, Yonezawa D. 1976. An attempt at preparing water-in-oil-in-water multiple phase emulsions. *J Colloid Interface Sci* 57(2): 353–361.

Matsumoto S, Kohda M, Murata S. 1977. Preparation of lipid vesicles on the basis of a technique for providing W/O/W emulsions. *J Colloid Interface Sci* 62(1): 149–157.

Matsumoto S. 1983. Development of W/O/W type dispersion during phase inversion of concentrated W/O emulsions. *J Colloid Interface Sci* 94(2): 362–368.

Matsumoto S, Kitayama T, Kohda M. 1985. Some trials in stabilizing W/O/W emulsions under the presence of electrolytes. *J Jap Oil Chem Soc* 34: 688–695.

Matsumoto S, Kohda M. 1980. The viscosity of W/O/W emulsions: An attempt to estimate the water permeation coefficient of the oil layer from the viscosity changes in diluted system on aging under osmotic pressure gradient. *J Colloid Interface Sci* 73(1): 13–20.

Matsumoto S. 1987. W/O/W type multiple emulsions. In: *Nonionic surfactant physical chemistry*, Schick MJ, eds. New York: Marcel Dekker, pp 549–600.

May SW, Li NN. 1974. Encapsulation of enzymes in liquid membrane emulsions. *Enzyme Eng* 2: 77–82.

Mikhin KV, Stepanow SV, Byakov VM. 2003. Formation of Ps bubble in liquid media. *Radiat Phys Chem* 68: 415–417.

Mohamed SM, Ghazy FS, Mahdy MA, Gad MA. 1989. Factors affecting the yield of multiple emulsion and drug release. Mans *J Pharm Sci* 5: 56–71.

Morimoto Y, Sugibayashi K, Yamagauchi Y, Kato Y. 1979. Detoxication capacity of a multiple (W/O/W) emulsion for the treatment of drug overdose: drug extraction into the emulsion in the gastro-intestinal tract of rabbits. *Chem Pharm Bull* 27(12): 3188–3192.

Morimoto Y, Yamagauchi Y, Sugibayashi K. 1979. Detoxication capacity of a multiple (W/O/W) emulsion for the treatment of drug overdose. II. Detoxication of quinine sulfate with the emulsion in the gastro-intestinal tract of rabbits. *Chem Pharm Bull* 27(12): 3188–3192.

Murray BS, Dickinson E. 1996. Interfacial rheology and the dynamic properties of adsorbed films of food proteins and surfactants. *Food Sci Technol Int* 2(3): 131–145.

Myers D. 1988. *Surfactant science and technology*. New York, NY: VCH Publishers, pp 209–253.

Myers SL, Shively ML. 1992. Preparation and characterization of emulsifiable glasses: Oil-in-water and water-in-oil-in-water emulsions. *J Colloid Interface Sci* 149(1): 271–278.

Nakhare S, Vyas SP. 1996. Preparation and characterization of multiple emulsion based systems for controlled diclofenac sodium release. *J Microencapsu* 13(3): 281–292.

Nakhare S, Vyas SP. 1995. Prolonged release of rifampicin from multiple W/O/W emulsion systems. *J Microencapsu* 12(4): 409–415.

Ng SM, Frank SG. 1982. Formation of multiple emulsions in a four-component system containing nonionic surfactants. *J Disper Sci Tech* 3: 217.

Ohwaki T. 1992. Improvement of the formation percentage of water-in-oil-in-water multiple emulsions by the addition of surfactants in the internal aqueous phase of the emulsion. *Int J Pharm* 85: 19–28.

Okada H. 1997. One-and three-month release injectable microspheres of the LH-RH superagonist leuprorelin acetate. *Adv Drug Deliv Rev* 28(1): 43–70.

Omotosho JA, Whateley TL, Law TK, Florence AT. 1986. The nature of the oil phase and the release of solutes from multiple (W/O/W) emulsions. *J Pharm Pharmacol* 38: 865–870.

Omotosho JA, Law TK, Whateley TL, Florence AT. 1986. The stabilization of W/O/W emulsions by interfacial interaction between albumin and non-ionic surfactants. *Colloids Surf* 20: 133–144.

Omotosho JA, Florence AT, Whateley TL. 1990. Absorption and lymphatic uptake of 5-fluorouracil in the rate following oral administration of W/O/W multiple emulsions. *Int J Pharm* 61: 51–56.

Omotosho JA. 1990. The effect of acacia, gelatin and polyvinylpyrrolidone on chloroquine transport from multiple W/O/W emulsions. *Int J Pharm* 62: 81–84.

Omotosho JA, Whateley TL, Law TK, Florence AT. 1989. Release of 5-fluorouracil from intramuscular W/O/W multiple emulsions. *Biopharm Drug Disp* 10: 257–268.

Opawale FO, Burgess DJ. 1998. Influence of interfacial rheological properties of mixed emulsifier films on the stability of water-in-oil-in-water emulsions. *J Pharm Pharmacol* 50: 965–973.

Opawale FO, Burgess DJ. 1998. Influence of interfacial properties of lipophilic surfactant on water-in-oil emulsion stability. *J Colloid Interface Sci* 197: 142–150.

Ostwald W. 1900. Uber die vermeintliche isomerie des rotten und gelben quecksilberoxyds und die oberflachenspannung fester korper. *Z Phys Chem* 34: 495.

Oza KP, Frank SG. 1989. Multiple emulsions stabilized by colloidal microcrystalline cellulose. *J Disper Sci and Tech* 10(2): 163–185.

Padday JF, Russel DR. 1960. The measurement of the surface tension of pure liquids and solutions. *J Colloid Sci* 15: 503–511.

Pal R. 1996. Multiple O/W/O emulsion rheology. *Langmuir* 12: 2220.

Panchal CJ, Zalic JE, Gerson DF. 1979. Multiple-phase emulsions using microbial emulsifiers. *J Colloid Interface Sci* 68: 295–307.

Pandit JK, Mishra B, Chand B. 1987. Drug release from multiple W/O/W emulsions. *Indian J Pharm Sci* 103–105.

Parkinson C, Matsumoto S, Sherman P. 1970. The influence of particle size distribution on the apparent viscosity of non-Newtonian dispersed systems. *J Colloid Interface Sci* 33(1): 150–160.

Raynal S, Grossiord JL, Seiller M, Clausse D. 1993. A topical W/O/W multiple emulsion containing several active substances: Formulation, characterization and study of release. *J Controlled Release* 26: 129–140.

Rosano HL, Gandolfo FG, Hidrot JP. 1998. Stability of $W_1/O/W_2$ multiple emulsions: Influence of ripening and interfacial interactions. *Colloids Surf A* 138: 109–121.

Seifriz W. 1925. Studies in emulsions. *J Phys Chem* 29: 738.

Sela Y, Magdassi S, Garti N. 1995. Release of markers from the inner water phase of W/O/W emulsions stabilized by silicone based polymeric surfactants. *J Controlled Release* 33: 1–12.

Sheriff M, Warburton B. 1974. Measurement of dynamic rheological properties using the principle of externally shifted and restored resonance. *Polymer* 15: 253–254.

Sheriff M, Warburton B. 1975. The theory of a universal oscillatory rheometer for the study of linear viscoelastic materials using the principle of normalized resonance. In: *Theoretical Rheology*. London: Applied Science Publishers, pp 299–316.

Shinoda K, Yoneyama T, Tsutsumi H. 1980. Evaluation of emulsifier blending. *J Disper Sci Technol* 1(1): 1–12.

Sherman P. 1953. Studies in water-in-oil emulsions. III. The properties of interfacial films of sorbitan sesquioleate. *J Colloid Sci* 8: 35–37.

Shimizu M, Nakane Y. 1995. Encapsulation of biologically active proteins in a multiple emulsion. *Biosci Biochem* 59(3): 492–496.

Silva-Cunha A, Cheron M, Grossiord JL, Puisieux F, Seiller M. 1998. W/O/W multiple emulsions of insulin containing a protease inhibitor and an absorption enhancer: Biological activity after oral administration to normal and diabetic rats. *Int J Pharm* 169: 33–44.

Singh S, Singh R, Vyas SP. 1995. Multiple emulsion based systems carrying insulin: Development and characterization. *J Microencapsul* 12(6): 609–615.

Smith A, Davis SS. 1973. The role of molecular diffusion in the bulk stability of O/W hydrocarbon emulsions. *J Pharm Sci* 25: 117.

Soma J, Papadopoulos KD. 1996. Ostwald ripening in sodium dodecyl sulfate-stabilized decane-in-water emulsions. *J Colloid Interface Sci* 181: 225.

Swarbrick J. 1997. Coarse dispersions. In: *Remington: The Science and Practice of Pharmacy*, 20th ed. Gennaro AR, ed. Mack Publishing Company, Easton, PA.

Taylor P, Ottewill RH. 1994. The formation and aging rates of oil-in-water miniemulsions. *Colloids Surf A* 88: 303.

Taylor P. 1995. Ostwald ripening in emulsions. *Colloids Surf A* 99: 175.

Tokgoz NS, Grossiord JL, Fructus A, Seiller M, Prognon P. 1996. Evaluation of two fluorescent probes for the characterization of W/O/W emulsions. *Int J Pharm* 141: 27–37.

Trevino L, Sole-Violan L, Daumur P, Devallez B, Postel M, Riess JG. 1993. Molecular diffusion in concentrated fluorocarbon emulsions and its effect on emulsion stability. *New J Chem* 17: 275.

Tsonopoulos C. 1999. Thermodynamic analysis of the mutual solubilities of normal alkanes and water. *Fluid Phase Equilib* 156: 21.

Tsonopoulos C. 2001. Thermodynamic analysis of the mutual solubilities of hydrocarbons and water. *Fluid Phase Equilib* 186: 185.

Ugelstad J, Mork PC, Kaggerud KH, Ellingsen T, Berge A. 1980. Swelling of oligomer-polymer particles: New methods of preparation of emulsion and polymer dispersions. *Adv Colloid Interface Sci* 13: 101.

Vaziri A, Warburton B. 1994. Slow release of chloroquine phosphate from multiple taste-masked W/O/W multiple emulsions. *J Microencapsul* 11(6): 641–648.

Walstra P. 1996. Emulsion stability. In: *Encyclopedia of Emulsion Technology*, Vol. 4, Becher P, ed. New York: Marcel Dekker, pp 1–62.

Warburton B. 1993. In: *Techniques in rheological measurement*, Collyer AA, (Ed). London: Chapman and Hall, p 55.

Welin-Berger K, Bergenstahl B. 2000. Inhibition of Ostwald ripening in local anesthetic emulsions by using hydrophobic excipients in the disperse phase. *Int J Pharm* 200: 249–260.

Weiss J, Herrmann N, McClements DJ. 1999. Ostwald ripening of hydrocarbon emulsion droplets in surfactant solutions. *Langmuir* 15: 6652.

Weiss J, Cancelliere C, McClements DJ. 2000. Mass transport phenomena in oil-in-water emulsions containing surfactant micelles: Ostwald ripening. *Langmuir* 16: 6833.

Wilke CR, Chang D. 1955. Correlation of diffusion coefficients in dilute solutions. *Am Inst Chem Eng J* 1: 264.

Yoshioka T, Ikeuchi K, Hashida M, Muranishi S, Sezaki H. 1982. Prolonged release of bleomycin from parenteral gelatin sphere-in-oil-in-water multiple emulsion. *Chem Pharm Bull* 30(4): 1408–1415.

Yoshioka T, Sternberg B, Florence AT. 1994. Preparation and properties of vesicle (niosomes) of sorbitan monoesters (Span 20, 40, 60, 80) and a sorbitan triester (Span 85). *Int J Pharm* 105: 1–6.

Zheng S, Zheng Y, Beissinger RL, Wasan DT, McCormick DL. 1993. Hemoglobin multiple emulsion as an oxygen delivery system. *Biochimica et Biophysica Acta* 1158: 65–74.

CHAPTER 2

Structure and Rheology of Stable Multiple Emulsions

P. PERRIN, F. PRIGENT and P. HÉBRAUD

Contents

2.1 INTRODUCTION

Emulsions are dispersions of two immiscible fluids, such as water and oil. Simple emulsions are of two different kinds: direct emulsions (O/W) are dispersions of oil into water, whereas inverse emulsions (W/O) are dispersions of water into an oil continuous phase. Multiple emulsions of water in oil in water ($W_1/O/W_2$) are direct emulsions (oil in W_2) where the dispersed oil phase is replaced by an inverse emulsion (W_1/O) (Garti, 1997). One thus obtains a dispersion of water droplets inside oil globules.

Amphiphilic molecules adsorbed at oil water interfaces are used to improve emulsion metastability. Several empirical rules—Bancroft rule (Bancroft,

Multiple Emulsions: Technology and Applications, Edited by Abraham Aserin

1913), hydrophilic lipophilic balance (HLB) rule (Tadros et al., 1997; Grossiord, 1998), for instance—aim at predicting the nature (direct or inverse) and stability of an emulsion from the thermodynamic equilibrium properties of solutions of surfactants used to stabilize their interfaces. Thus, according to Bancroft's rule, stabilization of direct and inverse emulsions requires the use of a surfactant with high and low HLB, respectively. Such systems are characterized by the existence of two opposite interfaces, of different radii of curvature. These interfaces are not thermodynamically stable. Control of the metastability of multiple emulsions is a key parameter in their applications.

In general, two surfactants are used, one of low HLB, and the other one of high HLB, to stabilize the internal (W_1/O) and external (W_2/O) interfaces, respectively (Florence et al., 1981). The presence of these two different surfactants is a major source of instability of multiple emulsions (Ficheux et al., 1998; Wen and Papadopoulos, 2000; Pays et al., 2001). Indeed the lifetime of the emulsions is considerably shortened by the rapid diffusion of the more water-soluble small-molecule surfactants toward the droplets interface. Among other mechanisms, the formation of inverse micelles allows the diffusion of molecules through the oil phase (Pays et al., 2000). Nevertheless, a general understanding of emulsion metastability is still lacking. It requires precise understanding of the droplets properties, such as their dynamics and their interaction with globules' interfaces. Such a study is inherently difficult, due to the finite lifetime of multiple emulsions.

In a first step toward a more general understanding of multiple emulsions stability and dynamics, in this chapter we study extremely stable multiple emulsions. We show how to conveniently and precisely probe the droplets' properties. Additionally we describe the structural and rheological properties of such stable multiple emulsions.

To prepare stable multiple emulsions that can be studied over a long time, we use amphiphilic copolymers instead of small molecular surfactants as the globule interface stabilizers (Sela et al., 1994; Garti, 1998; Benichou et al. 2004; Michaut et al. 2003, 2004a). However, the interface of inverse droplets is stabilized either with a small tensioactive molecule or with a polymeric surfactant.

We are interested in exploring two questions:

- What are the interactions between the inner surfaces of large globules and the outer surface of small inverse droplets,
- What is the composition of both interfaces? Do the surfactants migrate from one to the other (Pays et al., 2000)?

We can answer the first question by direct observation of the inner droplets using confocal microscopy. The dynamics of the inner droplets can be observed by a light diffusion technique suited to turbid media, diffusing wave spectroscopy (DWS). Both observations show that attractive interactions—either depletion or van der Waals interactions—occur between the droplets and the inner surfaces of the globules, and lead to an increase of the local concentration of droplets close to the globules' interface. The consequence is a slowdown of droplets' dynamics, as can be readily observed.

To answer the second question, we use a rheological means of observation. We prepare concentrated multiple emulsions, in which both globules and droplets are more concentrated than a random close-packing concentration. The rheological properties of the system are controlled by both interfaces, as it is well known in the case of simple emulsions (Princen and Kiss, 1986). Nevertheless, the quantity of interfaces of small droplets is much higher than that of large globules, and rheological properties are dominated by the interfacial properties of small droplets. Thus rheology is a unique in situ probe of the interfacial properties and likewise, as we will see, of the chemical composition of inner droplets. In our case, diffusion of the polymeric surfactant from the globules interface toward the droplets surface does not occur, so the stability of the multiple emulsions is enhanced.

2.2 PREPARATION OF MULTIPLE EMULSIONS

The amphiphilic polyelectrolyte used to stabilize the globules interfaces is a hydrophobically modified poly(sodium acrylate) having the chemical structure shown in Figure 2.1*a*. The molecular weight of the polymer is $50{,}000\,\mathrm{g \cdot mol^{-1}}$. The hydrophobic side chains are randomly distributed along the negatively charged backbone. Details of the synthesis have been reported in Wang et al. (1988). This polymer will be referred to as 10*Cn*, meaning 10% mol of *n*-alkyl chains onto the poly (sodium acrylate) backbone. The length of grafted alkyl chains is varied between 8 and 16. In this range the copolymer is soluble in water and exhibits surfactant properties. The inner droplets are stabilized either by the commercial sorbitan monooleate (Span 80) or by a polymeric surfactant, Admul Wol 1407 (Quest International); see Figure 2.1*b*.

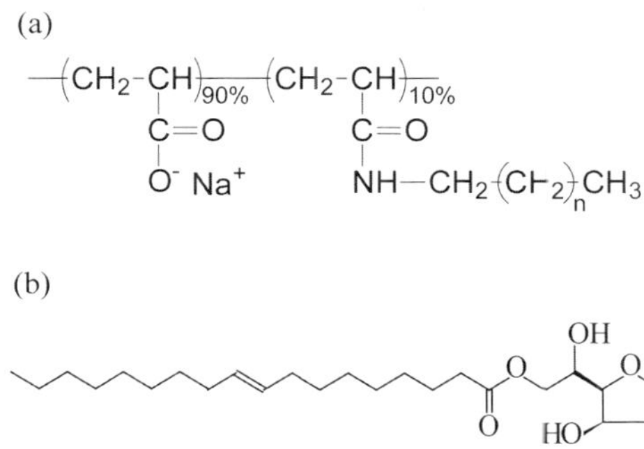

Figure 2.1 Chemical structure of (*a*) the hydrophobically modified poly(sodium acrylate) and of the sorbitan monooleate, Span 80, used as a stabilizer of (*b*) the inverse droplets.

Multiple emulsions of $W_1/O/W_2$ are prepared following a two-step emulsification process that allows the control of the concentration of both the primary (W_1/O) and the multiple emulsions. Concentrated emulsions are obtained by swelling the dispersed water phase. In the following discussion, ϕ_d refers to the volume fraction of the droplets relative to the oil phase, ϕ_g is the volume fraction of the globule phase, and ϕ that of the droplet phase, relative to the total sample volume. We thus have the relationship $\phi = \phi_g \phi_d$.

In the first step of sample preparation, water is dispersed in a 15% (w/v) Span 80/dodecane solution using a rotor-stator homogenizer (Heidolph DIAX 900) at 26,000 rpm for 5 minutes. An inverse emulsion with a dispersed phase volume fraction ϕ_d up to 68% is obtained. The second step consists in the dispersion of the inverse emulsion in an aqueous phase containing 1% (w/v) of 10*Cn*, using the same homogenizer at 8000 rpm for 10 seconds. The volume fraction of the direct emulsion, ϕ_g may be varied from 6% to 25%. Indeed we observed that following this two-step procedure, multiple emulsions with ϕ_g higher than 25% are destroyed during the emulsification process. Nevertheless, as we added NaCl in the aqueous phase, in the first preparation step, an osmotic stress was created between the dispersed and the continuous aqueous phases. Thus, once the multiple emulsion was prepared, the inverse droplets swelled to reach a pseudoequilibrium. The concentrated multiple emulsions then showed total droplet volume fractions ranging from $\phi = 60\%$ to 80%, and globule volume fraction ϕ_g from 70% to 90%. By this protocol, and by varying the amount of salt introduced, we could prepare multiple emulsions at any concentration, below or above random close packing, of droplets, globules, or both.

2.3 STRUCTURE OF STABLE MULTIPLE EMULSIONS

2.3.1 Static Structure: Direct Microscopic Observations

To begin, let us consider the spatial distribution of the inner droplets inside globules. As we did in the first step, we can doserve this distribution using confocal microscopy. Confocal microscopy allows us to image slightly turbid systems with a very good axial resolution (Pawley, 1995). It has been used to observe the structure of simple emulsions by rendering the interfaces fluorescent (Brujic et al. 2003; Zhou et al., 2006). In the subsequent experiments we will make droplets themselves visible by dissolving a fluorescent probe in the droplets. In order to avoid migration of this marker from the droplets toward the outer continuous aqueous phase, we have to use a neutral polysaccharide (dextran), of molecular weight 10,000, grafted with rhodamine B. This fluorescent probe is inserted inside the inner droplets with a concentration of $1\,\mathrm{mg \cdot ml^{-1}}$. With confocal microscopy, a succession of slices is obtained inside the sample. The radius of the globules is of the order of 15 μm, which is much larger than the optical thickness of the slices (of the order of 500 nm). We can then prepare multiple emulsions in which the globules' interface is stabilized

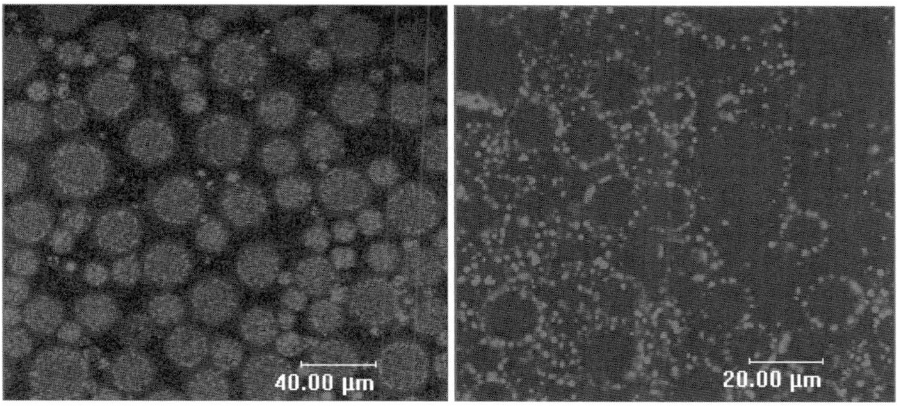

(a) (b)

Figure 2.2 Confocal slices of two multiple emulsions. (*a*) The globules' interfaces are stabilized by a modified polyacrylic acid, 10*C*12, and the inner droplets' interfaces are stabilized by a polymeric surfactant, Admul. (*b*) The globules' interfaces are also stabilized by the grafted polyacrylic acid, 10*C*12, whereas the inner droplets' interfaces are stabilized by a small surfactant molecule, Span 80.

by a grafted polyacrylic acid polymer, 10*C*12. In the first system, the inner droplets are stabilized by a small surfactant (Span 80), and in the second system, by a surfactant polymer (Admul). As shown in Figure 2.2, the inner droplets distributions are very different:

- When the inner droplets are stabilized by a small surfactant molecule, the droplets are adsorbed against the inner interface of the globules.
- In the case of a polymeric stabilization, the volume distribution of the droplets inside the globules is homogeneous.

The observations in Figure 2.2 can be made quantitatively by averaging the droplets' density as a function of their distance from the center of a globule, over the large number of globules (Figure 2.3). It clearly appears that, in the case of Span 80 stabilized molecules, droplets concentration exhibit a marked peak close to the inner globule interfaces. This attractive interaction between the globules interface and the inverse droplets may be attributed to depletion interactions of inverse micelles of Span 80 in the oil phase.

2.3.2 Dynamical Properties: DWS Probe of the Dynamics of Inner Droplets

Diffusing wave spectroscopy (DWS) is a light-scattering technique that allows the measurement of the dynamics of turbid systems (Weitz and Pine, 1993; Pine et al., 1990). A thick sample of multiple emulsions is illuminated with an enlarged laser beam. This sample diffuses light, and both backscattered and

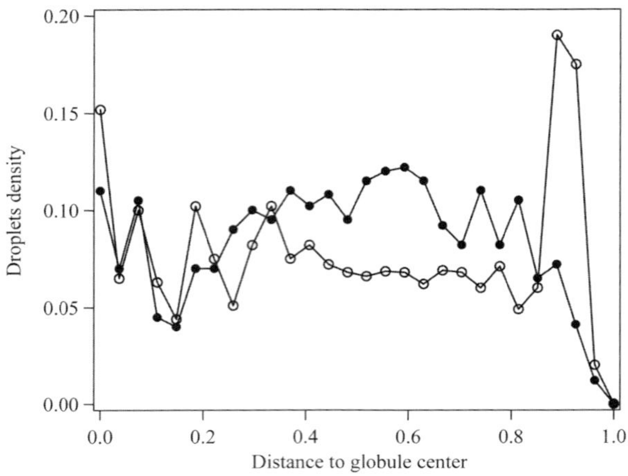

Figure 2.3 Radial distribution of the inverse droplets inside the globules, deduced from averaging over 100 globule slices observed by confocal microscopy. *Hollow circles*: The droplets are stabilized with Span 80. *Filled circles*: The droplets are stabilized with Admul. In both cases the globules are stabilized with the modified polyacrylic acid 10*C*12. The distance to a globule's center is given in a globule's radius unit.

transmitted light are collected through monomode fibers. Each signal is amplified by two photomultipliers to get rid of the high-frequency noise from each photomultiplier. Then the cross-correlation of these two amplified signals is computed with the real-time correlator, which obtains

$$g_2(t) = \frac{\langle I(\tau)I(\tau+t)\rangle_\tau}{\langle I(\tau)\rangle_\tau^2}, \tag{2.1}$$

the correlation function of the diffuse light. $g_2(t)$ is a function of the dynamics of the diffusing particles and of the photon transport mean free path (l^*) in the sample.

We independently measured the transmittance of the sample that is directly related to the photon transport mean free path when the diffusing particles are not interacting. From the field autocorrelation function, $g_1(t) = \sqrt{g_2(t) - 1}$, we were able to deduce the mean square displacement $\Delta r^2(t)$ of the diffusing particles. In the case of backscattering geometry, the field autocorrelation function is linked to the mean square displacement of the particles through

$$g_1(t) = \frac{\frac{L}{l^*} + \frac{4}{3}\sinh\left[\sqrt{\langle \Delta r^2(t)\rangle}\left(\frac{L}{l^*} - 1\right)\right] + \frac{2}{3}\sqrt{\langle \Delta r^2(t)\rangle}\cosh\left[\sqrt{\langle \Delta r^2(t)\rangle}\left(\frac{L}{l^*} - 1\right)\right]}{\frac{L}{l^*} - \frac{1}{3}\left(1 + \frac{8t}{3\tau_B}\right)\sinh\left[\frac{L}{l^*}\sqrt{\langle \Delta r^2(t)\rangle}\right] + \frac{4}{3}\sqrt{\langle \Delta r^2(t)\rangle}\cosh\left[\frac{L}{l^*}\sqrt{\langle \Delta r^2(t)\rangle}\right]} \tag{2.2}$$

Before measuring the droplet dynamics, the multiple emulsions were centrifuged to immobilize the globules. Whereas the concentration of the inner droplets relative to the globules' volume was 20%, the globules' volume fraction was much higher than the random close packing, such that their motion was extremely slow. More precisely, we checked that diffuse light was completely decorrelated by the rapid motion of small droplets, and that we were not sensitive to the slow motion of large globules. Under these experimental conditions, the DWS technique is only sensitive to the motion of the droplets. As we noted above, as viewed by direct confocal microscopy, they are located in the vicinity of the globules' interfaces. We next probe the dynamics of the inner droplets close to immobilized globules interfaces, and compare this dynamics to that of an inverse *O/W* emulsion.

In the inverse emulsion, the mean square displacement of the droplets is proportional to time ($\langle \Delta r^2(t) \rangle = 6Dt$, where D is the diffusion coefficient) over the entire time scale, which conforms with Brownian motion. In multiple emulsions, the dynamics of the droplets inside the globule are *sub*diffusive ($\langle \Delta r^2(t) \rangle \propto t^{\alpha}$ with $\alpha < 1$) over almost three decades of time (from $t < 1$ ms to 0.1 ms) before returning to the Brownian motion at larger time scales with a smaller diffusion coefficient than that of the inverse emulsion (Figure 2.4).

By changing the length of grafted alkyl chains of amphiphilic polymeric surfactants, we could fine-tune the interactions between the globules interface and the droplets. We thus varied the number of carbon of the grafted chain between 8 and 16, in a range in which the polymer exhibit surfactant properties but remains soluble in the aqueous phase. We could observe that the shape of the diffusing curves $\langle \Delta r^2(t) \rangle$ remains the same whatever the alkyl chain's length (Figure 2.5). With either polymer stabilizing the globule, a subdiffusive motion

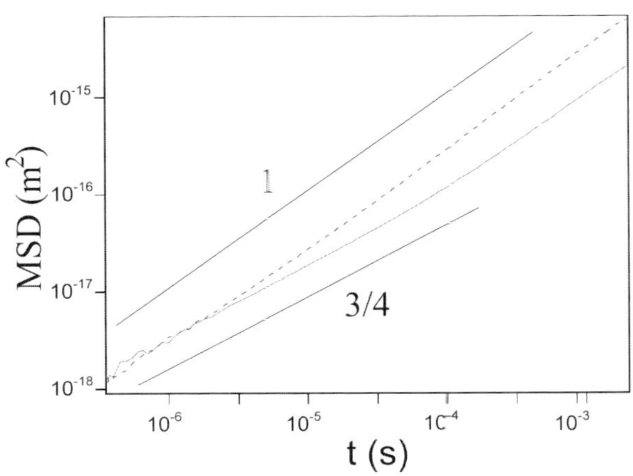

Figure 2.4 Mean square displacement of inverse droplets: In an inverse emulsion (*dashed line*); in globules of a multiple emulsion (*continuous line*). Straight lines, of slopes 1 and 3/4, are guides for the eye.

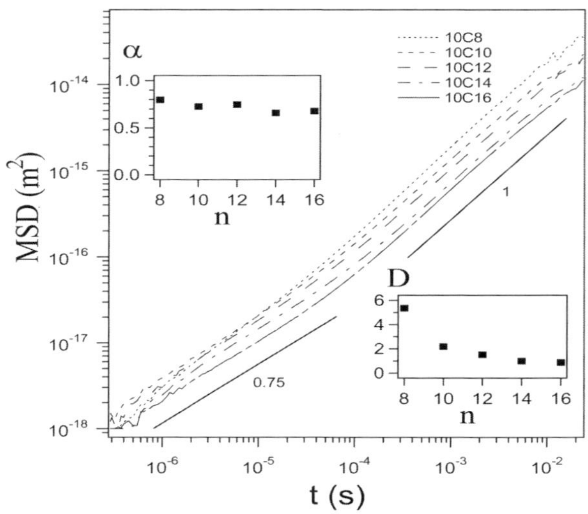

Figure 2.5 Mean square displacement of inverse droplets stabilized by a modified polyacrylic acid with grafted alkyl chains of lengths varying from $n = 8$ to $n = 16$. The continuous lines, of slopes 1 and 3/4, are guides for the eye.

was observed at short times, followed by a Brownian dynamics at longer times. As the pendant chain length increased, the diffusion coefficient of the Brownian dynamics decreased (Figure 2.5, bottom right insert). The droplets are submitted to a greater friction coefficient when longer chains cover the interface. In the subdiffusive regime the mean square displacement $\langle \Delta r^2(t) \rangle$ is always written $\langle \Delta r^2(t) \rangle \propto t^{\alpha}$, where $\alpha < 1$. For the different polymers studied, the value of α (Figure 2.5, top left insert) was always 0.75, which shows that changing the polymer hydrophobic length (and hence the nature of the interactions) does not qualitatively affect the dynamics in the subdiffusive regime.

From these structural studies we can conclude that when a small molecular surfactant is used to stabilize inverse droplets' inside the globule phase, the inverse droplets are submitted to attractive interaction toward the droplets' inner interface, and slowly diffuse along the surface of the globules. A single droplet thus spends an increased amount of time close to the globules' interface. This phenomenon appears to decrease the stability of multiple emulsions.

2.4 RHEOLOGY OF CONCENTRATED MULTIPLE EMULSIONS

At concentrations lower than random close packing of hard spheres (63%), simple emulsions are viscous liquids. Multiple emulsions behave in the same way. Nevertheless, it has been observed that attractive interactions between the globules lead to a shear-thinning behavior at volume fractions close to (but lower than) the random close-packing concentration (Pal, 1996). Then, at volume fractions larger than the random close-packing concentration, emulsions become mainly elastic systems. In the following, we analyze the origin of this elastic behavior in multiple emulsions.

2.4.1 Concentrated Inverse Emulsions

Before discussing the rheological properties of concentrated multiple emulsions, let us recall some fundamentals. Simple emulsions exhibit a transition from a viscous fluid to an elastic solid at dispersed phase volume fractions close to that of the random close-packing volume fraction, ϕ_c. At droplet concentrations lower than ϕ_c, there is almost no effect of the interfacial films on the rheological properties of the fluid, which is essentially viscous. However, at concentrations higher than ϕ_c, emulsions become elastic. The energy required to deform the system is stored within the films separating the close-packed droplets (Princen, 1983; Princen and Kiss, 1986). Thus the elasticity scales are measured as γ/R_d, where γ is the interfacial tension and R_d the droplet radius. Mason et al. (1995, 1998) empirically determined the following dependence of the elastic modulus, G', on the droplet volume fraction, ϕ:

$$G' = \alpha \frac{\gamma}{R_d} \phi(\phi - \phi_c), \tag{2.3}$$

where ϕ_c is close to the random close-packing volume fraction and α is an empirical coefficient.

We will consider first the dynamical rheological properties of concentrated inverse emulsions (stabilized by Span 80 in dodecane) for various volume fractions ϕ. The experimental results are given in the form of the plot $G'R_d/\phi$ in relation to ϕ (Figure 2.8, hollow symbols). From the value of the interfacial tension, $\gamma = 3.5\,\mathrm{mN \cdot m^{-1}}$, measured by tensiometry, and using 3, we calculate values for α and ϕ_c equal to 1 and 0.66, respectively. We will therefore use this measured value of α in the subsequent analysis of concentrated multiple emulsions.

2.4.2 Concentrated Multiple Emulsions

Once highly swollen multiple emulsions have reached pseudo equilibrium between salted droplets and the continuous aqueous phase is achieved, as detailed in Section 2.1, the amount of work W, needed to strain the multiple emulsion is the amount of work needed to increase the surface of both droplets and globules (Michaut et al., 2004b). The elastic modulus as a function of frequency is displayed in Figure 2.6.

The work needed to strain a droplet of surface tension γ with a shear amplitude ε is $W \approx \varepsilon \Pi V \approx \varepsilon \gamma R^2$, where Π is the Laplace pressure of the droplet, V

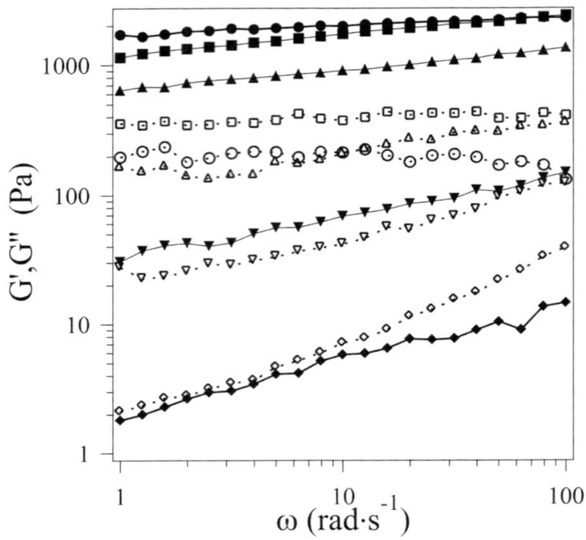

Figure 2.6 Variation of the elastic plateau G' (*filled symbols*) and loss G'' (*open symbols*) moduli as a function of frequency, measured at a shear rate of 0.2% for multiple emulsions stabilized by Span 80 (15% w/v in dodecane) and 10*C*12 (1% w/v in the aqueous phase). The volume fraction of the globules ϕ_g is 69% (♦), 72% (▼), 78% (▲), 82% (■), and 90% (●). In that state, both the droplets and globules volume fractions are larger than ϕ_c.

is its volume, and R is its radius. Then the work needed to shear a globule of radius R_g and surface tension γ_g containing N droplets of radius R_d and surface tension γ_d, is

$$W_{tot} = \varepsilon(\gamma_g R_g^2 + N\gamma_d R_d^2). \tag{2.4}$$

Because the droplets are close packed inside the globule, the total number of droplets in a globule is of the order $N(R_g/R_d)^3$. This means that the ratio of the work W_d needed to deform the droplets over the work W_g needed to deform the globule is the product of two terms:

$$\frac{W_d}{W_g} = \left(\frac{R_g}{R_d}\right)\left(\frac{\gamma_d}{\gamma_g}\right). \tag{2.5}$$

In our experiments, the radii ratio, R_g/R_d, is kept constant and equal to 60. The value of the second term depends on the surfactant system. The surface tension of the droplets is larger than that of the globule. We find that the globule interface separates an aqueous phase with 1% 10*C*12 surfactant and an oil phase with 15% Span 80 surfactant. At this concentration, which is nec-

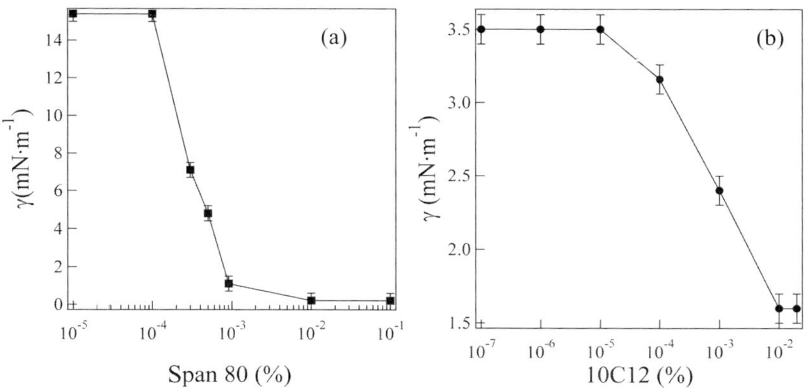

Figure 2.7 Gibbs isotherm of water/dodecane interface. (*a*) The water phase contains 1% 10C12, and the Span 80 concentration in the oil phase is increased from 10^{-5}% to 10^{-1}%. (*b*) The oil phase contains 2% Span 80, and 10C12 in the water phase is increased from 10^{-7}% to 2.10^{-2}%.

essary to stabilize the inverse droplets, the tension of the globule interface reached its plateau value, lower than $0.1\,\text{mN}\cdot\text{m}^{-1}$ (Figure 2.7*a*).

We do not know a priori the surface tension of the droplets interface, since we do not know the exact composition of the dispersed aqueous phase. We thus measure the interfacial tension of water (10C12) in relation to that of the interface (dodecane/Span 80) as a function of the 10C12 concentration. Whatever the 10C12 concentration, we find that the interfacial tension is comprised between 1 and $3.5\,\text{mN}\cdot\text{m}^{-1}$ (Figure 2.7*b*). Thus, in our system, the interfacial tension of the globule interface is smaller than the interfacial tension of the droplets' interface. As a consequence the work needed to strain a concentrated emulsion is stored by the droplets' interfaces. Rheology then becomes a tool to probe the interfacial properties of droplets' interfaces. Using Mason's equation, let us therefore consider $G'R_d/\phi$ in terms of ϕ (Figure 2.8, open symbols), where $\phi = \phi_g\phi_d$ is the droplet volume fraction calculated over the total sample volume and R_d the droplet radius.

We observe a sharp linear increase of $G'R_d/\phi$ above a volume fraction close to random close-packing concentration, in agreement with the Mason equation. From the linear regression, we get the value of the interfacial tension, $\gamma_d = 3.6\,\text{mN}\cdot\text{m}^{-1}$, between droplet and oil. The verification of the Mason equation for the multiple emulsions has several consequences:

- It confirms that the elasticity of concentrated multiple emulsions is governed by the internal droplets solely.
- The values of $G'R_d/\phi$ for inverse emulsions and for multiple emulsions superimpose (Figure 2.8). Thus the interfaces between the droplets and

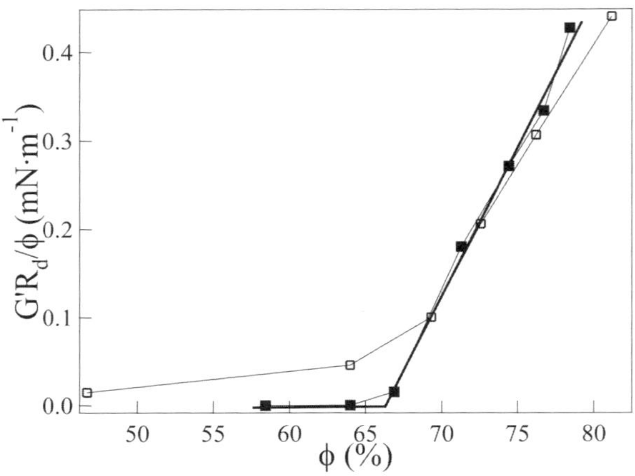

Figure 2.8 Mason's plot of the rescaled elastic modulus of concentrated emulsions as a function of the volume fraction: □: inverse dodecane in water emulsion stabilized with Span 80; ■: multiple emulsion *W*/*O*/*W* stabilized by Span 80 and modified polyacrylic acid 10*C*12. The straight line is a linear fit to the multiple emulsions' data.

globules, in multiple emulsions, have the same interfacial tension as the interface between droplets and continuous phase in the corresponding inverse emulsion.

We proceed next to study the implications of this result on the composition of the inner droplet/globule interface.

2.4.3 Characterization of the Interfaces

Let us consider the droplets interface in the concentrated multiple emulsion. A priori we do not know the quantity of 10*C*12 that migrated from the continuous water phase into the dispersed water. We studied the interface by tensiometry (2% Span 80 in dodecane)/(10*C*12 in water). For a detailed description of the tensiometry measurements, as a function of salt concentration the reader is referred to Michaut et al. (2004b). We varied the concentration of 10*C*12 in the aqueous compartment. When it does not contain any 10*C*12, the measured value of the interfacial tension is $3.5\,\mathrm{mN\cdot m^{-1}}$. Upon addition of 10*C*12, the interfacial tension decreases. Eventually it drops and reaches a plateau value of $1.6\,\mathrm{mN\cdot m^{-1}}$ (Figure 2.7*b*).

The value of the interfacial tension ($\gamma = 3.6\,\mathrm{mN\cdot m^{-1}}$) deduced from rheological experiments performed on concentrated swollen emulsions is very close to that measured by tensiometry, at zero concentration of 10*C*12. As a consequence the surface of the droplets is essentially covered with Span 80 molecules when 10*C*12 is used to cover the external interface.

In conclusion, rheological measurements show that the two interfaces in multiple emulsions are different in composition: the interface between the droplets and the oil phase is mostly covered with Span 80 molecules while the outer interface is a mixed 10*C*12/Span 80 interface. Indeed Span 80 molecules dispersed in the oil phase can easily diffuse toward the globule interface. As we mentioned in the introduction to this chapter, the migration of the hydrophilic emulsifier from the outer interface to the inner one is an important source of destabilization of multiple emulsions. It is in fact the adsorption of an increasing number of hydrophilic emulsifier molecules at the surface of the droplets that increases the value of the spontaneous curvature of the (mixed) amphiphile layer and creates a condition favorable to the coalescence of both the droplet/droplet and droplet/globule (Ficheux et al., 1998).

2.5 CONCLUSION

The use of an amphiphilic copolymer as a surfactant allows for an increase of the lifetime of multiple emulsions. The structural properties emulsions should thus be precisely studied. We performed direct confocal microscopy observations as well as light-scattering experiments in order to investigate the dynamics of the inverse droplets inside oil globules. Depending on the emulsion formulation, we showed that droplets are either uniformly distributed inside direct globules, or they are depleted toward the inner surface of the globules. These structural differences appear to play a key role in the destabilization and release processes.

Moreover, we used rheology to determine the interface composition of inverse droplets, which is difficult to measure otherwise. The elastic modulus of a concentrated emulsion is directly linked to the interfacial tension of the droplets interfaces. We showed that the measurement of the elastic modulus of a concentrated multiple emulsion leads to the value of the interfacial tension of the droplets interfaces. We were thus able to access the composition of these interfaces in situ.

When applied to multiple emulsions during their destabilization these techniques should lead to a better understanding and control of the destabilization and release mechanisms.

REFERENCES

Bancroft WD. 1913. The theory of emulsification. V. *Journal of Physical Chemistry* 17: 501–519.

Benichou M, Aserin A, Garti N. 2004. Double emulsions stabilized with hybrids of natural polymers for entrapment and slow release of active matters. *Adv Colloid Interface Sci* 108–109: 19–41.

Brujic J, Edwards SF, Hopkinson I, Makse HA. 2003. Measuring the distribution of interdroplet forces in a compressed emulsion system. *Physica A*, 3–4: 201–212.

Ficheux MF, Bonakdar L, Leal-Calderon F, Bibette J. 1998. Some stability criteria for double emulsions. *Langmuir* 14: 2702–2706.

Florence AT, Whitehill D. 1981. Some features of breakdown in water-in-oil-in-water multiple emulsions. *J Colloid Interface Sci*, 79: 1, 243–256.

Garti N. 1997. Double emulsions—Scope, limitations and new achievements. *Colloids Surf A* 123–124: 233–246.

Garti N. 1998 A new approach to improved stability and controlled release in double emulsions, by the use of graft-comb polymeric amphiphiles. *Acta Polym* 49: 606–616.

Grossiord JL, Seiller M. 1998. *Multiple Emulsions: Structure, Properties and Applications*. Editions de Santé. Paris.

Mason TG, Bibette J, Weitz DA. 1998. Yielding and flow of monodisperse emulsions. *J Colloids Interface Sci* 179: 439–448.

Mason TG, Bibette J, Weitz DA. 1995. Elasticity of compressed emulsions *Phys Rev Lett* 75: 2051–2054.

Michaut F, Hébraud P, Perrin P. 2003. Amphiphilic polyelectrolyte for stabilization of multiple emulsions. *Polym Int* 52: 594–601.

Michaut F, Hébraud P, Lafuma F, Perrin P. 2004a. A new oil-associative polymer for stabilizing inverse emulsions: strategy, synthesis, and physicochemical properties. *Langmuir* 19: 10086–10094.

Michaut F, Perrin P, Hébraud P. 2004b. Interface composition of multiple emulsions: Rheology as probe. *Langmuir* 20: 8576–8581.

Pal R. 1996. Multiple O/W/O emulsion rheology. *Langmuir* 12: 2220–2225.

Pawley J. 1995. *Handbook of Confocal Microscopy*. New York: Plenum Press.

Pays K, Giermanska-Kahn J, Pouligny B, Bibette J, Leal-Calderon F. 2000. Double emulsions: How does release occur? *J Control Release* 79: 193–205.

Pays K, Giermanska-Kahn J, Pouligny B, Bibette J, Leal-Calderon F. 2001. Double emulsions: A tool for probing thin-film metastability. *Phys Rev Lett* 87: article 178304.

Pine DJ, Weitz DA, Maret G, Wolf PE, Herbolzheimer E, Chaikin PM. 1990. Dynamical correlations of multiply scattered light. In: Light Scattering and its Applications, P. Sheng, ed. Singapore: World Scientific, pp 312–372.

Princen HM. 1983. Rheology of foams and highly concentrated emulsions. 1. Elastic properties and yield stress of cylindrical model system. *J Colloid Interface Sci* 91:160–175.

Princen HM, Kiss AD. 1986. Rheology of foams and highly concentrated emulsions: 3. Static shear modulus. *J Colloid Interface Sci*, 112: 427–437.

Sela Y, Magdassi S, Garti N. 1994. Polymeric surfactants based on polysiloxanes—Graft-poly (oxyethylene) for stabilization of multiple emulsions. *Colloids Surf A* 83: 99.

Tadros TF, Dederen C, Taelman MC. 1997. Preparation of double emulsions. *Cosme Toilet* 112: 75–85.

Wang KT, Iliopoulos I, Audebert R. 1988. Viscometric behavior of hydrophobically modified poly(sodium acrylate). *Polym Bull* 20: 577–582.

Weitz DA, Pine DJ. 1993. In: *Diffusing-wave Spectroscopy*, Brown W, ed. Oxford: Oxford University Press, pp 652–720.

Wen L, Papadopoulos KD. 2000. Visualization of water transport in $W_1/O/W_2$ emulsions. *Colloids Surf A* 174: 159–167.

Zhou J, Long S, Wang Q, Dinsmore AD. 2006. Measurements of forces inside a three-dimensional pile of frictionless droplets. *Science* 312: 1631–1633.

CHAPTER 3

Visualization of Stability and Transport in Double Emulsions

LOUISE BRAUD LAWSON and KYRIAKOS PAPADOPOULOS

Contents

3.1 INTRODUCTION

Double emulsions show potential for a variety of industrial, environmental, and medical applications. For practical design of these systems, a thorough understanding of their structure, response to external stimuli and long-term storage, the role of various components on this response, and many other factors is essential. A number of experimental techniques are useful for assessing these features in double-emulsion systems. Changes in rheological properties can indicate changes in the relative volume of the external phase due to release of the internal droplets. Conductivity measurements are commonly used to detect release of electrolytes from the globules' internal phase. Light-scattering and granulometric analyses can quantify changes and/or variations

Multiple Emulsions: Technology and Applications, Edited by Abraham Aserin

in droplet-size distributions due to different preparation conditions, external stimuli, or inherent instabilities over time.

In many investigations of the properties of double-emulsion systems, the use of microscopic visualization in conjunction with the experimental techniques mentioned above has correlated the information gained and in many cases enhanced the understanding of detected mechanisms. In other investigations, visualization techniques alone provide direct detection of changes in the stability and properties of double-emulsion systems. This chapter provides an overview of microscopy techniques applied in investigations of stability and transport in double-emulsion systems.

3.2 MICROGRAPHY OF DOUBLE EMULSIONS

3.2.1 Photomicrography

The ability to capture direct images of double-emulsion globules enables verification of the actual existence of the complex, three-phase system. The importance of visualization of double emulsions is seen in the work of Matsumoto and coworkers (1976) and their early success in developing a technique for preparation of stable double-emulsion systems. Similarly Kavaliunas and Frank (1978) used photomicrography to detect the structure of emulsions prepared with a water, *p*-xylene, and nonylphenol diethylene glycol ether system. As a result the authors were able to determine conditions favorable for the preparation of multiple emulsions, as shown in Figure 3.1. With polarized microscopy, the authors detected diffuse birefringence in multiple emul-

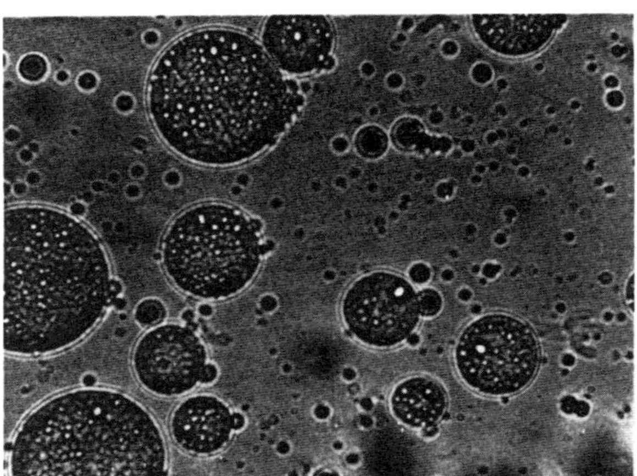

Figure 3.1 Photomicrograph of a *p*-xylene, water, nonylphenol diethylene glycol ether w/o/w emulsion (450×). Reprinted from Kavaliunas and Frank (1978) with permission from Elsevier.

sion systems, which indicated the existence of a liquid crystalline phase at both the internal and external oil/water interfaces of these emulsions.

In addition to verifying the formation of double emulsions, much information about their preparation and the role of different components in the preparation of double emulsions was gained through microscopic investigations since the 1980s. Visualization of different double-emulsion systems yielded information about the role of different components of the system in promoting emulsion stability. Garti and coworkers (Frenkel et al., 1983) employed optical microscopy to determine droplet size of emulsions prepared with various surfactants. From these observations, the investigators correlated droplet size with the likelihood of a water/oil/water emulsion inverting over time to form a simple oil/water emulsion as well as the surfactant properties that favored inversion. Larger droplet size and appropriate weighted HLB surfactants, which accounts for the HLB of the surfactants used in each phase of the emulsion and the relative concentration of each, promoted stability in the emulsion systems. With microscopic observations, Omotosho et al. (1986) detected the difference in droplet size when various hydrocarbon oils; the variation in droplet size was due to the differences in interfacial tensions with each hydrocarbon.

Photomicrography has continued to be used in understanding multiple-emulsion systems. Detection of an interesting, complex four-phase water/oil/water/oil system (Figure 3.2) was detected at specific mixing conditions (Goubault et al., 2001). Pal (1996) investigated oil/water/oil emulsions and their stability over a period of more than two days. Changes in droplet size

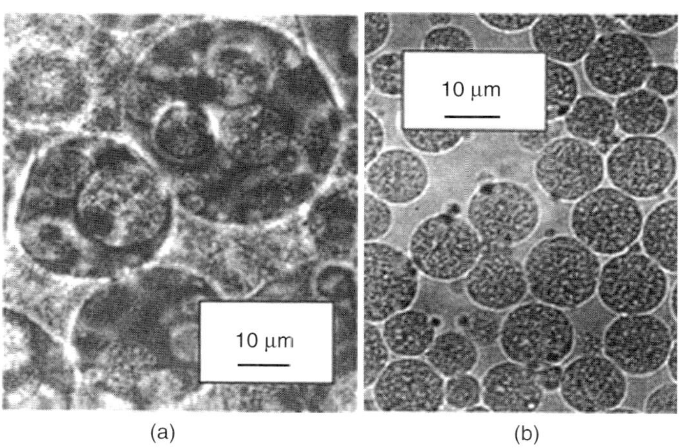

Figure 3.2 (*a*) Complex structures obtained when the incorporation of the primary inverted water/oil emulsion in the external water phase is too fast giving a W/O/W/O emulsion composed of water and dodecane. (*b*) Microscopic image of a typical W/O/W emulsion. Reprinted from Goubault et al. (2001) with permission from American Chemical Society.

due to coalescence of the emulsion globules were detected microscopically; these changes were reflected in an increase in the viscosity of the dispersion. Laugel et al. (1998) used optical microscopy to ensure multiplicity of oil/water/oil emulsions developed for potential use in drug delivery. The destabilization of water/oil/water double-emulsions in response to shear force was also assessed microscopically (Oliveiri et al., 2003). Comparison at different temperatures provided information for developing a system that exhibits enhanced release properties at body temperature for potential applications with cosmetics or therapeutic drugs.

Different techniques have been applied for observing double emulsions with photomicrography. For example, immersion microscopy allowed direct detection of the structure of emulsion globules (Cournarie et al., 2004). Matsumoto and coworkers [Matsumoto et al., 1976] used a glass slide having a small depression at its center to contain the liquid dispersion for microscopic observation. An example of a micrograph obtained with this method is shown in Figure 3.3. Zheng et al. (1991, 1993) worked with double emulsions for potential use as a red-blood cell substitute in which the internal phase consisted of an aqueous solution of human hemoglobin. Emulsion samples were diluted into a gelatin solution, yielding a homogeneous dispersion for detection of the internal droplet size and stability of multiple emulsions with phase-contrast microscopy. Phase-contrast optical microscopy of liquid dispersions has been used in recent investigations as well to visualize double-emulsion globules for droplet-size and globule stability determinations over long-term and short-term time periods (Doucet et al., 1998; Pays et al., 2001, 2002), as seen in Figures 3.4 and 3.5. Phase-contrast and dark-field microscopy enhanced detection of very thin oil films separating the inner and outer aqueous phases

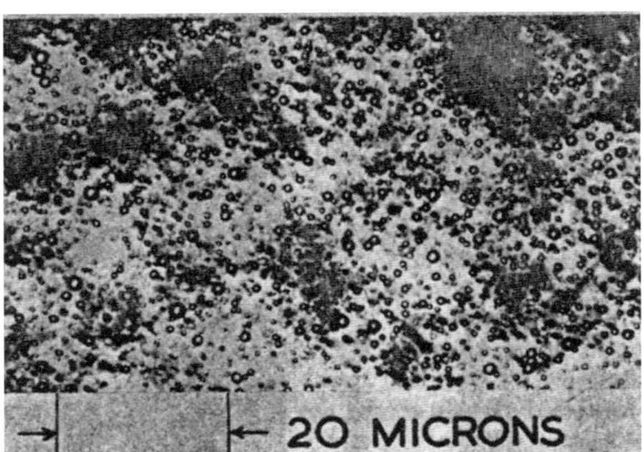

Figure 3.3 Photomicrograph of a freshly prepared W/O/W emulsion with liquid paraffin as the oil phase. Reprinted from Matsumoto et al. (1976) with permission from Elsevier.

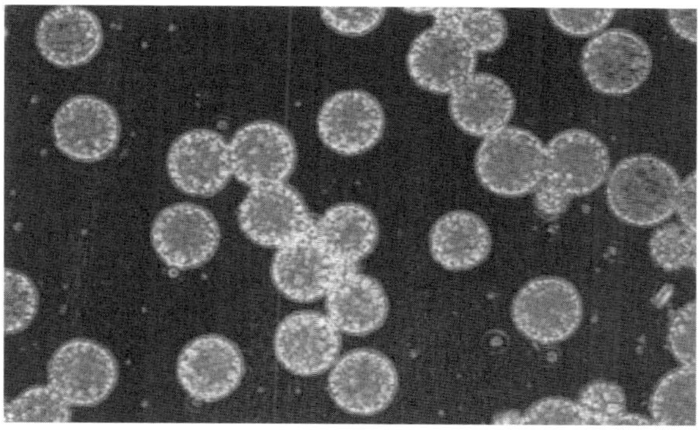

Figure 3.4 Phase-contrast optical microscopy of a water/oil/water emulsion. The oil phase is composed of vaseline oil and cetearyl octanoate. Reprinted from Doucet et al. (1998) with permission from Blackwell.

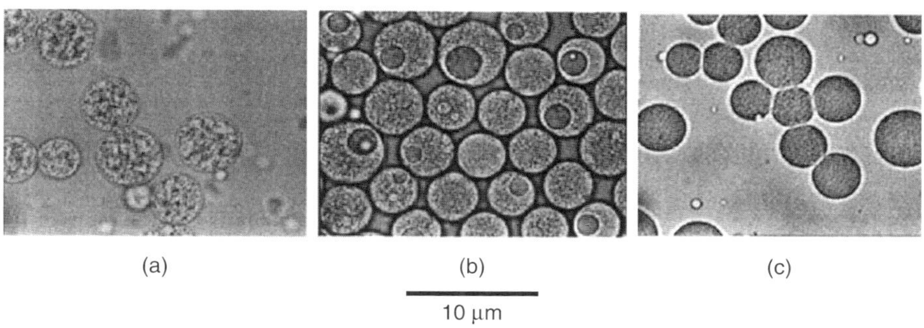

Figure 3.5 Progression from a W/O/W emulsion to a simple O/W emulsion detected with phase-contrast microscopy over time (*a*) 60 minutes after preparation, (*b*) 250 minntes after preparation, and (*c*) 850 minutes after preparation. Reprinted from Pays et al. (2001) with permission from American Chemical Society.

when the internal droplets exhibited extreme swelling (Florence et al., 1985).

Burgess and coworkers (Jiao et al., 2002) placed double-emulsion samples onto a glass slide. Subsequent pressure created with the application of a coverslip led to deformation or disruption of the globule structure, depending on the composition of the system. The double-emulsion system's ability to withstand the pressure induced by the coverslip indicated stability in the interfacial film and correlated with long-term stability. Sealing the edges of the coverslip prevented liquid evaporation over time. In instances where pressure from a coverslip is not desirable, this deformation or disruption is avoidable, as shown by Kanouni and coworkers (2002).

3.2.2 Video Micrography

Video micrography allows observation of emulsion systems with images of the same dispersion sample captured at regular time intervals. Sequential microscopic imaging was applied in the early 1980s by Florence and Whitehill (1981) to study double-emulsion stability using cinemicrography at various capture rates. This early investigation provided direct evidence of the occurrence of instability by mechanisms that had previously been proposed but not detected or differentiated. Micrography enabled the detection of changes in the mean droplet diameter and droplet size distribution, changes in ratio of single- to double-emulsion globules, and changes in the average number of internal droplets over periods of several weeks in addition to dynamic changes within individual globules in spans as short as 1/18 second. Video-enhanced microscopy and confocal microscopy verified the stability of double-emulsions encapsulating fluorescent markers within the internal phase and to determine the globule size of different formulations (Hai and Magdassi, 2004).

An immersion microscope equipped with a video camera allows direct visualization of double emulsions in a bulk system immediately after preparation and over time without having to transfer or otherwise disrupt the dispersion (Tedajo et al., 2001). Observation of the response to experimental changes in the system is also feasible with immersion micrography (Jager-Lezer et al., 1997). Figure 3.6 shows the swelling in a double-emulsion system with a higher concentration of $MgSO_4$ in the internal phase than the external phase conditions captured with immersion microscopy. For long-term observation of emulsion systems, video micrography with a rectangular cuvette with height of 200 μm and length and width of 3 cm and 4 mm was used. The top view and side view images of the emulsion globules were captured over periods of typically several hours (Gonzalez-Ochoa et al., 2003). Video tracking of emulsion globules was used to determine the type of instability caused by mechanical

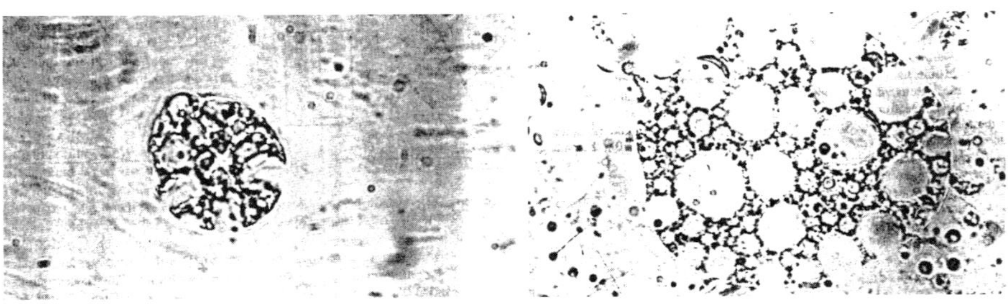

(a) (b)

Figure 3.6 Micrograph of a W/O/W emulsion (*a*) immediately after dilution of the continuous phase to create an osmotic pressure gradient between the inner and outer aqueous phases with a higher concentration of $MgSO_4$ in the internal phase and (*b*) one hour after dilution. Reprinted from Jager-Lezer et al. (1997) with permission from Elsevier.

agitation of double-emulsion systems containing various surfactants. Microscopic video observation along with macroscopic visual detection of phase separation in water/oil/water systems upon stirring allowed detection of two types of instability. These were either breakup of internal aqueous droplets within the oil phase or breakup of the water/oil globules dispersed in the continuous aqueous phase, leading to formation of a single oil/water emulsion or separation of the oil and water continuous phases (Karouni et al., 2002).

3.2.3 Capillary Microscopy

While many of visualization techniques assess the structure of emulsion globules in a bulk system, a technique developed by Papadopoulos and coworkers in the 1990s offers the capability of observing the dynamic behavior over time of an individual emulsion globule from the moment of its preparation. The formation of individual double-emulsion globules within thin-walled capillaries of diameters of 150 to 200 μm enabled direct observation of their behavior in a series of investigations (e.g., Hou and Papadopoulos, 1997; Wen and Papadopoulos, 2000b; Villa et al., 2003). Globules are prepared directly within a capillary filled with the continuous outer phase upon injection of each the oil phase and the internal water droplets with separate pipettes inserted into the capillary as shown in Figure 3.7. Capturing images at a high frame rate and over extended periods of time (over 24 hours in some cases) of an individual globule is possible due to the limited mobility of the globule within the confines of the capillary that would not exist in a bulk system. To study the mechanism of transport of water in these emulsions, micropipettes were inserted within the capillary, and this facilitated precise control of the location, size, and number of internal droplets. Different transport mechanisms of water from the internal phase of water/oil/water systems were distinguishable when the inner droplet was in contact or not with the outer phase. These mechanisms

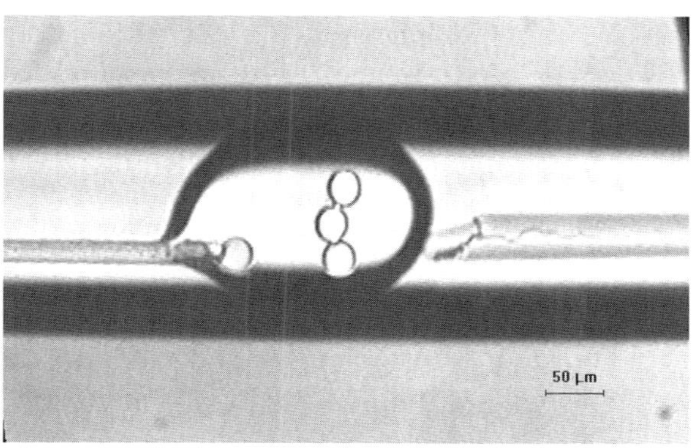

Figure 3.7 Preparation of a double-emulsion globule recorded with capillary microscopy.

were detectable within the thin-walled capillary as the droplets' position within the oil phase was manipulable (Wen and Papadopoulos, 2000b). To better understand the conditions that favor a stable system, the interactions between the inner droplets and the external phase and among internal droplets were enhanced with video observation of globules prepared within a capillary. Events leading to instability immediately upon preparation were also detected (Hou and Papadopoulos, 1996; Lawson and Papadopoulos, 2004)

3.2.4 Electron Microscopy

Electron micrography provides a method for visualizing droplets and capturing the fine details a double emulsion. One of the limitations of optical micrography is the visualization of fine droplets smaller than a few micrometers in diameter, as noted by Zheng et al. (1991, 1993). Another limitation is the inability to discriminate between water droplets covered by an oil globule and those dispersed within the oil phase (Matsumoto et al., 1980). DiStefano et al. (1983) captured micrographs of oil/water/oil emulsion globules as small as 0.05 μm in diameter and internal oil droplets 0.02 μm in diameter with transmission electron micrography. Also noted were the variations in number of internal droplets at different mixing speeds during emulsion preparation. Freeze-fracturing of water/oil/water emulsion globules provides a technique for droplet-size determination without distortion by the oil phase and with higher resolution of small droplets (Davis and Burbage, 1977). An example of the micrographs obtained is shown in Figure 3.8. More recently Okochi and Nakano (2000) employed scanning electron micrography to confirm the preservation of W/O/W structure after one week of storage.

3.3 INSTABILITY IN DOUBLE EMULSIONS

Visualization techniques have not only been useful in confirming the existence of double emulsions but also in improving our understanding of various phenomena occurring in these complex systems. Microscopic techniques have enabled identification of mechanisms that were more difficult to detect with other indirect methods of investigation, yielding information that is beneficial in the design of double-emulsion systems for practical applications.

3.3.1 Visualization of Transport

The use of microscopy techniques to investigate transport in double emulsions has brought some insight about mechanisms that can lead to migration of water and/or solutes within these systems. Optical microscopy was used to assess swelling in the internal aqueous phase of W/O/W emulsions due to migration of water in the presence of an osmotic pressure gradient (Matsumoto et al., 1980). Quantification of the rate of water migration in dif-

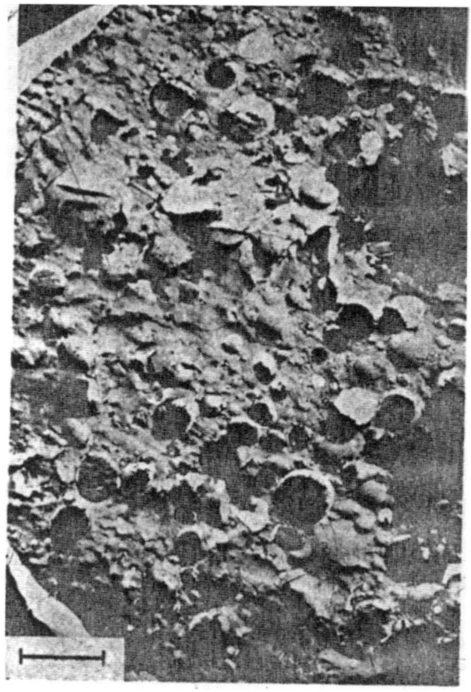

Figure 3.8 Electron micrograph of a W/O/W emulsion with light liquid paraffin in the oil phase. Bar = 1 μm. Reprinted from Davis and Burbage (1977) with permission from Elsevier.

ferent oils based on droplet-size measurements taken as a function of time was used to determine the permeation coefficient of different oils. The concentration of surfactant in the oil phase also plays a role in the degree of swelling in a water/oil/water emulsion. The degree of swelling varies, but can increase the size of emulsion globules significantly, as seen in Figure 3.6. In some cases the extent of swelling leads to eventual breakdown of the oil phase of a water/oil/water emulsion (Jager-Lezer et al., 1997). Such extreme swelling causes extensive thinning in the oil film separating the aqueous phases. Without adequate stabilizers, this can lead to rupture of the thin film and thus release of the swollen droplets to the external phase (Florence et al., 1985).

One of the mechanisms by which water is transported across the oil phase is by reverse micellar transport. Sela et al. (1995) provides evidence of this mechanism in assessing water solubility by reverse micelles in the oil phase of water/oil/water systems. Microscopic observation of collected oil-phase samples ensured that no water droplets were present in the oil phase and that any water detected via titration methods was solubilized in the oil. The migration of water from an internal droplet to the continuous aqueous phase, resulting in a decrease in the droplet size, was visualized and quantified in a series of investigations. Capillary microscopy was used in determining the effects of the osmotic pressure gradient (Wen and Papadopoulos, 2001) and surfactant concentrations in each phase of the emulsion (Wen and Papadopoulos, 2000a) on the rate of water transport (see Figure 3.9). Transport of water molecules to the outer saline aqueous phase by spontaneous emulsification of the internal pure water droplets, a previously unobserved mechanism, was detected (Wen and Papadopoulos, 2000b) and is shown in Figure 3.10. Spontaneous emulsification was evident by the formation of many tiny aqueous droplets dispersed within the oil phase. The decrease in the diameter of the internal aqueous droplet indicated the overall migration of water molecules to the outer aqueous phase.

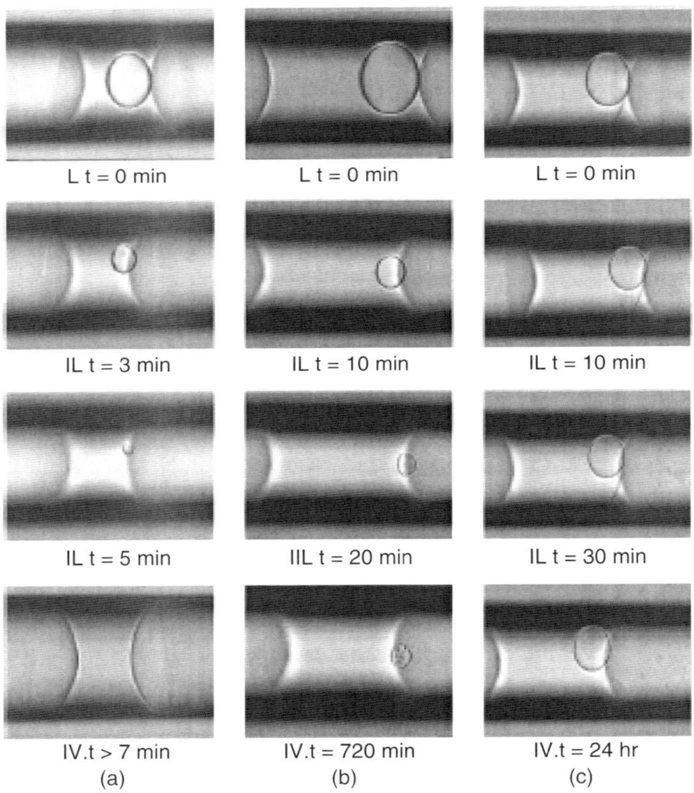

Figure 3.9 Water migration from the internal aqueous droplet to the outer continuous aqueous phase with different salt concentrations in the internal phase and *n*-hexadecane in the oil phase and 5.0 M NaCl in the external aqueous droplet. (*a*) Pure water in the internal aqueous phase; (*b*) 0.1 M NaCl in the internal aqueous phase; (*c*) 2.5 M NaCl in the internal aqueous phase. Reprinted from Wen and Papadopoulos (2001) with permission from Elsevier.

Figure 3.10 Spontaneous emulsification in a W/O/W system with pure water in the internal water phase, *n*-hexadecane in the oil phase, and 5 M NaCl in the outer aqueous phase. Reprinted from Wen and Papadopoulos (2000a) with permission from Elsevier.

The effect of Ostwald ripening as a factor in emulsion instability in water/oil/water systems was noted through the use of video microscopy (Rosano et al., 1998). The curvature at the internal water/oil interface is greater than that at the interface separating the outer aqueous phase from the oil phase. Thus the solubility of water is greater at the interface of the internal droplets, leading to a net transfer of water from the internal aqueous phase to the external phase. Transport due to Laplace pressure differences between the internal and external aqueous phases of water/oil/water systems can be balanced by inducing an osmotic pressure gradient in the opposite direction. Emulsions containing NaCl in the internal aqueous phase but not the external phase were more stable than those with no electrolyte in the internal phase.

In other investigations of molecular transport from the internal phase of a double-emulsion system, video-recorded behavior of double-emulsion systems over time support the occurrence of molecular transport across the middle phase of the emulsion without release of internal droplets through coalescence between the internal and external phases. Magdassi and Garti (1984) noted this type of release from an otherwise stable emulsion: factors that affected the rate of release of electrolytes from the internal phase included the concentration of electrolyte and its hydrophobicity. In a later investigation, microscopic observation indicated that there was no detectable change in internal droplet size or number within the system despite the release of various salts and active agents from the inner phase; this indicated the ability of the molecules to migrate through the oil phase without any disruption in the structure of double emulsions (Sela et al., 1995). In this study emulsion samples were assessed periodically over a period of one month. Similarly Hai and Magdassi (2004) visualized double-emulsion globules over time and confirmed that release of fluorescent markers from the internal phase was not due to release of the entire droplets; stable emulsions were detected after 24 hours, while significant release of the fluorescent marker was detected within 2 hours of preparation. Rojas et al. (2006) also noted release of a fluorescent dye dissolved in the internal aqueous phase of an otherwise stable double-emulsion system. By contrast, when the dye was encapsulated within nanohydrogel particles in the internal aqueous phase, it was not released to the external phase. Release of dye from these particles, however, could be triggered by in situ addition of hydrochloric acid to the emulsion system. The nanohydrogel particles remained encapsulated within the emulsion globule as the stability of the globule was maintained. Addition of HCl to either the internal or external aqueous phase prompted release of the dye from the internal phase particles (Figure 3.11), indicating ionic transport of HCl through the oil phase; transport of the dye to the external phase was indicated by an increased fluorescence in the external phase.

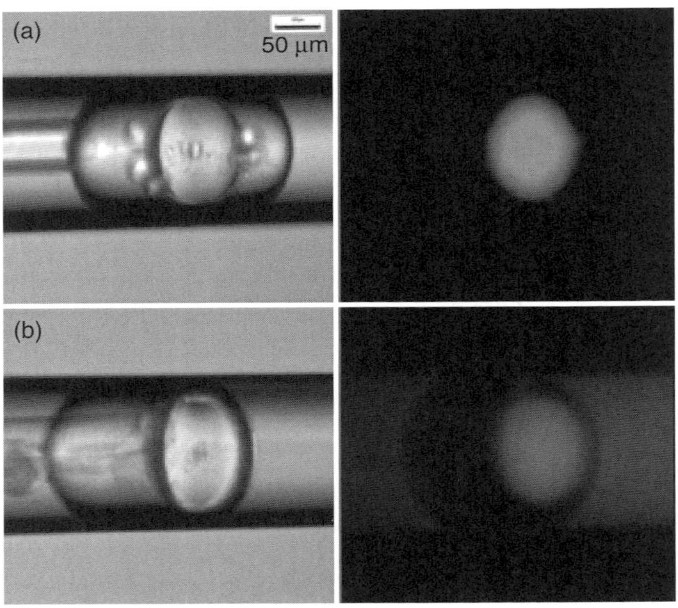

Figure 3.11 Activation of fluorescent dye release from nanohydrogel particles within the internal aqueous phase of a W/O/W globule prompted by continuous injection of a 0.2 M HCl aqueous solution into the external aqueous phase. The oil phase is composed of *n*-hexadecane and Span 80. (*a*) Initially after globule preparation; (*b*) 90 seconds after the onset of acid injection. Bright field microscopy images are in the left column while fluorescence microscopy images are in the right column. Reprinted from Rojas et al. (2006) with permission from Elsevier.

In an investigation by Garti et al. (1994), the release of NaCl from a water/oil/water emulsion was affected by the presence of protein in the inner phase of the emulsion but not by the addition of protein to the external phase. However, the addition of protein to either phase enhanced emulsion stability, as indicated by the visualization of stable emulsions containing protein in the outer aqueous phase after seven days of storage (Figure 3.12). Similarly the addition of a polysaccharide, along with protein, in the external aqueous phase enhanced globule stability, as was indicated by confocal scanning laser microscopy over time and the long-term (seven-day) encapsulation of vitamin B1 from the internal aqueous phase over time. Encapsulation or release of vitamin B1 was noted with confocal scanning laser microscopy due to its fluorescence properties (Benichou et al., 2007). A recent study compared the rate of transport of different ions across the oil phase of a water/oil/water double emulsion. NaCl and $AgNO_3$ were added to separate aqueous phases of the emulsion system. AgCl deposition observed within an individual globule through the use of capillary video microscopy enabled detection of the ionic transport rate (see Figure 3.13), which was a result of reverse micellar transport and was faster for Ag^+ than for Cl^- (Cheng et al., 2007).

In the investigations mentioned here, the visualization of stable emulsion globules over time indicated that release from the internal phase could occur without coalescence of the internal droplets to the external phase but rather by diffusion or facilitated transport across the middle phase. These findings are useful in design of controlled-release systems in that the choice of surfactants, oil, and other components can be used to tailor the rate at which release occurs.

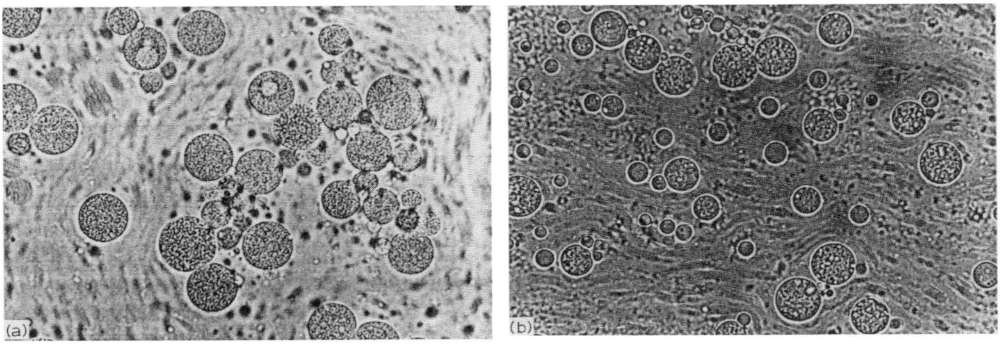

Figure 3.12 Microscopic observation of a W/O/W containing albumin protein in the external aqueous phase and light mineral oil in the oil phase (*a*) immediately after preparation and (*b*) seven days after preparation. Reprinted from Garti et al. (1994) with permission from Elsevier.

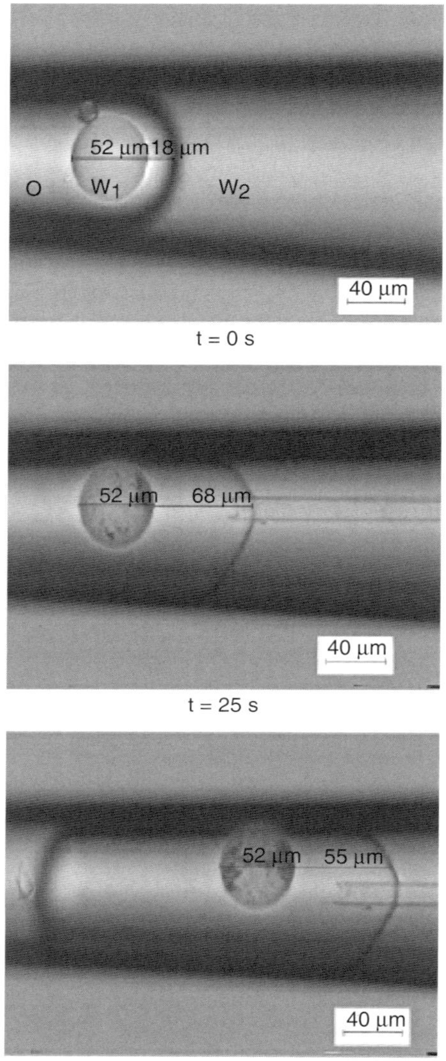

Figure 3.13 Transport of Ag^{+} through the oil film of a W/O/W globule indicated by AgCl deposition in the internal aqueous phase. The internal aqueous phase is a 2.5 M NaCl solution, the oil phase contains hexadecane and Span 80 while the external aqueous phase contains $AgNO_3$. Reprinted from Cheng et al. (2007) with permission from Elsevier.

3.3.2 Coalescence Events

In the past decade microscopic detection of specific types of coalescence within double-emulsion droplets has contributed to a better understanding of these mechanisms in order to control their occurrence. Three types of coalescence events have been identified in double-emulsion systems. These include coalescence between contacting inner droplets (internal coalescence), coalescence of interior droplets with the outer continuous phase (external coalescence), and coalescence between contacting two-phase emulsion globules. Direct evidence of each type of coalescence has been collected for double emulsions with microscopic visualization. The type and concentrations of surfactants in each phase of the water/oil/water double emulsions are critical in controlling the occurrence and type of coalescence (Ficheux et al., 1998; Villa et al., 2003; Hou and Papadopoulos, 1996; Hou and Papadopoulos, 1997; Pays et al., 2001).

Coalescence between contacting multiple emulsion globules was seen within the first weeks after preparation of a bulk water/oil/water system and was directly detected with high-speed cinemicrography (Florence and Whitehill, 1981). Within minutes or hours of preparation, internal coalescence has also been detected microscopically (Ficheux et al., 1998; Pays et al., 2001). A clear picture of coalescence between contacting internal droplets, namely internal coalescence, was captured with capillary video microscopy, as seen in Figure 3.14. Over time internal coalescence leads to a reduction in the number of internal droplets and an increase in their size, as seen in the sequence in Figure 3.15 (Villa et al., 2003). In these investigations the influence of surfactants in preventing or promoting internal coalescence has become more clear.

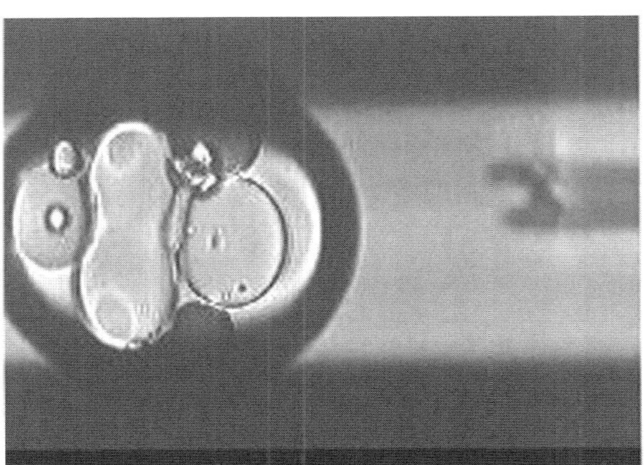

Figure 3.14 Micrograph capturing the occurrence of internal coalescence in a W/O/W globule with *n*-hexadecane as the oil phase. Reprinted from Villa et al. (2003) with permission from American Chemical Society.

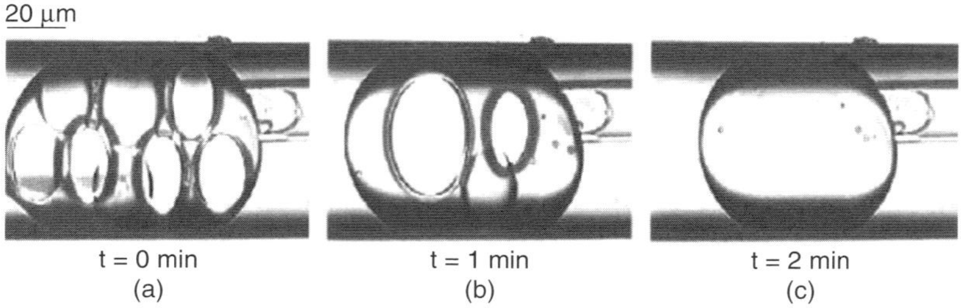

Figure 3.15 Internal coalescence followed by external coalescence in a W/O/W globule with *n*-hexadecane as the oil phase. Reprinted from Villa et al. (2003) with permission from American Chemical Society.

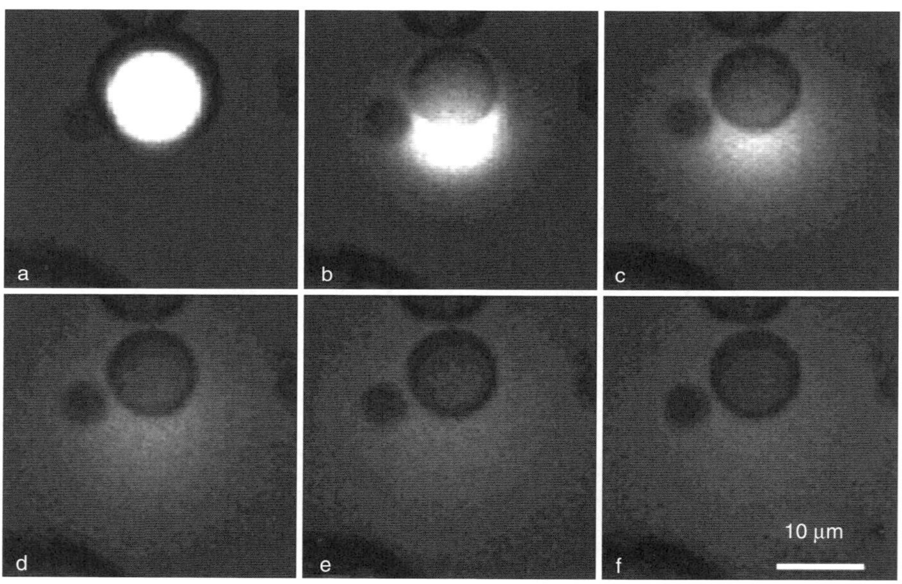

Figure 3.16 External coalescence in a W/O/W emulsion with fluorescein in the internal aqueous phase and dodecane in the oil phase. Images were captured at a rate of 30 fps. Reprinted from Gonzalez-Ochoa et al. (2003) with permission from American Chemical Society.

External coalescence leads to release of internal droplets and their contents to the outer continuous phase. The release of a fluorescent label from the internal aqueous phase of a water/oil/water double emulsion was visualized with fluorescence microscopy and indicated the occurrence of external coalescence (Figure 3.16) (Gonzalez-Ochoa et al., 2003). External coalescence was directly detected in bulk water/oil/water systems (see Figure 3.17) (Pays et al.,

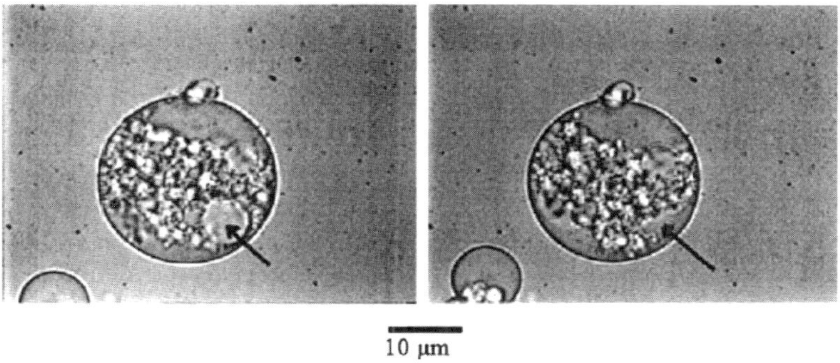

Figure 3.17 External coalescence in a W/O/W system. Time interval between images is 1 second. Reprinted from Pays et al. (2001) with permission from American Chemical Society.

2001). This study provides information about the effects of different surfactant concentrations and volume fractions of each phase on the progression from a double emulsion to a simple emulsion due to coalescence. With direct visualization of individual water/oil/water double-emulsion globules containing various concentrations of surfactants in the oil and external aqueous phases, the differing rate at which external coalescence occurred was quantified (Lawson and Papadopoulos, 2004). An example sequence of images indicating progressive release of internal droplets is shown in Figure 3.18. In emulsions containing chemical substances for controlled release, the concentration and type of surfactant dictates whether release will be dominated by external coalescence or by transport of the substance across the middle phase as discussed above. Visualization of the emulsion systems over time is a useful technique for differentiating between these two release mechanisms (Pays et al., 2002). Through direct visualization the effect of internal droplet size on instability has also been explored (Ficheux et al., 1998). Similarly Hou and Papadopoulos (1997) systematically assessed the effects of pH, ionic strength, and surfactant type and concentration on instability in double emulsions. From the experimental results collected with capillary microscopy, a theoretical model of the interactions among internal water droplets and between internal water droplets and the external phase was developed (Hou and Papadopoulos, 1996).

3.4 CONCLUSION

Various microscopic techniques have been used to explore double emulsions, including electron microscopy, video microscopy, capillary microscopy, among many others. At the most fundamental level the use of such techniques

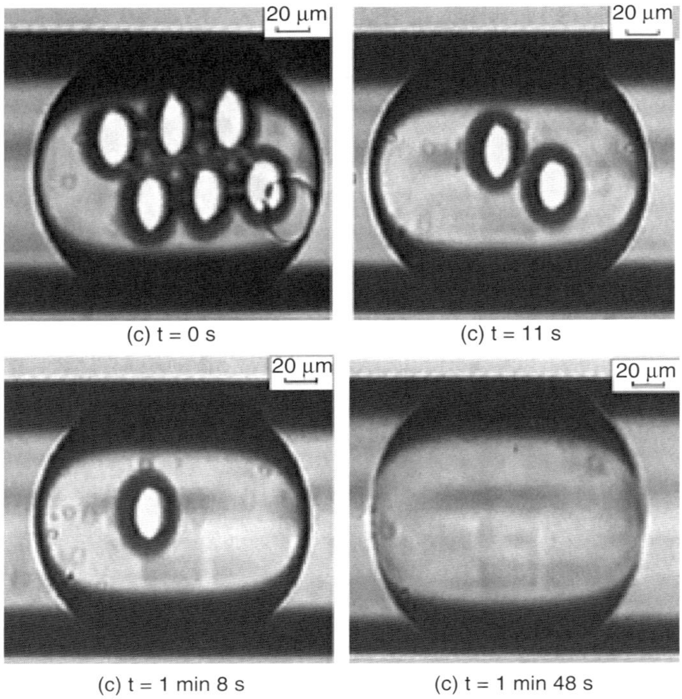

Figure 3.18 External coalescence in a W/O/W emulsion with *n*-hexadecane in the oil phase. Reprinted from Lawson and Papadopoulos (2004) with permission from Elsevier.

provides direct evidence of the existence of the complex, multi-phase double-emulsion globules. However, dynamic mechanisms leading to instability and/or release have been identified and better understood through microscopic visualization. In many cases the mechanisms recorded visually would be difficult or nearly impossible to detect with other more indirect methods. One such example is distinguishing transport through the oil phase of a water/oil/water system with or without release of the entire internal droplets. When used in conjunction with other analysis tools, such as conductivity or rheological measurements, visualization of the behavior of multiple-emulsion systems provides unique information and, in many cases, a more thorough description of the mechanisms taking place.

REFERENCES

Benichou A, Aserin A, Garti N. 2007. W/O/W double emulsions stabilized with WPI-polysaccharide complexes. *Colloids Surf A* 294(1–3): 20–32.

Cheng J, Chen J-F, Zhao M, Luo Q, Wen L-W, Papadopoulos KD. 2007. Transport of ions through the oil phase of $W_1/O/W_2$ double emulsions. *J Colloid Interface Sci* 305: 175–182.

Cournarie F, Savelli M-P, Rosilio V, Bretez F, Vauthier C, Grossiord J-L, Seiller M. 2004. Insulin-loaded W/O/W multiple emulsions: Comparison of the performances of systems prepared with medium-chain-triglycerides and fish oil. *Eur J Pharmaceut Biopharm* 58: 477–482.

Davis SS, Burbage AS. 1977. Electron micrography of water-in-oil-in-water emulsions. *J Colloid Interface Sci* 62(2): 361–363.

DiStefano FV, Shaffer OM, El-Aasser MA, Vanderhoff JW. 1983. Multiple oil-in-water-in-oil emulsions of extremely fine droplet size. *J Colloid Interface Sci* 92(1): 269–272.

Doucet O, Ferrero L, Garcia N, Zastrow L. 1998. O/W emulsion and W/O/W multiple emulsion: physical characterization and skin pharmacokinetic comparison in the delivery process of caffeine. *Int J Cosmet Sci* 20: 283–295.

Ficheux M-F, Bonakdar L, Leal-Calderon F, Bibette J. 1998. Some stability criteria for double emulsions. *Langmuir* 14(10): 2702–2706.

Florence AT, Law TK, Whateley TL. 1985. Nonaqueous foam structures from osmotically swollen W/O/W emulsion droplets. *J Colloid Interface Sci* 107(2): 584–588.

Florence AT, Whitehill D. 1981. Some features of breakdown in water-in-oil-in-water multiple emulsions. *J Colloid Interface Sci* 79(1): 243–257.

Frenkel M, Shwartz R, Garti N. 1983. Multiple emulsions I. Stability: Inversion, apparent and weighted HLB. *J Colloid Interface Sci* 94(1): 174–178.

Garti N, Aserin A, Cohen Y. 1994. Mechanistic considerations on the release of electrolytes from multiple emulsions stabilized by BSA and nonionic surfactants. *J Controlled Release* 29: 41–51.

Gonzalez-Ochoa H, Ibarra-Bracamontes L, Arauz-Lara JL. 2003. Two-stage coalescence in double emulsions. *Langmuir* 19(19): 7837–7840.

Goubault C, Pays K, Olea D, Gorria P, Bibette J, Schmitt V, Leal-Calderon F. 2001. Shear rupturing of complex fluids: Application to the preparation of quasi-monodisperse water-in-oil-in-water double emulsions. *Langmuir* 17(17): 5184–5188.

Hai M, Magdassi S. 2004. Investigation on the release of fluorescent markers from w/o/w emulsions by fluorescence-activated cell sorter. *J Controlled Release* 96: 393–402.

Hou W, Papadopoulos KD. 1996. Stability of water-in-oil-in-water type globules. *Chem Eng Sci* 51(22): 5043–5051.

Hou W, Papadopoulos KD. 1997. $W_1/O/W_2$ and $O_1/W/O_2$ globules stabilized with Span 80 and Tween 80. *Colloids Surf A* 125: 181–187.

Jager-Lezer N, Terrisse L, Bruneau F, Tokgoz S, Ferreira L, Clausse D, Seiller M, Grossiord J-L. 1997. Influence of lipophilic surfactant on the release kinetics of water-soluble molecules entrapped in a W/O/W multiple emulsion. *J Controlled Release* 45: 1–13.

Jiao J, Rhodes DG, Burgess DJ. 2002. Multiple emulsion stability: Pressure balance and interfacial film strength. *J Colloid Interface Sci* 250: 444–450.

Kanouni M, Rosano HL, Naouli N. 2002. Preparation of a stable double emulsion ($W_1/O/W_2$): Role of the interfacial films on the stability of the system. *Adv Colloid Interface Sci* 99: 229–254.

Kavaliunas DR, Frank SG. 1978. Liquid crystal stabilization of multiple emulsions. *J Colloid Interface Sci* 66(3): 586–588.

Laugel C, Baillet A, Piemi MPY, Marty JP, Ferrier D. 1998. Oil-water-oil multiple emulsions for prolonged delivery of hydrocortisone after topical application: Comparison with simple emulsions. *Int J Pharmaceut* 160: 109–117.

Lawson LB, Papadopoulos KD. 2004. Effects of a phospholipid cosurfactant on external coalescence in water-in-oil-in-water double-emulsion globules. *Colloids Surfaces A* 250: 337–342.

Magdassi S, Garti N. 1984. Release of electrolytes in multiple emulsions: Coalescence and breakdown or diffusion through oil phase? *Colloids Surf* 12: 367–373.

Matsumoto S, Inoue T, Kohda M, Ikura K. 1980. Water permeability of oil layers in W/O/W emulsions under osmotic pressure gradients. *J Colloid Interface Sci* 77(2): 555–563.

Matsumoto S, Kita Y, Yonezawa D. 1976. An attempt at preparing water-in-oil-in-water multiple-phase emulsions. *J Colloid Interface Sci* 57(2): 353–361.

Okochi H, Nakano M. 2000. Preparation and evaluation of w/o/w type emulsions containing vancomycin. *Adv Drug Deliv Rev* 45: 5–26.

Oliveiri L, Seiller M, Bromberg L, Besnard M, Duong T-N-L, Grossiord J-L. 2003. Optimization of a thermally reversible W/O/W multiple emulsion for shear-induced drug release. *J Controlled Release* 88: 401–412.

Omotosho JA, Whateley TL, Law TK, Florence AT. 1986. The nature of the oil phase and the release of solutes from multiple (w/o/w) emulsions. *J Pharm Pharmacol* 38: 865–870.

Pal R. 1996. Multiple O/W/O emulsion rheology. *Langmuir* 12(9): 2220–2225.

Pays K, Giermanska-Kahn J, Pouligny B, Bibette J, Leal-Calderon F. 2001. Coalescence in surfactant-stabilized double emulsions. *Langmuir* 17(25): 7758–7769.

Pays K, Giermanska-Kahn J, Pouligny B, Bibette J, Leal-Calderon F. 2002. Double emulsions: How does release occur? *J Controlled Release* 79: 193–205.

Rojas EC, Sahiner N, Lawson LB, John VT, Papadopoulos KD. 2006. Controlled release from a nanocarrier entrapped within a microcarrier. *J Colloid Interface Sci* 301: 617–623.

Rosano HL, Gandolfo FG, Hidrot JP. 1998. Stability of $W_1/O/W_2$ multiple emulsions: Influence of ripening and interfacial interactions. *Colloids Surf A* 138: 109–121.

Sela Y, Magdassi S, Garti N. 1995. Release of markers from the inner water phase of W/O/W emulsions stabilized by silicone based polymeric surfactants. *J Controlled Release* 33: 1–12.

Tedajo GM, Seiller M, Prognon P, Grossiord JL. 2001. pH compartmented W/O/W multiple emulsions: A diffusion study. *J Controlled Release* 75: 45–53.

Villa CH, Lawson LB, Liu Y, Papadopoulos KD. 2003. Internal coalescence as a mechanism of instability in water-in-oil-in-water emulsions. *Langmuir* 19(2): 244–249.

Wen L, Papadopoulos KD. 2000a. Effects of surfactants on water transport in $W_1/O/W_2$ emulsions. *Langmuir* 16(20): 7612–7617.

Wen L, Papadopoulos KD. 2000b. Visualization of water transport in $W_1/O/W_2$ emulsions. *Colloids Surf A* 174: 159–167.

Wen L, Papadopoulos KD. 2001. Effects of osmotic pressure on water transport in $W_1/O/W_2$ emulsions. *J Colloid Interface Sci* 235: 1–7.

Zheng S, Beissinger RL, Wasan DT. 1991. The stabilization of hemoglobin multiple emulsion for use as a red blood cell substitute. *J Colloid Interface Sci* 144(1): 72–85.

Zheng S, Zheng Y, Beissinger RL, Wasan DT, McCormick DL. 1993. Hemoglobin multiple emulsions as an oxygen delivery system. *Biochim Biophys Acta* 1158: 65–74.

CHAPTER 4

Effect of an Oil-Insoluble Solute on the Stability of Multiple Water-Oil-Water Emulsions

MOUHCINE KANOUNI and HENRI ROSANO

Contents

4.1 MODEL FOR BEHAVIOR OF TWO W/O DROPLETS, OR G/W BUBBLES CONTAINING A SOLUTE

This model covers both the case of two water-in-oil droplets (W/O) and that of two gas-in-water bubbles (G/W) containing a solute (or a gas) insoluble in

---

Multiple Emulsions: Technology and Applications, Edited by Abraham Aserin

the continuous phase. For simplicity, we present the model as applied to the case of two droplets of saline solution dispersed in a solvent.

4.1.1 Flux of Water due to Laplace Pressure between Two Droplets in a Solvent

We consider two water droplets of radii r_1 and r_2, $r_1 < r_2$, dispersed in a solvent (the continuous phase), and we assume that water is only sparingly soluble in this solvent. Let γ equal the interfacial tension (assumed to be constant) and C_m the concentration of solvent in the continuous phase.

Kabalnov and Shchukin (1965) used a Kelvin equation to express the increase in solubility of a substance inside spherical particles as the size of these particles diminishes:

$$C(r) = C(\infty)e^{2\gamma V_m/RTr} \approx C(\infty)\left(1 + \frac{2\gamma V_m}{RTr}\right), \tag{4.1}$$

where $C(r)$ is the solubility of the dispersed solvent in the continuous phase at the surface of a droplet of radius r, $C(\infty)$ is the solubility of this same solvent in the continuous phase but with a flat interface, and V_m is the molar volume of the solvent (water).

Due to the Laplace pressure, the solubility of water in the interfacial region around the smaller droplet (r_1) is larger than the solubility of water in the solvent between the two droplets. A water gradient of concentration is created in the continuous phase. After a certain time the concentration of water in the continuous phase is larger than the concentration of the water at the interface of the larger droplet (r_2), and water enters the larger droplet.

After a while the flux of water leaving the smaller droplet equals the flux entering the larger one. Using Fick's law at the interface of each droplet, as Princen (1965) and Kralchevsky, Danov, and Denkov (2002) did for the diffusion of a gas in a soap film, we can express the relation between the gradient of concentration at the surface of each droplet and the radii:

$$J_1 = J_2 \quad \text{with} \quad J_i = -4\pi r_i^2 D \left.\frac{\partial c}{\partial r}\right|_{r=r_i} \qquad \text{(Fick's law)}, \tag{4.2}$$

$$-4\pi r_1^2 D \left.\frac{\partial c}{\partial r}\right|_{r=r_1} = -4\pi r_2^2 D \left.\frac{\partial c}{\partial r}\right|_{r=r_2},$$

$$r_1^2 \left.\frac{\partial c}{\partial r}\right|_{r=r_1} = r_2^2 \left.\frac{\partial c}{\partial r}\right|_{r=r_2}, \tag{4.3}$$

where r_1 and r_2 represent the radii of the two droplets and D is the water diffusion coefficient in the continuous medium.

If the gradient of the water concentration in the interfacial region (of thickness δ) decreases linearly. We can assume the same value of δ for the two droplets, so the gradient can be expressed as follows:

$$\left.\frac{\partial C}{\partial r}\right|_{r=r_1} = \left(\frac{C_m - C(r_1)}{\delta}\right) \quad \text{and} \quad \left.\frac{\partial C}{\partial r}\right|_{r=r_2} = \left(\frac{C(r_2) - C_m}{\delta}\right). \tag{4.4}$$

From equations (4.3) and (4.4), the equality of the two fluxes may be expressed as

$$r_1^2\left(\frac{C(r_1) - C_m}{\delta}\right) = r_2^2\left(\frac{C_m - C(r_2)}{\delta}\right), \tag{4.5}$$

and C_m can be expressed as a function of the two radii r_1 and r_2:

$$C_m = \frac{r_1^2 C(r_1) + r_2^2 C(r_2)}{r_1^2 + r_2^2}. \tag{4.6}$$

Therefore the flux of water going through the interfacial regions surrounding the two droplets can be expressed as

$$J_L = -4\pi r_1^2 D \frac{(C_m - C(r_1))}{\delta} \tag{4.7}$$

using equation (4.6). After $C(r_1)$ and $C(r_2)$ are replaced by their values found through Kabalnov's equation (1.1), the Laplace flux becomes

$$J_L = \frac{4\pi r_1^2 r_2^2 D\alpha C(\infty)}{\delta(r_1^2 + r_2^2)}\left[\frac{1}{r_1} - \frac{1}{r_2}\right] \tag{4.8}$$

with $\alpha = 2\gamma V_m / RT$ (which has the dimension of length). Because the radius r_1 is smaller than the radius r_2, J_L, as written above, takes a positive value.

4.1.2 Flux of Water between Two Water Droplets due to Osmotic Pressure

If the two droplets contain, in a concentration of m_0 (moles/liter), a solute soluble in the dispersed phase but totally insoluble in the continuous phase (in the present case NaCl), we can conceive a difference in osmotic pressure between the two droplets. We know that the smaller will decrease in size and the larger will increase because of the Laplace pressure effect.

The solvent separating the two droplets can be considered to be a homogeneous liquid membrane. The flux, per unit of interface, due to the osmotic pressure can be expressed as follows:

$$J_0 = -L_0(\Delta P) = -L_0 RT(\Delta C), \tag{4.9}$$

where L_0 is an overall permeability constant (consistent with the definition given by Lemlich (1978), of the permeability of the continuous phase). L_0 is a result of the resistance across the two interfaces and across the liquid lamella separating the two droplets (we assume, as in Section 4.1.1 above, that the resistance to the migration of water of the membrane of the liquid lamella, compared to that of the interfaces, is negligible, because of the greater gradient of concentration in the interfaces). ΔP is the difference between the osmotic pressures of the two droplets.

Replacing ΔP by the Van't Hoff expression for the osmotic pressure, we find that the concentration of the solute in a droplet is equal to the initial number of moles present divided by the volume of the droplet at a time t:

$$C = \frac{m_0 \times (4\pi r_0^3/3)}{4\pi r_t^3/3} = m_0 \times \frac{r_0^3}{r_t^3}, \tag{4.10}$$

where r_0 is the initial radius of a droplet and r_t is the radius at a later time t.

From equations (4.9) and (4.10) we can express the flux between the two droplets of radius r_1 and r_2, $r_1 < r_2$, as

$$J_0 = -L_0 RT(C_1 - C_2) = -L_0 RTm_0\left(\frac{r_{01}^3}{r_1^3} - \frac{r_{02}^3}{r_2^3}\right),$$

$$J_0 = -L_0 RTm_0 r_{02}^3\left[\left(\frac{r_{01}}{r_{02}}\right)^3 \frac{1}{r_1^3} - \frac{1}{r_2^3}\right],$$

where r_{01} and r_{02} are the initial radii of the droplets 1 and 2. If we set θ equal to the ratio of the radii of the two droplets, we obtain a final expression:

$$J_0 = -L_0 RTm_0 r_{02}^3\left(\theta^3 \frac{1}{r_1^3} - \frac{1}{r_2^3}\right) \quad \text{with} \quad \theta = \frac{r_{01}}{r_{02}} \leq 1. \tag{4.11}$$

We can consider the surface of the smaller droplet to be the limiting surface for the osmotic flux. Then we can express the total flux due to the osmotic pressure as

$$J_{0T} = 4\pi r_1^2 J_0 = -4\pi L_0 RTm_0 r_{02}^3 r_1^2\left(\theta^3 \frac{1}{r_1^3} - \frac{1}{r_2^3}\right). \tag{4.12}$$

As the size of the smaller droplet decreases and the size of the bigger one increases, C_1 is superior to C_2, and the osmotic flux, as written, takes a negative value.

4.1.3 Total Flux of Water between Two Droplets Containing an Insoluble Solute in Terms of Laplace and Osmotic Fluxes

Using the separate expressions for Laplace and osmotic fluxes that we established in equations (4.8) and (4.12), we can write the total flux of water containing an insoluble solute (NaCl) dispersed in a continuous phase as follows:

$$J_T = J_L + J_{0T},$$

$$J_T = \frac{4\pi r_1^2 r_2^2 D\alpha C(\infty)}{\delta(r_1^2 + r_2^2)}\left(\frac{1}{r_1} - \frac{1}{r_2}\right) - 4\pi L_0 RTm_0 r_{02}^3 r_1^2\left(\theta^3\frac{1}{r_1^3} - \frac{1}{r_2^3}\right). \tag{4.13}$$

We can abbreviate the coefficient of each term in (4.13) in order to simplify the writing of this equation:

$$K_L = \frac{4\pi D\alpha C(\infty)}{\delta} \quad \text{and} \quad K_0 = 4\pi L_0 RTm_0 r_{02}^3.$$

Because of conservation of matter, the quantity $C_m + V_0$, where V_0 is the total initial volume, is constant.

Assuming that the quantity of water dissolved in the continuous phase is very small and varies little, we can express the conservation of matter for the system by considering only the total volume of the two droplets. So, at any instant, the state of the system is

$$V_0 = \frac{4\pi r_{01}^3}{3} + \frac{4\pi r_{02}^3}{3} = \frac{4\pi r_1^3}{3} + \frac{4\pi r_2^3}{3}.$$

The radius r_2 can thus be expressed as a function of r_1 only:

$$r_2 = \sqrt[3]{\Omega - r_1^3} \quad \text{with} \quad \Omega = \frac{3V_0}{4\pi}.$$

And equation (4.13) becomes

$$J_T = K_L \frac{r_1^2(\Omega - r_1^3)^{2/3}}{r_1^2 + (\Omega - r_1^3)^{2/3}}\left[\frac{1}{r_1} - \frac{1}{(\Omega - r_1^3)^{1/3}}\right] - K_0 r_1^2\left(\theta^3\frac{1}{r_1^3} - \frac{1}{(\Omega - r_1^3)}\right). \tag{4.14}$$

4.1.4 Role of the Concentration of the Insoluble Substance on the Equilibrium between Droplets

***Determining the Parameters* K_L, K_0** We have $K_L = 4\pi D\alpha C(\infty)/\delta$, or about $1.25 * 10^{-20}\,\text{mol}\cdot\text{cm}^{-1}\cdot\text{s}^{-1}$ if we assign the following values to the different constants:

D = the diffusion coefficient at the interfaces, = $10^{-5}\,cm^2 \cdot s^{-1}$.

α = 10^{-7} cm (2).

$C(\infty)$ = the solubility of the dispersed phase in the continuous one, = $10^{-15}\,mol \cdot cm^{-3}$.

δ = the length of the diffusion layer, = 10^{-6} cm.

K_0 = $4\pi L_0 RT m_0 r_{02}^3$, or about $1.25 * 10^{-20}\,mol \cdot cm \cdot s^{-1}$.

In calculating L_0, the overall permeability of the interbubble medium (diffusion coefficient divided by a thickness and divided by RT), we set the diffusion coefficient D to about $10^{-5}\,cm^2 \cdot t^{-1}$ and use a thickness of about 10^{-5} cm. Then we have L_0 = about 10^{-2}. We set the variable m_0 to about $10^{-6}\,mol \cdot cm^{-3}$ and r_{02} to about 10^{-5} cm.

We cannot, of course, know the exact values of these two parameters, but by setting the approximations above, we can assign very similar values to them. We end up treating the evolution of the total flux of water between the two droplets as a function solely of the radius of the smaller droplet. It is obvious that equilibrium will be reached when the total flux is equal to zero. In turn the zero value of the flux gives us the radius of the smaller droplet at equilibrium.

Evolution of the Total Flux between the Two Droplets as the Radius* r_1 *Diminishes The three curves plotted in Figure 4.1 correspond to three different initial droplet–radius ratios. As the initial difference between the sizes of the droplets increases, the initial flux—essentially the resulting difference in Laplace pressure—increases. We find that the greater the initial difference,

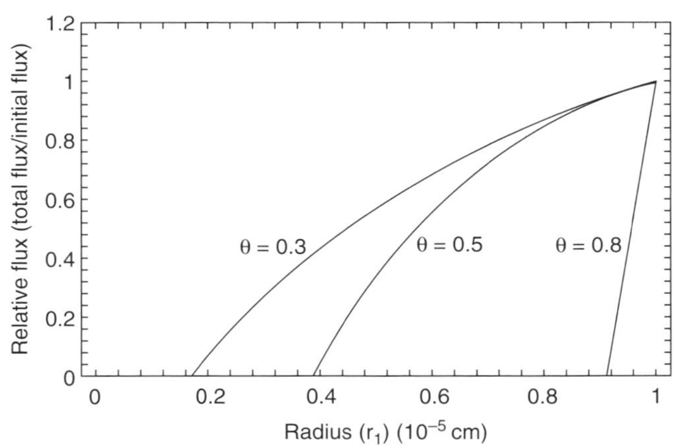

Figure 4.1 Evolution of the relative flux of water (total flux of water divided by the initial flux) versus the radius (r_1), for different initial radius ratios. K_0 is taken to be equal to K_l.

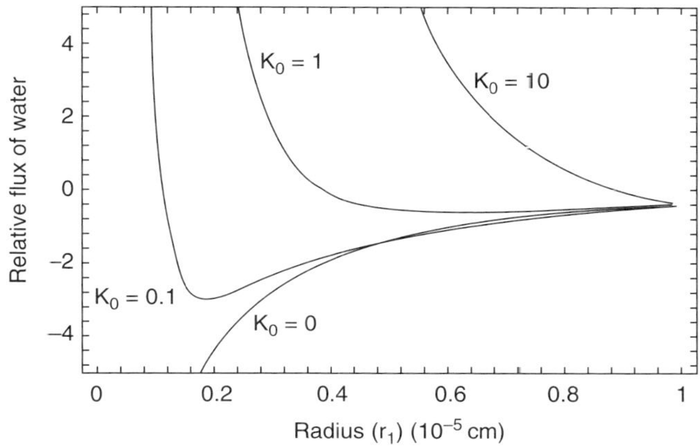

Figure 4.2 Evolution of the relative flux of water (as in Figure 4.1) versus the radius r_1, for different values of K_0, that is, different concentrations of insoluble solute.

the smaller the droplet will be at equilibrium. This is because the system requires a greater evolution in size to permit the osmotic flux to counterbalance the initial Laplace flux.

Because K_0 is proportional to the concentration of the solute in the dispersed phase, we can visualize the effect of the concentration on the evolution of the system just by considering the flux as a function of r_1, for different values of K_0. When K_0 is equal to zero, we have the disappearance of the smaller bubble or droplet due to the Laplace pressure difference. When we take a nonzero value for K_0, we find a zero flux for a nonzero radius r_1, the equilibrium radius.

The plots for $K_0 = 1$ and $K_0 = 10$ (Figure 4.3) show the tremendous effect of the concentration on the final size of the droplet. Indeed the equilibrium radius will be $0.4r_{01}$ for $K_0 = 1$ and $0.9r_{01}$ for $K_0 = 10$.

Limits of the Model For the simulation we assume that the system is in steady state. This implies that the variation of C_m is not important and does not create a large gradient of concentration in the continuous medium.

Figure 4.3 allows us to verify that the steady state is valid only in cases where the droplet does not disappear. The exponential included in expression (4.6) for the solubility at the interface of the smaller droplet is now predominant and induces a rapid increase in C_m. However, equation (4.14) cannot be used to compute the disappearance of the droplet because, in this case, the exponential of the Kelvin equation cannot be given a first-order approximation. Such other physical events as the desorption of the surfactant from the interface must be taken into account.

To sum up, as shown in Figure 4.3, we can apply the model if we consider a diminution in size down to, but not beyond, a radius of a few nanometers.

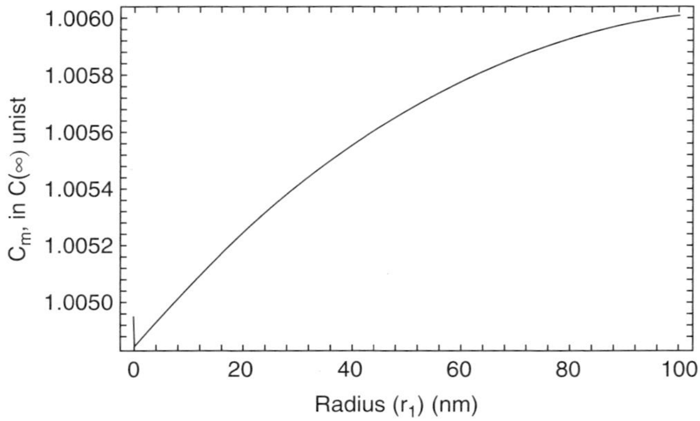

Figure 4.3 Evolution of the mean concentration of the dispersed solvent in the continuous phase, with a diminution of the size of the smaller droplet.

The model then corresponds to the evolution of the system balanced by osmotic pressure as shown in Figures 4.2 and 4.3.

4.2 PREDICTING CHANGE IN BUBBLE-SIZE DISTRIBUTION DUE TO INTERBUBBLE GAS DIFFUSION IN FOAMS

4.2.1 Lemlich's Theory

At least two distinct spontaneous phenomena can change an existing distribution of bubble sizes in a liquid foam: the rupture of the lamellae between bubbles and the transfer of gas between bubbles by diffusion. While some liquid foams are extremely resistant to rupture, none are resistant to diffusion of gas. The following treatment of interbubble gas diffusion and the resultant change in bubble-size distribution builds on the work of Lemlich (1978). Lemlich's approach, instead of viewing the gas as diffusing directly from bubble to bubble, interprets the phenomenon as first diffusing into the liquid region midway between the bubbles. The concentration of gas in this liquid can be considered as being equivalent (by Henry's law) to a gas pressure in the liquid. Then, by virtue of the law of Laplace and Young, this gas-pressure equivalent can be considered to be the gas pressure that would exist within a fictitious spherical bubble of radius ρ. Thus the pressure difference ΔP_{Lap} between a bubble of any radius r and the liquid is

$$\Delta P_{\mathrm{Lap}} = 2\gamma\left(\frac{1}{\rho} - \frac{1}{r}\right), \tag{4.15}$$

where γ is the surface tension.

The molar rate of gas transfer $S(r, t)$ from a bubble to the liquid is assumed to be given by

$$S(r, t) = -JA\Delta P_{\mathrm{Lap}}, \tag{4.16}$$

where J is the effective permeability of the continuous phase to the transfer and A is the surface area through which the transfer takes place. If n is the number of moles of gas in the bubble, by the conservation of moles we have $S(r, t) = -dn/dt$. If we approximate A by $4\pi r^2$, equations (4.15) and (4.16) yield

$$\frac{dn}{dt} = 8\pi J\gamma\left(\frac{r^2}{\rho} - r\right). \tag{4.17}$$

We can assume that the conservation of gaseous moles occurs throughout the foam as a whole, $\sum_i dn/dt = 0$. The summation can then be taken over all the bubbles. So, using (4.17), we get

$$\sum_i n_i\left(\frac{r_i^2}{\rho} - r_i\right) = 0, \tag{4.18}$$

where n_i is the number of bubbles of radius r_i.

Lemlich carried this calculation further and considered an average effective ρ independent of r but varying with time. Thus, by Lemlich's approach, (4.18) gives

$$\rho = \frac{\sum_i n_i r_i^2}{\sum_i r_i} \quad \text{or} \quad \rho = \frac{\int_0^\infty r^2 F(r, t)dt}{\int_0^\infty rF(r, t)dt}, \tag{4.19}$$

where $F(r, t)$ is the frequency distribution function of r at time t.

Furthermore, as the pressure in a bubble is only slightly higher than the surrounding pressure P_a, which is typically the atmospheric pressure, from the ideal gas law and the formula for the volume of a sphere, we have

$$n = \frac{4\pi P_a r^3}{3R_g T}, \tag{4.20}$$

where R_g is the ideal gas constant and T is the absolute temperature.

Combining Equations (4.17), (4.19), and (4.20), Lemlich deduced that the rate of change of the radius of a bubble over time is given by

$$\frac{dr}{dt} = \frac{2J\gamma R_g T}{P_a}\left(\frac{\int_0^\infty rF(r, t)dt}{\int_0^\infty r^2 F(r, t)dt} - \frac{1}{r}\right), \tag{4.21}$$

or more succinctly,

$$\frac{dr}{dt} = K\left(\frac{1}{r_{21}} - \frac{1}{r}\right), \tag{4.22}$$

where $K = 2J\gamma R_g T/P_a$ and $r_{21} = \int_0^\infty r^2 F(r,t)dt \Big/ \int_0^\infty rF(r,t)dt$. Thus bubbles with $r > r_{21}$ grow in size, whereas bubbles with $r < r_{21}$ shrink in size and eventually disappear.

For the sake of generality and utility in simulations, Equation (4.22) is recast in a dimensionless form as

$$\frac{dR}{dY} = \left(\frac{1}{R_{21}} - \frac{1}{R}\right) \quad \text{with} \quad R = \frac{r}{r_c},\ Y = \frac{Kt}{r_c^2},$$

where r_c is some convenient characteristic radius such as the average initial radius of the distribution.

4.2.2 Lemlich's Theory Applied to Bubble Containing a Insoluble Gas

As the bubbles change in size, osmotic pressure will develop between gas bubbles that contain a certain amount of gas that is totally insoluble in the continuous phase. If we consider the phenomenon with two bubbles of different initial radii, initially the concentration of the insoluble vapor is constant and only the Laplace pressure acts on the bubbles. The smaller bubble shrinks as a result, and the larger one increases in size. With the changes in size, the concentration of insoluble gas increases in the smaller bubble and decreases in the larger one, giving rise to an osmotic pressure that opposes the Laplace pressure.

The osmotic pressure difference ΔP_{osm} that drives the flux of the soluble gas opposing the flux driven by the Laplace pressure difference can be written as

$$\Delta P_{\text{osm}} = R_g T \Delta m,$$

where R_g is the gas constant, T the absolute temperature, and Δm the difference in concentration of the insoluble gas between the fictitious bubble of radius ρ and a bubble of radius r.

Using m_0 as the initial concentration of insoluble gas, and r_0 and ρ_0 as the initial radii of the bubble and the fictitious bubble, respectively, we get

$$\Delta P_{\text{osm}} = m_0 R_g T\left[\left(\frac{r_0}{r}\right)^3 - \left(\frac{\rho_0}{\rho}\right)^3\right]. \tag{4.23}$$

Recalling Equation (4.16), we have

$$S(r,t) = -JA(\Delta P_{\text{Lap}} + \Delta P_{\text{osm}}). \tag{4.24}$$

The conservation of the total number of moles of the soluble gas gives

$$\sum_i n_i\left(\frac{r_i^2}{\rho} - r_i\right) + \frac{m_0 R_g T}{2\gamma}\sum_i n_i\left(\frac{r_{i0}^3}{r_i} - \frac{\rho_0^3}{\rho^2}r_i^2\right) = 0. \tag{4.25}$$

In the case where ρ is independent of r but varies with time, (4.25) would yield a cubic equation for ρ as a function of the initial radius and at time t. Without pursuing this line of reasoning, we note only that ρ is initially equal to r_{21}, and for the purpose of the following discussion we will assume that ρ has this value.

From (4.23) and (4.24), following Lemlich, we derive the rate of change of the radius time:

$$\frac{dr}{dt} = \frac{2J\gamma R_g T}{P_a}\left\{\frac{1}{r_{21}} - \frac{1}{r} + \frac{m_0 R_g T}{2\gamma}\left[\left(\frac{r_0}{r}\right)^3 - \left(\frac{r_{21_0}}{r_{21}}\right)^3\right]\right\}. \tag{4.26}$$

For the sake of generality and utility in simulations, Equation (4.26) is recast in a dimensionless form as

$$\frac{dR}{dY} = \frac{1}{R_{21}} - \frac{1}{R} + \Phi\left[\left(\frac{R_0}{R}\right)^3 - \left(\frac{R_{21_0}}{R_{21}}\right)^3\right] \tag{4.27}$$

with

$$R = \frac{r}{r_c}, \quad Y = \frac{2J\gamma R_g T}{P_a}\frac{t}{r_c^2}, \quad \Phi = \frac{m_0 R_g T r_c}{2\gamma},$$

where r_c is some convenient characteristic radius, such as the average initial radius of the distribution.

4.2.3 Numerical Simulation

Values for R_{21} ***and*** Φ For values typically found in familiar liquid foams—$m_0 = 0.01\,\text{mol/L}$, $\gamma = 30\,\text{dynes/cm}$, $T = 298\,\text{K}$, and $r_c = 1\,\mu\text{m}$—we have $\Phi = 0.4$. With this convention in mind, we see that a reasonable range of values to investigate for Φ is between 0.01 and 1.

Beginning with the empirical distribution of de Vries and the Maxwell-Boltzmann distribution of Bayens, we can determine the following initial values of R_{21}: 1.48 for the distribution of de Vries and 1.18 for the distribution of Bayens (see Figures 4.4 and 4.5).

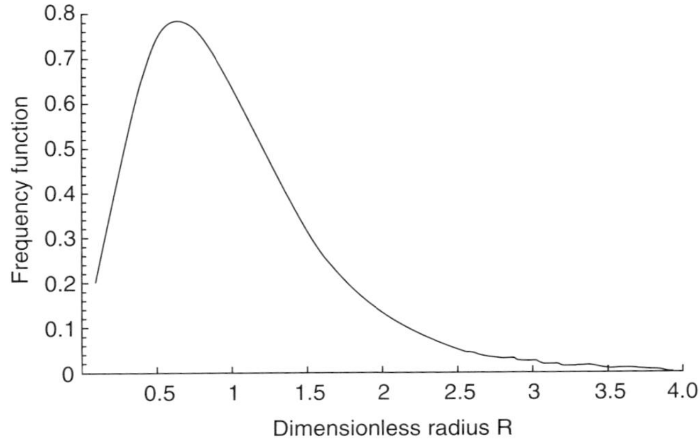

Figure 4.4 Distribution of de Vries.

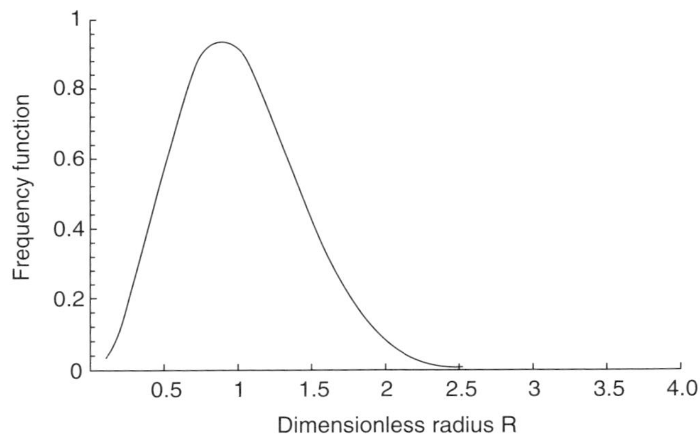

Figure 4.5 Distribution of Bayens.

***Curves of* dR/dY *as a Function of* R** Using equation (4.27), and assuming that R_{21} is a constant equal to its initial value, we can plot dR/dY as a function of R if we know the initial value of the radius of a bubble. Figure 4.6 shows one such plot for $R_0 < R_{21}$. We start with an initial radius (R_0) of 0.6. At this radius dR/dY is negative, so the radius will decrease. This initial decrease is due to the Laplace pressure. As the minimum value of dR/dY is reached, the osmotic pressure becomes more significant, equaling and then exceeding the Laplace pressure, with a resultant rapid decrease in dR/dY to a value of 0. This equilibrium is a stable one, since, if the bubble continues to shrink, dR/dY will become positive and the radius will increase again.

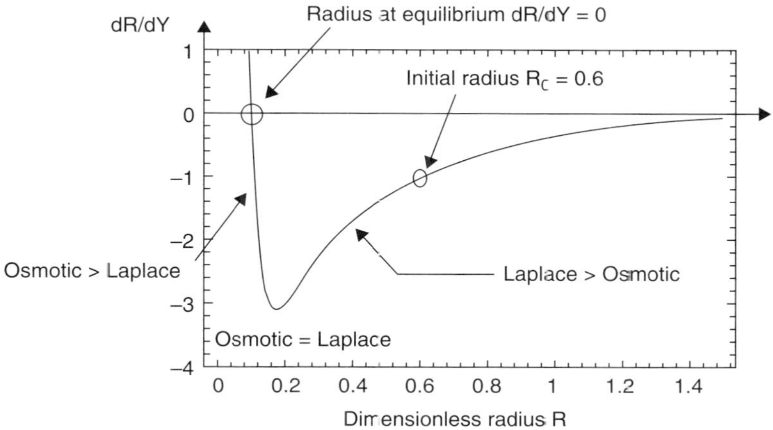

Figure 4.6 Plot of dR/dY as a function of the radius R for $\Phi = 0.05$, and $R_{21} = 1.5$, and $R_0 = 0.6$.

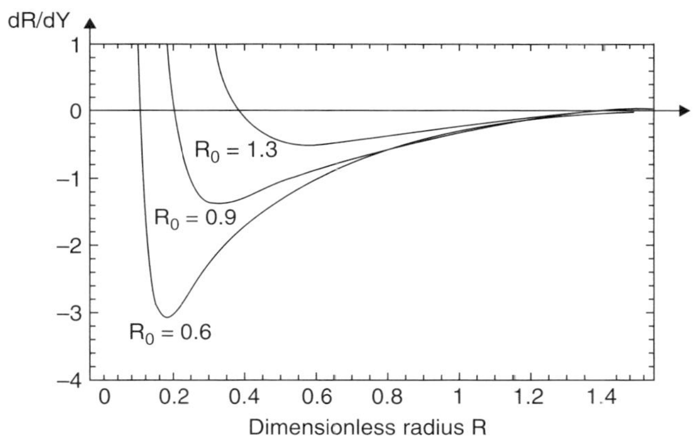

Figure 4.7 Plot of dr/dt for several initial radii (R_0) and with $\Phi = 0.05$ and $R_{21} = 1.5$.

Figure 4.7 shows the difference in the initial rate of shrinkage for bubbles with different initial radii. We see that the smaller the bubble, the greater is the Laplace pressure. For initial radii of 0.6, 0.9, and 1.3, the equilibrium radii are 0.107, 0.204, and 0.378, respectively.

Influence of the Amount of Insoluble Gas Figure 4.8 shows the effect of increasing the osmotic pressure on dr/dt. As can be seen, an increase of Φ reduces the effect of the Laplace pressure, and the equilibrium radius increases as well. Figure 4.9 shows that an increase in Φ leads to an equilibrium radius closer to the initial radius.

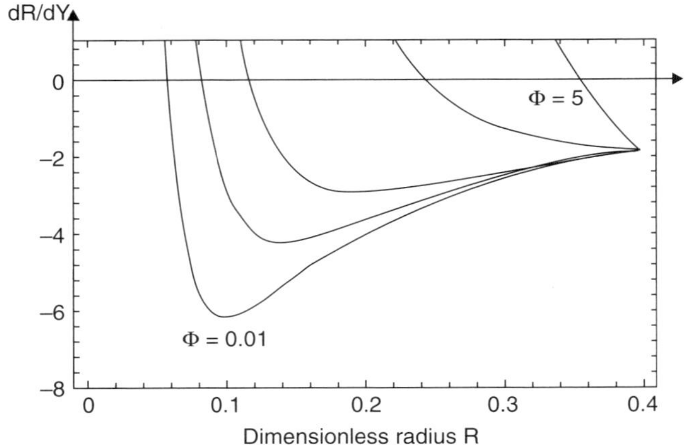

Figure 4.8 Effect of increasing the osmotic pressure (Φ = 0.05, 0.1, 0.2, 1, 5) for R_{21} = 1.5 and R_0 = 0.4.

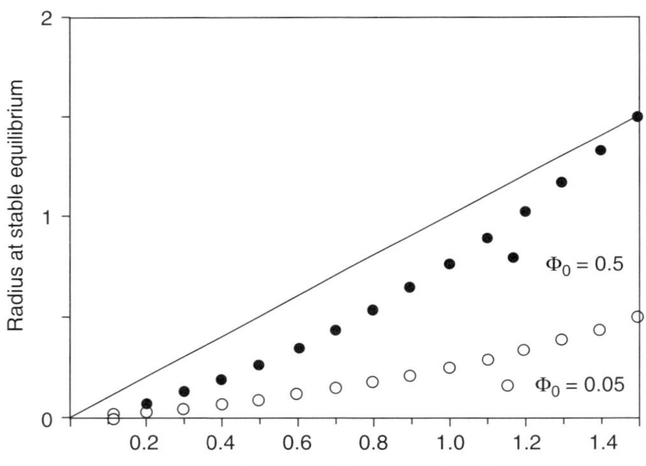

Figure 4.9 Radius at equilibrium versus initial radius R_0.

Variation of the Radius as a Function of Time As Monsalve and Schechter (1984) showed, if we assume R_{21} to be constant, then the equation of Lemlich (4.22), which is equation (4.26) with = 0, can be integrated to give

$$Y = R_{21}\left[R - R_0 + R_{21}\ln\left(\frac{R - R_{21}}{R_0 - R_{21}}\right)\right].$$

We can then plot Y as a function of R (and vice versa, R as a function of Y) as shown in Figure 4.10.

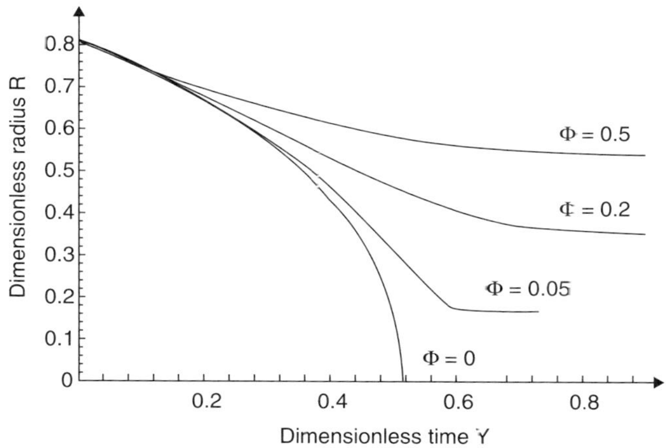

Figure 4.10 Curves of radius versus time for a bubble with and without an osmotic pressure, with $R_0 = 0.8$, $R_{21} = 1.5$, and $\Phi = 0, 0.05, 0.2, 0.5$.

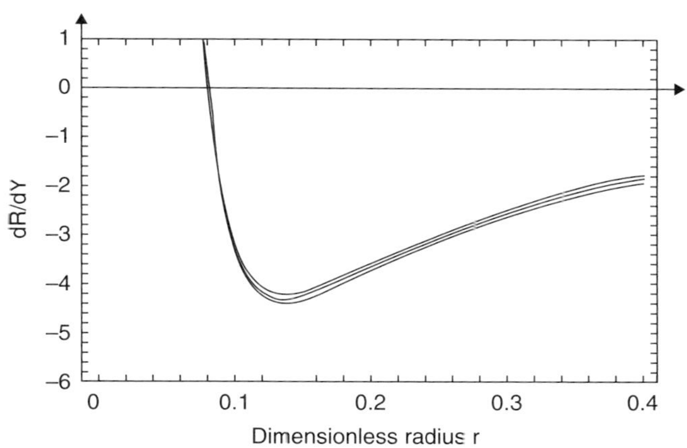

Figure 4.11 Plot of dR/dY for different R_{21} (1.5, 1.7, 2) and $\Phi = 0.05$, $R_0 = 0.5$, and $R_{21_0} = 1.5$.

If take into consideration the contribution of an osmotic pressure (i.e., $\Phi \neq 0$), we integrate the equation numerically (using Mathematica) and plot R as a function of time; see Figure 4.10. As the figure shows, adding a small amount of insoluble compound prevents the bubble from disappearing, and an equilibrium radius is attained.

Is the Hypothesis of Constant* R_{21} *Reasonable? As can be seen from Figure 4.11, a change in R_{21} has very little effect on the curves for dR/dY. Thus the hypothesis of constant R_{21} is tenable, at least for $R_0 < R_{21}$.

4.2.4 Stable Equilibrium Condition

The conditions necessary for one bubble of radius r to reach a stable equilibrium are (i) that $dr/dt = 0$ and (ii) that $d^2r/dt^2 \leq 0$. Let us expand the second condition, taking into consideration the fact that r_{21} is a function of time.

$$\frac{d^2r}{dt^2} = K\left(-\frac{1}{r_{21}^2} + 3\frac{m_0 R_g T}{2\gamma}\left(\frac{r_{21_0}}{r_{21}}\right)^3 \frac{1}{r_{21}}\right)\frac{dr_{21}}{dt}.$$

As $dr_{21}/dt \geq 0$, the condition $3m_0 R_g T r_{21_0}^2 / 2\gamma \leq r_{21}^2$ must be fulfilled.

4.3 EQUILIBRIUM OF W/O DROPLET CONTAINING SALT

We consider a water droplet of radii r containing a certain amount of salt, in a solvent (oil phase). We assume that the film around this water droplet prevents coalescence but is not impermeable to water.

The Laplace pressure $2\gamma/r$ of a droplet containing a salt, dispersed in a solvent, will increase the shrinkage of the droplet, while the osmotic pressure will increase as well. In turn the diffusible solvent (water in the present case) slows in activity, and thus stimulates the backdiffusion of the solvent. In the ideal case the osmotic pressure is given by

$$\Pi_{\text{osm}} = mRT = m_0\left(\frac{r_0}{r}\right)^3 RT,$$

where m is the molar concentration of salt and m_0 refers to the original droplet.

We can write the excess pressure ΔP in the droplet as

$$\Delta P = \frac{2\gamma}{r} - m_0\left(\frac{r_0}{r}\right)^3 RT$$

At equilibrium, we should have

$$\frac{d(\Delta P)}{dr} = -\frac{2\gamma}{r^2} - 3m_0\left(\frac{r_0}{r}\right)^3 \frac{1}{r} RT = 0.$$

Thus

$$2\gamma \leq 3\,mRTr. \tag{4.28}$$

Using equation (4.28), which was proposed by Walstra (1996), we can calculate the minimum percentages difference of NaCl required (for various

TABLE 4.1 Estimates of the minimum salt concentration required for various $\gamma = 10\,mN/m$ and r to prevent Ostwald ripening in a water/oil emulsion

Radius (r) nm	γ (mN/m)	m(mol/m^3)	%NaCl
10	10	2.69	0.016
100	10	0.269	0.0016
200	10	0.134	0.0008
400	10	0.067	0.0004

values of the radius of the droplet and a surface tension of 10 mN/m.) to counterbalance the Ostwald ripening between two droplets; these values are listed in Table 4.1. Another condition is that m has to be greater than $2\gamma/3RTr$. It is apparent that very small quantities of a suitable solute will suffice to stop Ostwald ripening.

4.4 CONCLUSION

From of our model, which is greatly simplified, it is clear that the presence of a small quantity of an insoluble solute in a water-in-oil dispersion (or an insoluble vapor in the case of a foam) can prevent the destabilizing effect of the Ostwald ripening phenomenon. It appears that with changes to the radii of two bubbles, a difference in salt concentration on the order of 0.0016% to 0.0004% for a radii of 100 to 400 nm is sufficient to stop Oswald ripening (when $\gamma = 10\,mN/m$).

REFERENCES

Kabalnov AS, Shchukin ED. 1965. Ostwald ripening theory: Applications to fluorocarbon emulsion stability. *Adv Colloid Interface Sci* 38: 69–97.

Kralchevsky PA, Danov KD, Denkov ND. 2002. Chemical physics of colloid systems and interfaces. In: *Handbook of Surface and Colloid Chemistry*, Birdi KS ed. New York: CRC Press, pp 166–243.

Lemlich P. 1978. Prediction of changes in bubble size distribution due to interbubble gas diffusion in foam. *Ind Eng Chem Fundam* 17(2): 89–93.

Monsalve A, Schechter RS. 1984. The stability of foams: dependence of observation on the bubble size distribution. *J Colloid Interface Sci* 97(2): 327.

Princen HM, Mason SG. 1965. The permeability of soap films to gases. *J Colloid Interface Sci* 20: 353–375.

Walstra P. 1996. *Encyclopedia of Emulsion Technology*, Vol. 4, Becher P, ed. New York: Marcel Dekker, pp 1–62.

CHAPTER 5

Multiple Emulsions Stabilized by Biopolymers

RACHEL LUTZ and ABRAHAM ASERIN

Contents

5.1 INTRODUCTION

Multiple (or double) emulsions are complex liquid dispersion systems known also as emulsions of emulsions, in which the droplets of one dispersed liquid (water in oil or oil in water) are further dispersed in another liquid (water or

Multiple Emulsions: Technology and Applications, Edited by Abraham Aserin

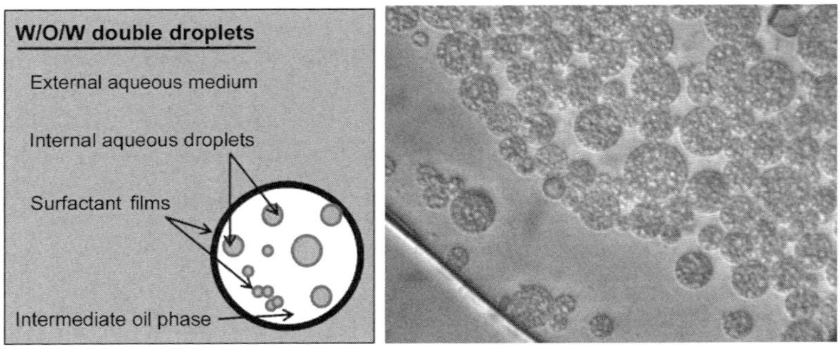

Figure 5.1 Schematic representation and microscopic image of W/O/W multiple emulsion.

oil, respectively), producing W/O/W or O/W/O. The inner dispersed droplets (hereafter called inner droplets or just droplets, while the droplets of the multiple emulsion will be termed, for simplicity, the globules) in the multiple emulsion are separated from the outer liquid phase by a layer of another phase. A schematic representation and microscopic image of a W/O/W multiple emulsion globule is shown in Figure 5.1. W/O/W multiple emulsions are the most studied system because they have the potential to become commercial products—mainly in food systems but also in cosmetics and pharmaceuticals (Garti and Benichou, 2001; Garti and Lutz, 2004).

For close to 60 years, multiple emulsions were considered an emerging technology and a major scientific challenge (see Seifriz, 1925, for the first report), but in practice, only a few industrial products based on multiple emulsions exist in the marketplace. The main use of multiple emulsion systems is as confinement and protection systems for controlled release of active compounds.

In the food industry (Matsumoto, 1986; Yoshida et al., 1999) W/O/W multiple emulsions can improve the solubility of certain active matter, solubilize oil-insoluble materials, serve as protective liquid reservoirs for molecules sensitive to external environmental reactivity such as oxidation, light, and enzymes, and serve as entrapment reservoirs for masking undesired flavors and odorants. Applications in the cosmetics industry (Kim and Lee, 1999; Gallarate et al., 1999) include aqueous preparations that provide a good "feel" and slow release of active materials or flavors, deposition of water-soluble agents onto the skin from wash-off systems. Most applications are related to the pharmaceutical industry (Okochi and Nakano, 2000; Shima et al., 2006; Vasiljevic et al., 2006), such as enhancing the chemotherapeutic effect of anticancer drugs, drug immobilization, treatment of drug overdoses, and protecting insulin from enzymatic degradation. In the agriculture industry the potential applications (Versteeg, 1978) are sustained release systems for fertilizers and pesticides, and in the fuel industry (Lin and Wang, 2003) as alternative fuel for diesel engines. This chapter deals mainly with the food industry.

The formation of a multiple emulsion requires only two simple consecutive emulsification steps, yet close examination of the results of such simple preparations reveals that the unsolved technical problems are not easy to overcome. In recent decades researchers have contributed information on the selection of different ingredients such as various blends of emulsifiers (mostly nonionic), oils, and viscosity agents. These reviews provided new ideas on novel analytical tools and advanced preparation techniques to better understand the nature of the instability of multiple emulsions and to control release pathways (Garti and Benichou, 2001; Grossiord and Seiller, 2001; Bibette et al., 2002; Garti and Lutz, 2004).

5.2 STABILITY CONSIDERATIONS

Because of their large droplet sizes, multiple emulsions, even more than regular emulsions, are inherently thermodynamically unstable. In addition the release of active material from the inner phase to the outer phase is mostly uncontrolled. Many studies have dealt with these two major problems (Garti and Benichou, 2001; Jiao et al., 2002; Garti and Lutz, 2004; Wang et al., 2006).

The stability mechanisms have been extensively studied in empty multiple emulsions, but most recent investigations look at the emulsion containing entrapped solutes. In these systems it is very difficult to distinguish between stability and release mechanisms because the two phenomena occur in parallel and are sometimes derived from the same driving forces. The osmotic pressure gradient affects and motivates the release of ingredients from the core, and must be controlled. Many researchers loaded the multiple emulsions with active ingredients in the presence of salts, glucose, and other osmotic pressure regulators.

5.3 NONIONIC AND POLYMERIC SURFACTANTS

During years of investigation to improve stability and to control, sustain, and/or prolong the release of active materials, monomeric surfactants have been progressively replaced by polymeric emulsifiers as outer or inner interface stabilizers.

Low-molecular-weight emulsifiers (monomeric) can migrate from the W/O interface to the oil phase, and hydrophilic surfactants can be transported to the inner interface and alter the required hydrophilic/hydrophobic balance of each of the phases. Most of the studies between 1970 and 1985, searched for monomeric emulsifier blends of hydrophilic and hydrophobic surfactants with the best ratios for optimal stabilization. Matsumoto et al. (1976, 1985) established an empirical weight ratio of 10 for the internal hydrophobic to the external hydrophilic emulsifiers. Garti et al. (Magdassi and Garti, 1986; Benichou et al., 2002a) proved that free exchange between the internal and

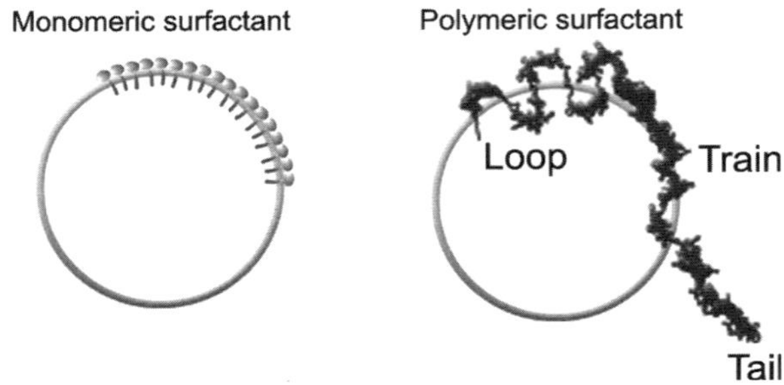

Figure 5.2 Schematic representation of monomeric and polymeric surfactant adsorption on interface.

external emulsifiers required calculation of an effective HLB value of emulsifiers to optimize the stabilization of the emulsion. In most cases the internal emulsifiers are used in great excess relative to the external emulsifiers. The nature of the emulsifiers also dictated the number of compartments and the internal volume that the inner phase occupies.

Among the most promising emulsifiers for making multiple emulsions are synthetic polymeric amphiphiles and naturally occurring biopolymers (proteins or hydrocolloids). Polymeric amphiphiles are known to be multi-anchoring amphiphiles with irreversible adsorption capabilities that can improve the droplets' interfacial coverage during emulsification since they provide strong steric stabilization capabilities (Figure 5.2). The gain in free energy by such adsorption is much greater than that of the absorbed monomeric surfactants (Ortega-Vinuesa et al., 1996). The polymeric amphiphiles form thick and flexible films that are strongly anchored into the oil-water interface in most cases.

5.4 STERIC STABILIZATION BY POLYMER

There are three main mechanisms of stabilization with polymeric amphiphiles:

1. Steric stabilization resulting from hydrophobic interactions among adsorbed polymers.
2. Depletion stabilization by nonadsorbing macromolecules that prevent collision between particles (droplets) and provide elasticity to the system.
3. Electrostatic repulsion between two droplets carrying the same charge.

5.4.1 Steric Stabilization

The main contribution to the stability of macromolecular-stabilized emulsions is related to droplets reaching a distance where compression due to polymer to polymer interactions occurs. Such polymer to polymer interactions are the basis for steric stabilization by two contributions: osmotic pressure effects and entropic or volume restriction. If there are polymeric chains covering the external surface of the droplets, (the average thickness of such polymeric layer is δ), then increasing in the local concentration will leads to different in the osmotic pressure between the solution and the overlap zone (Figure 5.3). Osmotic pressure effect will occur when the two particles are closer than a distance equal to 2δ (Figure 5.3*a*). The osmotic pressure of the solution in the overlap zone will be less than that in the regions external to it, leading to a driving force for the spontaneous flow of solution into the overlap zone; that force pushes the particles apart. However, if the two particles are closer than a distance equal to δ (Figure 5.3*b*), at least some of the polymer molecules will undergo elastic compression. Thermodynamically the compression corresponds to a net loss in configurational entropy, lose of certain degree of freedom. Hence, in order to regain the lost entropy, the droplets must move to obtain more freedom of movement.

The net energy (G_i) of the droplets interaction will be including the attractive van der Waals forces (ΔG_a) and the repulsive steric interaction (ΔG_s).

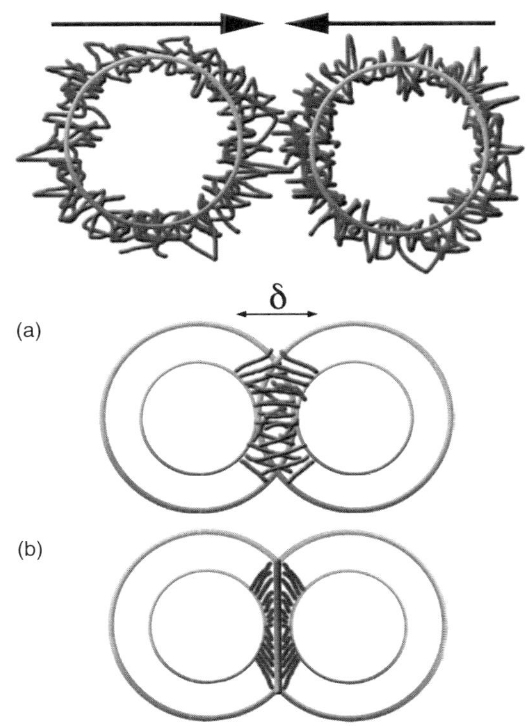

Figure 5.3 Schematic representation of steric effects: (*a*) Osmotic effect; (*b*) elastic repulsion.

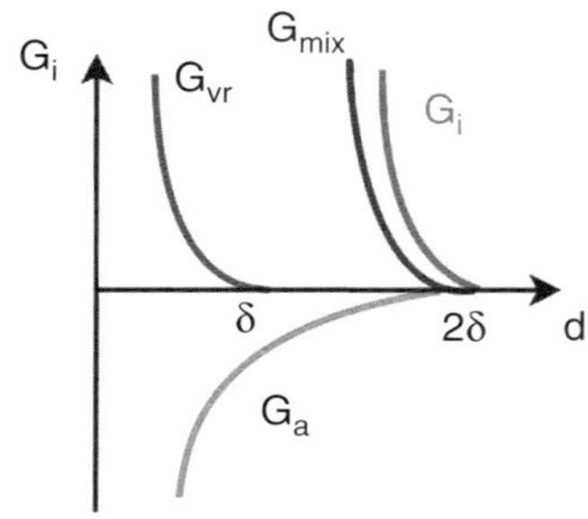

Figure 5.4 Different free energies involved in the steric stabilization mechanism as a function of distance (d).

$\Delta G_i = +\Delta G_s - \Delta G_a$. ΔG_s, is composed of volume restriction—termed G_{vr} and from the osmotic effect mixing—G_{osm}, thus the steric free energy is $\Delta G_{\mathrm{s}} = \Delta G_{\mathrm{osm}} + \Delta G_{\mathrm{vr}}$ (Figure 5.4).

The nature of the surfactant/droplet affinity and the nature of the continuous phase will dictate the overall repulsion or attraction interactions leading to droplet repulsion or coalescence. In cases of "poor" solvent or the bulk solvency decreasing toward the dispersed phase, a significant minimal value of the total free energy of interaction can be reached.

Steric stabilization becomes a dominant factor when polymeric amphiphiles are used as surfactants. The polymeric chains act to cover and stabilize mainly the globules by adsorbing onto the external interface (Ortega-Vinuesa et al., 1996; Myers, 1998a; Benichou et al., 2002a).

5.4.2 Depletion Stabilization (Myers, 1998a)

Depletion force is expected to occur whenever nonadsorbing polymer is added to a colloidal dispersion. A polymer chain in solution will keep, on average, a configuration that is entropically most favorable. The polymer may approach a surface to a distance such that its farthest segments just meet the surface. To approach more closely, the polymer must adopt a less favorable conformation with a resulting loss of configurational entropy and also loss of system stability.

The distance between two droplets or colloidal dispersion determines the depletion force as well. A gradient of polymer segmental concentration exists in the surface area. Depletion attraction arises when the depletion layers, each of thickness Δ (or R_g), associated with two surfaces, overlap. Theoretically, at large separations ($D > 2\Delta$), the segment concentration increases from zero at the surface to that of the bulk solution in the middle of the gap between the surfaces. However, at smaller separations ($D < 2\Delta$), the concentration at the midplane falls below that of the bulk. As a result the pressure in the bulk solution is greater than that in the gap, and there is an attractive osmotic force between the surfaces (Kuhl et al., 1998).

The first researchers to relate particle flocculation in the presence of non-adsorbing macromolecules were Asakura and Oosawa (1954). They proposed that the destabilization is due to a differential osmotic pressure setup once the macromolecules are excluded from the region between the particles. Other researchers contributed to understanding the depletion stability. Vincent et al. (Fleer et al., 1984) interpreted the destabilization of colloidal dispersions that occurs at high polymer concentrations to be due to a decrease in the depletion layer thickness, resulting in a weakened attraction. Some years earlier Feigin and Napper (1980) had attributed depletion stabilization to a repulsive force due to the high energy cost of demixing the solution prior to the onset of the attractive regime. Walz and Sharma (1994) later predicted a repulsive barrier at separations between one and two diameters of the excluded macromolecule due to excluded volume effects when second-order interactions are taken into account. Mao et al. (1995a, b) and more recently Mezzenga et al. (2004) have shown that excluded volume considerations can lead to a repulsive force regime between particles in semidilute concentrations of smaller nonabsorbent particles. When the excluded particles are rod shaped, the repulsive barrier is greatly enhanced, leading to kinetic stabilization.

5.4.3 Electrostatic Stabilization

The electrostatic stabilization mechanism is well documented for simple O/W emulsions (Myers, 1998a). Multiple emulsion droplets are much larger in size, and therefore the repulsive electrostatic forces are less pronounced (Figure 5.5).

5.5 INSTABILITY CONSIDERATIONS

In addition to the common instability of multiple emulsions, there are two instabilities related to polymeric surfactants. The common destabilization mechanisms are demonstrated in Figure 5.6 and include aggregation, flocculation, and coalescence (Mezzenga et al., 2004).

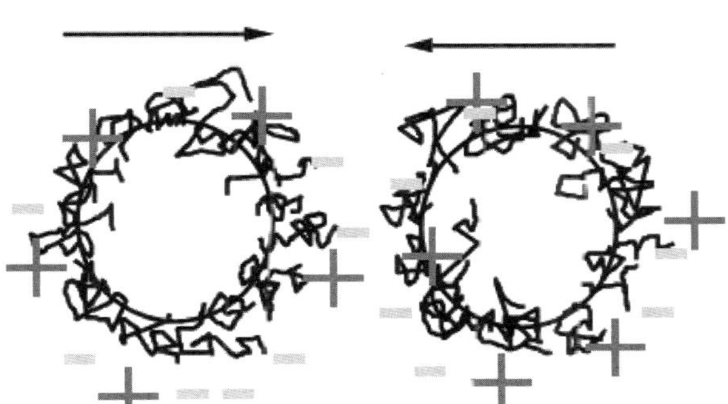

Figure 5.5 Multiple emulsion repulsive electrostatic forces.

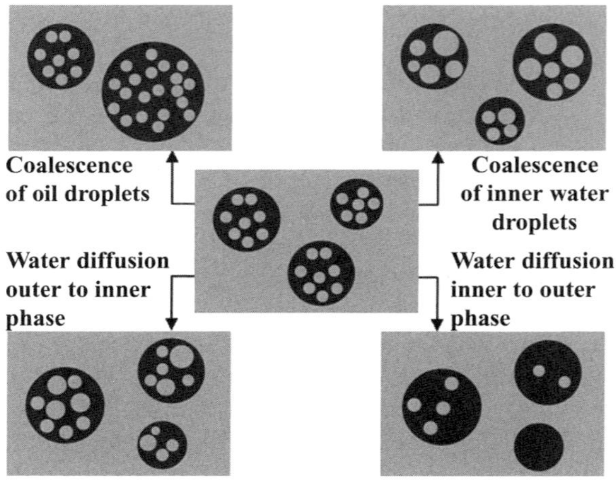

Figure 5.6 Schematic representation of possible instabilities occurring in W/O/W multiple emulsions (Mezzenga et al., 2004).

The other destabilization mechanisms are related, for example, to uncontrolled transport of the inner phase, the inner emulsifiers, and the inner addenda. Water as well as solubilized molecules can be transported across the oil layer (in W/O/W emulsions) from the inner phase to the outer phase or in the opposite direction, depending on the osmotic pressure gradient between the two aqueous phases (Wen and Papadopoulos, 2001). Water transported from the inner phase empties the inner droplets, and turns the W/O/W multiple emulsion globules into a simple O/W emulsion. Transport of water from the continuous phase to the inner phase may cause an internal "explosion" of the aqueous phase.

In addition, the use of polymeric amphiphiles may be beneficial but is mostly problematic because of two additional possible typical destabilization pathways that can occur in these systems, namely depletion and bridging flocculation.

5.5.1 Depletion Flocculation

Polymeric amphiphiles, even when they are not adsorbed to the droplet interfaces, provide an entropic effect that can expressed as a change in viscosity. Depletion flocculation occurs because the center of mass of the depleting agent (the polymers) will not approach the droplet more closely than its effective radius (as mentioned before). Hence there is an area around the droplet that is not accessible to the polymer because of its reduced entropy. If the total depletion volume of the droplets is reduced by the overlap of their depletion zones, the excluded volume is decreased, and the volume accessible to the polymer is increased, increasing its entropy. The attractive force that appears

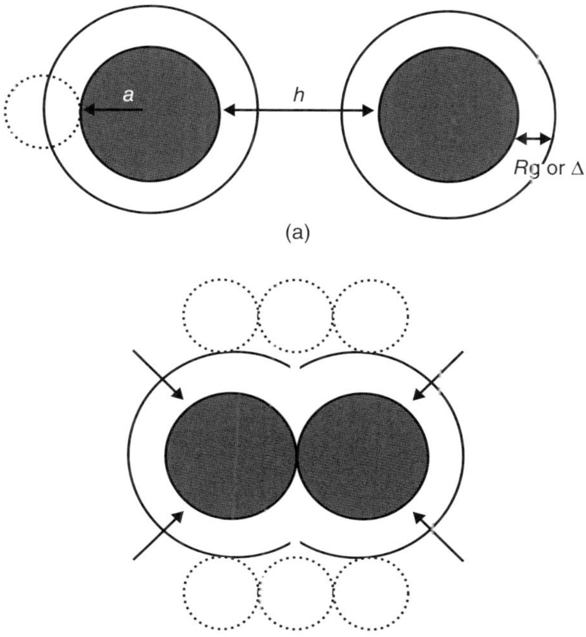

Figure 5.7 Schematic diagram illustrating depletion flocculation. (*a*) Two large spheres (radius *a*) separated by a surface. Surface distance *h* possesses a depletion layer (Δ) with a thickness of the order of the radius R_g of the nonadsorbed species (*dashed circle*). (*b*) Osmotic pressure of the external medium pushes the large spheres together in the directions of the arrows. For clarity, this diagram is not to scale: the size ratio of large spheres to nonadsorbed species is typically considerably greater (Radford and Dickinson, 2004).

between the droplets leads to their flocculation. The force holding the flocs together is weak and reversible, however, and the flocculation can be readily redispersed by mixing (Figure 5.7) (Jerdjev et al., 2006). Droplets that are flocculated will not coalesce but can aggregate into flocs and will have a strong tendency to cream.

5.5.2 Bridging Flocculation

If a very high molecular weight polymer is added to the dispersion system, the various possible points of attachment can encounter two different droplets rather than attaching themselves to the same droplet. That is especially the case where there is a large excess of droplets relative to the concentration of polymer. Attachment of the same polymer chain to two droplets essentially bridges them, resulting in droplet flocculation (Figure 5.8) (Myers, 1998b).

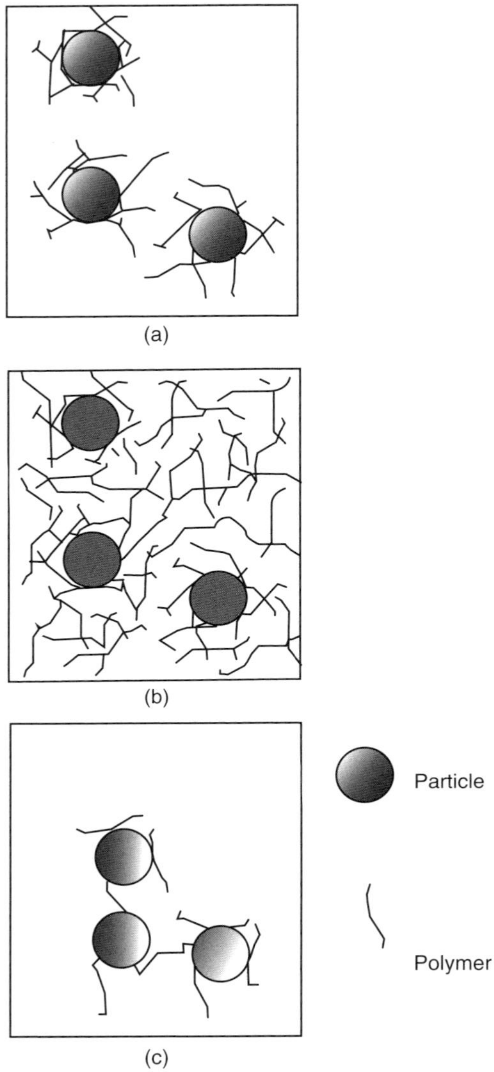

Figure 5.8 Schematic representation of three alternative effects of the adsorption of stiff hydrocolloid polymers on the surface of spherical emulsion droplets, depending on the hydrocolloid concentration and the nature of the hydrocolloid–protein interaction: (*a*) A sterically stabilized system, (*b*) an emulsion gel, and (*c*) a system flocculated by macromolecular bridging (Dickinson, 2003).

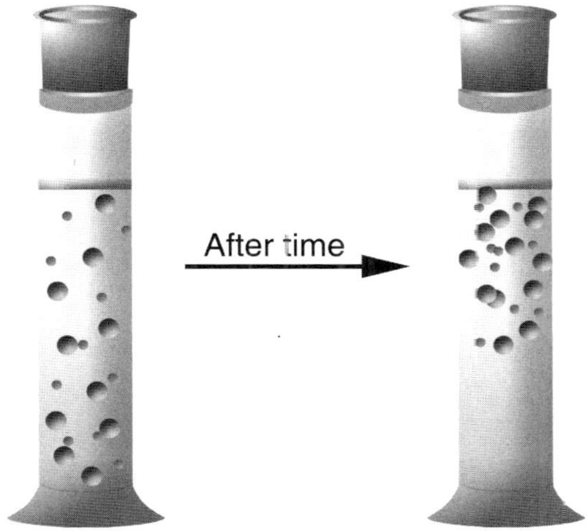

Figure 5.9 Illustration of creaming phenomenon.

5.6 CREAMING PHENOMENON

Other instabilities in W/O/W multiple emulsions are related to the creaming phenomenon—creaming of the oil globules in the continuous aqueous phase (Figure 5.9). The creaming phenomenon depends on oil droplet size, continuous aqueous phase viscosity, and density differences, according to Stocks law. For monodispersed droplets, creaming rate is depicted by $v = 2gr^2(\rho_1 - \rho_2)/9\eta$, and for polydispersed droplets $v = [\Sigma 8\pi/27\eta V]\ [gn_i r_i^5(\rho_1 - \rho_2)]$, where g is the gravimetric constant, r is the radius of the droplets, $\rho_1 - \rho_2$ is the density difference between the dispersed phase and the continuous phase, and η is the viscosity of the continuous (dispersion) phase (Shields et al., 2001). Reducing the oil globule size, increasing the continuous aqueous phase viscosity, and reducing density differences will minimize the creaming phenomenon.

The most common solution for creaming is the use of polymeric amphiphile or polymeric nonadsorbing compounds that increase the viscosity of the external aqueous phase and retard (or slow down) creaming of the W/O globules.

5.7 BIOPOLYMER AS STABILIZED MULTIPLE EMULSIONS

Synthetic amphiphilic polymers are not food-grade systems, and therefore naturally occurring biopolymers must be utilized. Proteins, lipoproteins, and polysaccharides are food-grade molecules and can be used as emulsifiers as well as depletion stabilizers.

Various proteins have been utilized for several decades as emulsifiers. Some of the more common are bovine serum albumin (BSA), casein, egg albumin,

TABLE 5.1 General characteristics and differences between proteins and polysaccharides as functional biopolymers in food systems

Protein	Polysaccharides
Similarities	
Natural polymers	
Widespread in food colloids	
Environmentally friendly polymer	
Used in pharmaceuticals, cosmetic, personal products	
Complicated structure	
Complex aggregation behavior	
Gelling/stabilizing agents	
Differences	
Wide-ranging structures	Similar structures
Reactive	Unreactive
Monodisperse	Polydisperse
Many segment types	Few segment types
Linear chain	Linear or branched
Flexible chain	Rigid chain
Medium molecular weight	High molecular weight
Small molecular volume	Large molecular volume
Amphiphilic	Hydrophilic
Surface active	Not surface active
Polyelectrolyte	Non-ionic or charged
Emulsifying/foaming	Thickening/water holding
Temperature sensitive	Temperature insensitive
Strong surfactant binding	Weak surfactant binding

Source: Dickinson (2003).

whey proteins, soy proteins, and lysozyme, most of which have been used in both simple O/W emulsions and in multiple emulsions.

Hydrocolloids are high-molecular-weight hydrophilic biopolymers used as functional ingredients in the food industry for the control of viscosity, gelation, microstructure, texture, flavor, and shelf-life. The term hydrocolloid encompasses all the polysaccharides that are extracted from plants, seaweeds, and microbial sources, as well as gums derived from plant exudates, and modified biopolymers made by chemical or enzymatic treatment to be soluble or dispersible in water. The general molecular and functional properties of proteins and polysaccharides are compared in Table 5.1.

In the formulation of emulsion systems, one normally distinguishes between two types of ingredients: proteins are considered to be the emulsifying agent (or emulsifier) and hydrocolloids are considered to be stabilizers. There is a debate in the literature as to whether hydrocolloids can serve as emulsifying agents. Several authors claim that except for some specific hydrocolloids (e.g., gum arabic, and acetylated pectin), most serve as stabilizers, and those that exhibit surface activity contain proteins at various levels, or even as traces, that are the hydrophobic part of the emulsifier (Bergenstähl, 1997). Others claim that some hydrocolloids, despite their hydrophilicity and structural rigidity, can under certain conditions adsorb onto oil droplets and serve as emulsifiers. These authors are genuinely convinced that this surface activity is an intrinsic property of the polysaccharide itself (Garti and Reichman, 1994; Garti et al., 1997).

5.7.1 Multiple Emulsion Stabilized by Proteins

The use of proteins as macromolecular amphiphiles was adopted long ago by scientists exploring the stability of multiple emulsions. Gelatin (Zhang et al., 1992), whey proteins (Cornec et al., 1998), bovine serum albumin (BSA) (Dickinson et al., 1991; Fredrokumbaradzi and Simov, 1992), human serum albumin (HSA), caseins, and other proteins have been evaluated. In many cases the proteins were used in combination with other monomeric emulsifiers, especially when proteins were used in the internal phase. However, whether the adsorption is competitive or complementary remains a controversial question (Wilde and Clark, 1993; Kerstens et al., 2006).

Research on the use of protein to improve the stability of multiple emulsions deals with proteins present (or adsorbed) in the inner phase as a surfactant or as encapsulating agent, or at the outer aqueous phase as a surfactant.

Omotosho et al. (1986) were the first to study the influence of BSA with a nonionic surfactant in the inner aqueous phase to stabilize W/O/W emulsions. They concluded that interfacial complexation between BSA and the non-ionic surfactant occurs at the inner W/O interface. This complex membrane has been found to enhance the stability of multiple emulsions and to slow down release of solute entrapped within the emulsion droplets. BSA has been investigated as a replacement for some of the monomeric surfactants in the inner phase and found to provide good stabilization for W/O/W multiple emulsions (Fredrokumbaradzi and Simov, 1992; Evison et al., 1995).

Koberstein-Hajda and Dickinson (1996) incorporated unmodified faba protein into the inner aqueous phase using sorbitan monooleate (Span 80) as a hydrophobic emulsifier and sodium caseinate as a hydrophilic emulsifier. The authors concluded that encapsulation of the inner phase with caseinate and a small quantity of Span 80 did not improve stability of the multiple emulsion, while increasing the protein and the Span 80 content improved both yield and stability.

Maydani (1994) have used lysozyme with monomeric emulsifiers, in both the inner and the outer interfaces. The multiple emulsions with lysozyme tend to coalescence and not create a sealed interfacial layer, probably because lysozyme is considered a small protein, and it can anchor well at the interface between oil and water (Maydani, 1994).

The authors of this review (Garti et al., 1994) also investigated the use of BSA along with monomeric emulsifiers, in both the inner and the outer interfaces (in low concentrations of up to 0.2 wt%). They found significant improvement in stability and a delay in release of markers, compared to multiple emulsions stabilized by BSA as an external phase emulsifier alone. It was postulated that although BSA has no stability effect at the inner phase, it has a strong effect on the release of the markers. On the other hand, BSA, when utilized as an external emulsifier together with small amounts of monomeric emulsifiers (or hydrocolloids), was found to be a good steric stabilizer and to improve the stability and shelf-life of the multiple emulsions, and to slow down the release of the markers through the external interface. The release of the marker was found to take place mainly via reverse micellar transport. It was concluded that BSA reduces the chance of reverse micelle formation in the oil phase, thus reducing the release rate. It was concluded that BSA can play a double role in multiple emulsions; it is a film-former, enhancing steric stabilization and providing a barrier to the release of small molecules from the internal interface, and a steric stabilizer at the external interface.

Su et al. (2006) recently investigated the influence of polyglycerol polyricinoleic acid (PGPR) alone and in combination with sodium caseinate as emulsifier for primary water-in-oil (W/O) emulsions and caseinate as the sole emulsifier for secondary W/O/W emulsion. It was proposed that casein forms either an extra film around the internal water droplets or a complex with the existing emulsifiers. An improved preparation yield, higher stability, and unchanged W/O/W emulsion droplet size were reported. It was suggested that synergistic effect could have stabilized the primary W/O emulsion interface against coalescence in the presence of both PGPR and sodium caseinate as hydrophobic emulsifiers. A reduction in droplet size was observed upon increasing the caseinate concentration from 0.03 to 1% (w/v) (Figure 5.10). At 0.03% (w/v) caseinate some of the oil droplets were outside the range of the Mastersizer, indicating coalesced oil droplets greater than 50 μm. Confocal microscopy showed the presence of several flocculated droplets in the W/O/W emulsion prepared at low concentrations of caseinate (Figure 5.11*a* and *b*), while flocculation was not observed in W/O/W emulsions prepared with 0.5% (w/v) caseinate in the external aqueous phase (Figure 5.11*c* and *d*). The droplet size of all W/O/W emulsions showed no significant change after storage for four weeks at 20°C. Inserting 0.5% (w/v) caseinate to the inner phase allowed reduction of the PGPR content from 8 to 2 wt% without affecting the yield or the stability of the multiple emulsion. The use of casein allowed replacement of part of the PGPR (which is nonfriendly in food applications) by caseinates. The combination of PGPR and caseinate may produce a more viscoelastic adsorbed layer, thereby preventing the release of the dye from the internal aqueous phase to the external aqueous phase, as investigated by the authors. This conclusion is in good agreement with Garti et al. in the BSA study (Su et al., 2006).

Dickinson et al. (1994) also prepared stable food grade W/O/W emulsions with, sodium caseinate and the polyglycerol ester of polyricinoleic acid (PGPR) as the hydrophobic emulsifier. They showed that when caseinate concentration

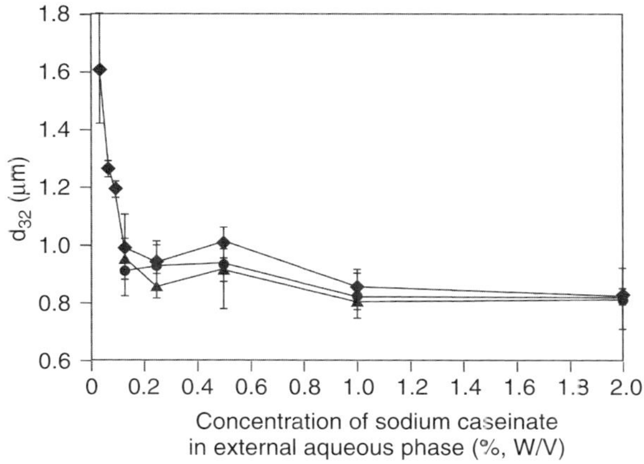

Figure 5.10 Effect of sodium caseinate concentration in external aqueous phase on droplet size (d_{32}) of W/O/W emulsions prepared with 2 (■), 4 (▲), and 6% w/v (◆) PGPR in the oil phase. Error bars represent standard deviation (Su et al., 2006).

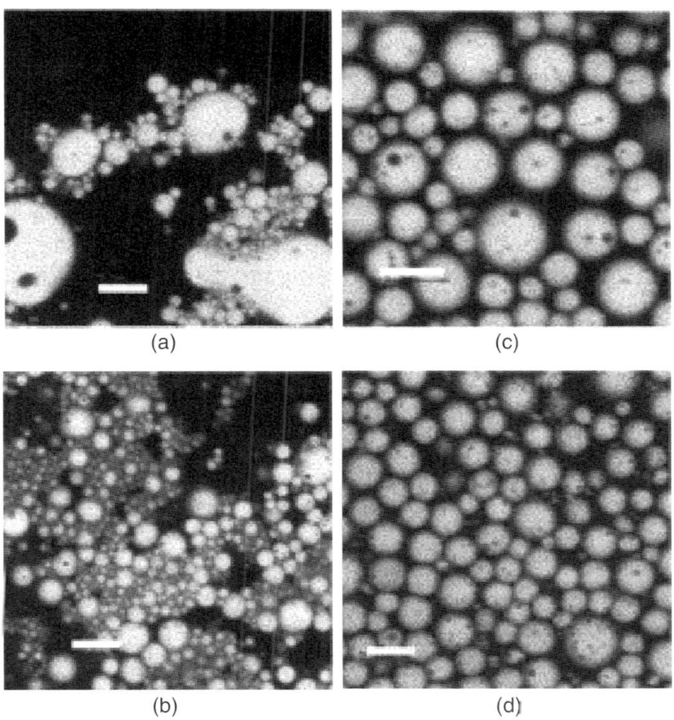

Figure 5.11 (*a, b*) Confocal microscopy images of freshly prepared W/O/W emulsions at 0.09% w/v caseinate at the external phase; (*c, d*) confocal microscopy images of freshly prepared W/O/W emulsions at 2% (w/v) caseinate at the external phase. Scale bar represents 5 µm (Su et al., 2006).

was increased from 0.125 to 1 wt%, the droplet size decreased from around 1.4 to around 0.85 μm, and resulted in a finer emulsion that was more stable against creaming. From the release curves it seems that the marker transport is more controlled.

Proteins cannot completely replace hydrophobic monomeric emulsifiers in the inner phase of multiple emulsions (Dickinson et al., 1994). However, proteins in combination with other stabilizers do have the capacity to enhance the stability of multiple emulsions, mainly by encapsulating the inner phase and thereby substantially reducing the hydrophobic emulsifier concentration. Protein in the external phase can enhance the multiple emulsion stability without the addition of any emulsifier or stabilizer.

Multiple emulsion stability was significantly improved in the presence of amphiphilic proteins because of two important factors. First, the multianchoring flexible macromolecules act to improve the steric stabilization by forming a thick multilayered coating on the droplets. Second, the proteins used in the inner phase create a mechanical film barrier that prevents uncontrolled release of the entrapped ingredients.

5.7.2 Multiple Emulsion Stabilized by Hydrocolloids

Most often hydrocolloids are added to the multiple emulsion in order to increase the viscosity of the inner or the outer phase. Florence and Whitehill (1985) suggested three pathways to reduce the mobility of ingredients in multiple emulsions:

1. Use of highly viscous oil to prevent diffusion of water/surfactant/active material from the inner phase.
2. Gelation of the oily or aqueous phase, also known as the encapsulated method.
3. Polymerization of the interfacial adsorbed surfactant molecules.

The two first pathways essentially involve depletion stabilization, whereas the interfacial polymerization of the third pathway involves some steric stabilization. All three options are found to form semisolid or gel-like formulations. The most common gel-like multiple emulsions are primarily those used in cosmetic creams, and are not applicable in nonviscous diluted emulsions such as beverages.

Use of High Viscosity Oil Recently Weiss et al. (2005) investigated the possibility of using a semicrystalline oil phase in W/O/gel to control the release of encapsulated hydrophilic compounds from polysaccharide gels with multiple emulsions. Various lipid phases prevent the diffusion of water/surfactant/active material from the inner phase and also gelation of the oily or aqueous phase. For example, MCT-oil and various vegetable fats were investigated for

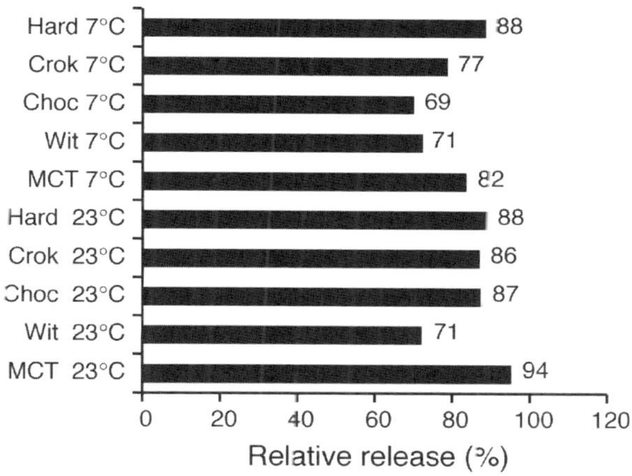

Figure 5.12 Relative release of tryptophan (%) from emulsion gels compared to gels prepared without multiple emulsions of Witocan (Wit), Chocosine (Choc), Crokcool (Crok), and Hardstock (Hard) (=100% during a period of 24 hours). Effects of the lipid phase were observed at environmental temperatures of 7°C and 23°C (Weiss et al., 2005).

their ability to act as a hydrophobic barrier in the oil phase. The content PGPR (hydrophobic surfactant) in the inner W/O emulsions was varied. In order to create W/O/gel, the W/O/W was homogenized in a Ca^{2+}-alginate gel containing maltodextrin as bulking agent.

An increase in encapsulation efficacy was observed when various lipids were used for multiple emulsion preparation, rather than pure water; even the system containing pure MCT-oil (including 2% PGPR) showed a comparably lower relative release of the marker (94%) (Figure 5.12). The cumulative release of tryptophan from multiple emulsion gels varying in composition, recorded over a storage period of 24 hours in pure water, is summarized in Figure 5.12. At 23°C, the lowest release rate was observed in Witocan (Wit) and the highest melting point in fat (Figure 5.12).

Reducing the temperature from 23°C to 7°C caused a reduction in the release rate of the marker. A significant reduction in encapsulation efficiency was obtained by increasing the oil phase content.

Gelation of the Outer Aqueous Phase (W/O/Gel) Most of the additives inserted in the multiple emulsions under study are not permitted in foods. Viscosities were found to be enhanced by cellulose derivatives such as hydroxypropyl cellulose (HPC), hydroxypropylmethyl cellulose (HPMC), hydroxyethyl cellulose (HEC), and the natural hydrocolloids, such as xanthan, guar gum, and carrageenan. The viscosity or gelling agents can affect some of the multiple emulsion properties such as entrapping yields, stability, droplet size, consistency, and skin feel (Ozer et al., 2000).

An interesting study by Muguet et al. (2001) in which multiple emulsions were stabilized by viscosity-increasing agents or gelling agents, such as hydroxypropyl cellulose in the outer aqueous phase, has shown that the release process is moderated by these agents but the globules will still break under shear. The emulsifiers were fit to a Taylor theoretical framework so that the bursting mechanisms of the globules under shear were the same for any composition of multiple emulsions. This way stability and release could be controlled by the shear.

In a recent study by Kanouni et al. (2002), Abil EM90 (polysiloxane, cetyldimethicone copolyol) was used in the inner water phase. The outer phase contained betaine and SDS in combination with hydrophilic surfactants, and xanthan gum served as the stabilizer (viscosity agent). From microscopic observations along with surface isotherms linked to rheology measurements, Kanouni and colleagues concluded that in the inner phase, it is essential to use a low-HLB surfactant so that the resulting expandable and compressible interfacial layers are irreversibly adsorbed. Long-term stability requires, in addition, a balance between Laplace and osmotic pressures between W_1 droplets and the oil phase. Small quantities of salt in the inner phase can balance the excess pressure. As for the outer phase, it was concluded that the presence of thickener in the W_2 phase is necessary in order to reach a viscosity ratio (permeability ≈ 1) between W_1/O and W_2 that allows dispersion of the viscous primary emulsion into the W_2 aqueous phase. Because the interactions between the low- and high-HLB emulsifiers at the O/W_2 interface do not destabilize the films, they suggested using the high-HLB surfactant (betaine derivative) in combination with an anionic surfactant (SLES, sodium lauryl ether sulfate) as the blend of emulsifiers for the external interface.

The thick interfacial layer and the viscous or gelled aqueous phase might have some advantages when protection of sensitive addenda is required. Kim et al. (2003) treated the internal aqueous phase with hydroxypropyl-beta-cyclodextrin (to make the internal aqueous phase more hydrophilic) and introduced kojic acid (5-hydroxy-2-(hydroxymethyl)-4-pyrone) as a model antioxidant. The multiple emulsions were also stabilized sterically by addition of xanthan gum into the external aqueous phase. The authors found improved stability against coalescence and improved chemical stability. The stability of kojic acid could be maintained at 90% for 10 weeks (Figure 5.13).

Polymerization of the Interfacial Adsorbed Surfactant Molecules
Chitosan has surface activity that seems to stabilize W/O/W emulsions. Chitosan reacts with anionic emulsifiers such as SDS at certain ratios to form a water-insoluble complex that has strong emulsification capabilities. Chitosan solution was used to form multiple emulsions of O/W/O as intermediates from which, by a simple procedure of striping the water, the authors formed interesting porous spherical particles of chitosan (Schulz et al., 1998; Adachi et al., 1999).

Cyclodextrins (α, β, and γ) (Figure 5.14) have been shown to be potential stabilizers for O/W/O emulsions (Yu et al., 1999). The advantages of the

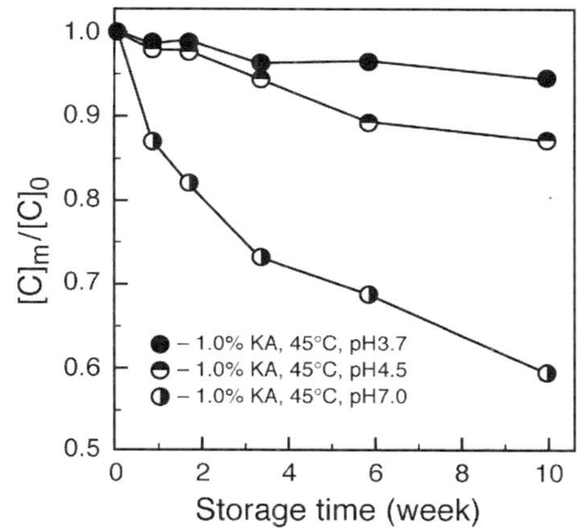

Figure 5.13 Stability of kojic acid over time under different conditions: Emulsion formulation, concentration, and storage temperature. The pH of the formulation was adjusted to 3.7 with lactic acid (Kim et al., 2003).

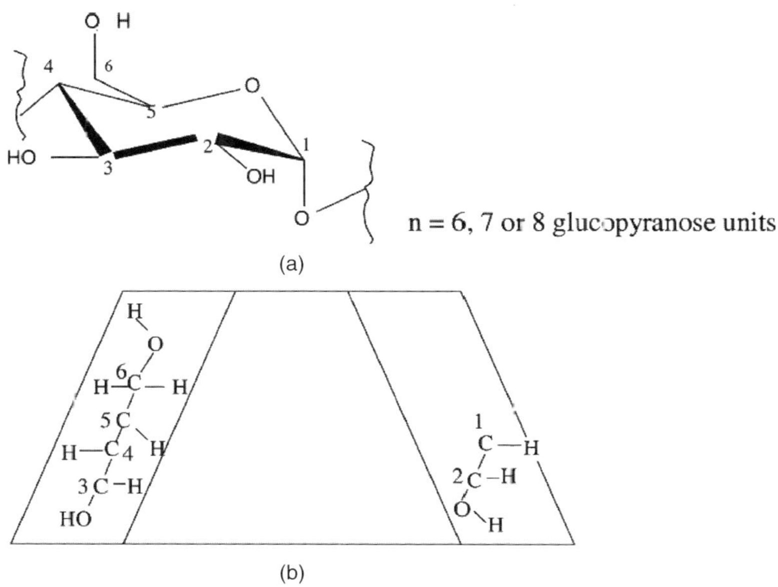

Figure 5.14 Chemical representation of a glucopyranose unit of a CD molecule. (*a*) The secondary hydroxyls at the 2- and 3-positions exist on the secondary face, and the primary hydroxyls at the 6-position exist on the primary face. (*b*) Schematic representation of a CD cross section with carbon atom numerotation (Shan-Chen et al., 2003).

cyclodextrins are their ability to complex with certain oil components at the oil/water interface, and so no additional surfactant is needed. It appears that the stabilization efficacy depends on the nature of the oil and the type of cyclodextrin ($\alpha > \beta > \gamma$). The presence of any active matter in the inner phase destabilizes the emulsion. However, only the α-cyclodextrin yielded stable emulsions because the interfacial interaction between the components present at the interface changed the HLB and caused a destabilization effect. So this elegant idea of interfacial complexation between the oil components and the surfactant cannot be a universal solution. The idea suffers from very severe intrinsic disadvantages as each different additive is included in the emulsion. For every additive at any concentration, an adjustment must be made; a given cyclodextrin or complexing agent is not necessarily totally fit.

In a more recent article, Shan-Chen et al. (2003) encapsulate camphor (CA) in oil/water/oil multiple emulsions prepared with cyclodextrin (CD). The O/W/O stabilized by CD/CA as an emulsifier was not perfectly stable. After storage of 20 days at 20°C, only the α-CD emulsion without CA was perfectly stable (Table 5.2). O/W/O multiple emulsions can be prepared in the absence of classical surfactants, and by using appropriate α and β CD. The conditions of emulsion stability in the presence of a lipophilic active ingredient, or any additive, depend on possible competition between the lipophilic molecule and the fatty acid residues of triglycerides to enter the CD cavity.

Vaziri and Warburton (1994) investigated the effects of different combinations of surfactants (Span 80 and Tween 80) and hydrocolloids (acacia and gelatin) on the release rate and stability of multiple emulsions. The combination of surfactants produced stable systems and more prolonged release because a better interfacial film was formed by the emulsifiers compared with that of the multiple emulsions in which each surfactant was employed individually (see Figure 5.15 and Table 5.3). The enhanced stabilization can also be attributed to the protein content in the acacia gum. The protein provided strong adsorption interaction with the oil-water interface during emulsification.

Mechanical Stabilization by Microcrystalline Cellulose Some attempts have been made to improve the shelf-life of multiple emulsions by incorporating small solid particles in the surfactant formulations. The idea was to allow the particles to adsorb onto the oil droplets so that they can provide a mechanical barrier against coalescence (Aveyard et al., 2003). Oza and Frank (1989) were the first to develop the concept by using colloidal microcrystalline cellulose (MCC) instead of water-soluble surfactant in W/O/W emulsions containing oil-soluble hydrophobic surfactants (Spans). Emulsions, stable for up to 1 month, contained a network of MCC particles adsorbed at the outer oil–water interface.

Kim et al. (2005) improved the storage stability of antagonistic microorganisms by encapsulating them in W/O/W multiple emulsions stabilized by MCC. MCC was very useful in obtaining a multiple emulsion with excellent stability, exhibiting better storage stability. The activity of the antagonistic microorganism was better than in trials without multiple emulsions.

TABLE 5.2 Multiple emulsion stability after storage for 20 days, at 4°C, 20°C, and 40°C

	α–CD Multiple Emulsion			β–CD Multiple Emulsion			γ–CD Multiple Emulsion		
	4°C	20°C	40°C	4°C	20°C	40°C	4°C	20°C	40°C
Multiple emulsion without CA	Stable	Stable	3% WPS	Stable	Some oil exuded	5% WPS*	Stable	Some oil exuded	2% WPS
Multiple emulsion with CA	Stable	3% oil phase exuded	11% WPS	Stable	Some oil exuded	9% WPS*	Stable	Some oil exuded	2% WPS

Source: Shan-Chen et al. (2003).

Note: At WPS (waxy phase separation) of the total multiple emulsions the liquid state at 40°C becomes solid at 20°C.

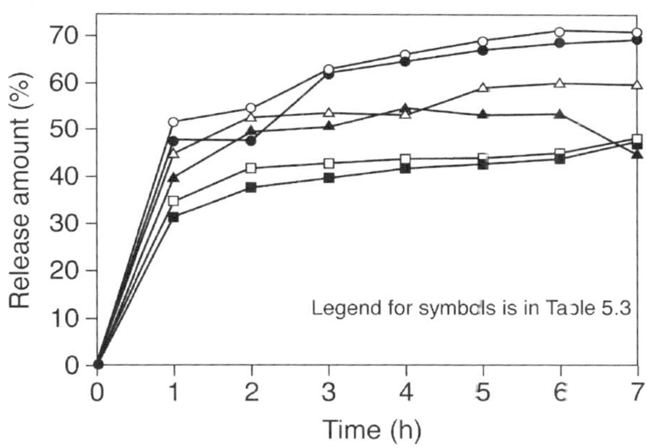

Figure 5.15 Release of chloroquine diphosphate from W/O/W emulsions (Vaziri and Warburton, 1994).

TABLE 5.3 Formulation of olive oil W/O/W multiple emulsion containing chloroquine diphosphate (Vaziri and Warburton, 1994)

	Inner Aqueous Phase			Oily Phase Surfactant	Outer Aqueous Phase		
Symbol	Gelatin	Acacia	Tween 80	Span 80	Gelatin	Acacia	Tween 80
○	+	–	–	+	+	–	–
●	+	–	+	+	+	–	+
△	–	+	–	+	–	+	–
▲	–	+	+	+	–	+	+
□	+	+	–	+	+	+	–
■	+	+	+	+	+	+	+

Garti et al. (1999) used the submicron crystalline fat particle α form of triglycerides as hydrophilic emulsifiers in the outer interface of W/O/W emulsions in combination with other surfactants and achieved a significant improvement in stability over systems that contained just the emulsifiers.

Hydrocolloids significantly improve the stability of multiple emulsions because they help encapsulate better the inner aqueous phase, preventing uncontrolled release of the entrapped ingredients. Stabilization of the multiple emulsion is achieved mainly by depletion stabilization. However, hydrocolloids significantly affect viscosity and the rheology of a multiple emulsion.

5.7.3 Multiple Emulsion Stabilized by Hybrid of Proteins and Hydrocolloids

W/O/W Under certain conditions some proteins and polysaccharides form hybrids (complexes) with enhanced functional properties compared to the proteins and polysaccharides alone. Conversely, electrostatic complexing between oppositely charged proteins and polysaccharides allows better anchoring of the newly formed macromolecular amphiphile onto the oil–water interfaces.

Interaction between Protein and Hydrocolloids in an Aqueous Solution Mixing two aqueous solutions of biopolymer may yield one of the structures depicted in Figure 5.16. Interactions between the two polymers can be segregative or associative. For instance, for very dilute solutions the proteins and polysaccharides form a solution-like aqueous system. With increasing concentrations of each component, the system becomes unstable; in some cases biopolymer mixtures tend to segregate. For polymers of similar and extended structures, this trend is usually attributed to repulsive interactions among the polymer segments. For polymers with dissimilar shape and structure, segregation leads to a local reduction in concentration of one of the polymers in the area of the second one, due to a decrease of conformational entropy of macromolecules at the interface boundary (depletion phenomenon). Segregation has also been observed under associative interaction between the two biopolymers. In this case the polysaccharide molecules can adsorb onto protein and even bridge several protein molecules (complex coacervation). In an aqueous solution, electrostatic attraction causes complex coacervation between two oppositely charged polymers. For instance, the complexation between the proteins and anionic polysaccharides occurs below the protein isoelectric point and at low ionic strengths. Factors that influence compatibility and complex formation are protein/polysaccharide ratio, pH, ionic strength, and the nature of the polymers (molecular weight, net charge, ternary structure, and flexibility of chains). Under certain conditions proteins and polysaccharides form hybrids (complexes) with enhanced interfacial properties compared to the proteins

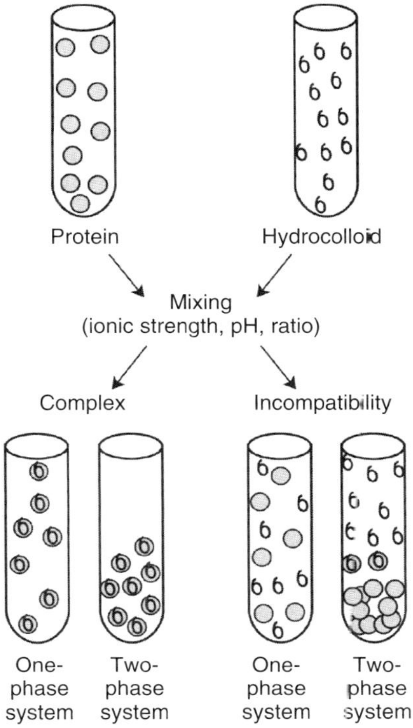

Figure 5.16 Illustration of water–protein–polysaccharide mixtures describing the different types of interactions between the protein and the polysaccharide in aqueous medium.

and polysaccharides alone, where uncomplexed blends of biopolymers are formed. Electrostatic complexation between oppositely charged proteins and polysaccharides allows better anchoring of the newly formed macromolecular amphiphile onto oil/water interfaces (Grinberg and Tolstoguzov, 1997; Tolstoguzov, 2003; Benichou et al., 2004).

Numerous articles have been written on the interaction between proteins and hydrocolloids in aqueous solution and research aspects of this technology (Grinberg and Tolstoguzov, 1997; Tolstoguzov, 2003; Benichou et al., 2004; Ibanoglu, 2005; Ribotta et al., 2005; Uruakpa and Arntfield, 2006; Nunes et al., 2006).

Multiple Emulsion Stabilized by Hybrid of Protein (WPI) and Hydrocolloids
In our most recent studies (Benichou et al., 2002b), we demonstrated that proteins (BSA, WPI, caseins) are capable of forming soluble complexes (hybrids) with certain hydrocolloids (xanthan, guar, LBG).

The complex is an amphiphilic macromolecule that strongly adsorbs onto the oil/water outer interface and upon severe shear forms smaller globules with improved stability over multiple emulsions stabilized by protein alone.

Benichou et al. (2002b) were the first to study the influence of WPI and xanthan hybrids on multiple emulsion stability. Additional information on the advantages of such systems will be published shortly.

Effect of WPI/Hydrocolloid Ratio on the Multiple Emulsion The ratio between proteins and hydrocolloids is one of the factors that influences the formation of the hybrid. Phase diagrams are constructed (data not shown) to determine various compositions along the isotropic clear regions; the best ratio for hybrid formation is 4 parts of WPI to 0.5 parts of hydrocolloid, by weight.

The effect of the WPI/xanthan or WPI/modified pectin (MP) (Garti and Wicker, 2005) ratio on stability of W/O/W multiple emulsions was examined (Tables 5.4 and 5.5) Addition of hydrocolloids caused significant improvement, mainly to the multiple emulsion droplet sizes. The droplets were smaller in comparison to droplets of multiple emulsions prepared solely with WPI. Multiple emulsion droplets stabilized with WPI and pectins are one-third the size

TABLE 5.4 Globule size distribution of multiple emulsions made with WPI (whey protein isolate), xanthan gum, LBG (locust bean gum), and blends (hybrids) of the gums with WPI at their optimum ratios

	WPI 4 wt%	Xanthan Gum 0.5 wt%	LBG 0.5 wt%	WPI/Xanthan 4/0.5	WPI/LBG 4/0.5
d_{43} (μm)	12.0	23.4	28.4	3.3	4.8
d_{90} (μm)	10.8	19.8	23.6	3.1	4.0

Source: Benichou et al. (2002b).

Note: *d_{90}: average droplet diameter of 90% of the droplets; d_{43}: average droplet diameter r*, which is calculated as $d(4,3) = \Sigma_i n_i d_i^4 / \Sigma_i n_i d_i^3$.

TABLE 5.5 Globule size distribution of multiple emulsions made with WPI (whey protein isolate), modified pectin, and blends (hybrids) of the gums with WPI at their optimum ratios

External Phase	WPI/MP 4:1 (wt/wt)	WPI/MP 4:0.5 (wt/wt)	WPI/MP 4:0.1 (wt/wt)	WPI/MP 4:0 (wt/wt)
$t = 0$ (μm)	4.6	4.6	4.7	14.5
$t = 20$ days (μm) 25°C	5.6	6.6	7.2	10.6

Source: Garti and Wicker (2005).

Note: The internal phase in all the systems consists of 30 wt% water (17 wt% glucose) at pH 6, 10% PGPR, and 60% R(+)-limonene.

of droplets stabilized with only WPI (ca. 5 µm for WPI/pectin and ca. 14.5 µm for WPI). Systems based on WPI/xanthan yielded multiple emulsions with droplets one-fourth smaller in comparison to droplets derived from WPI alone and one-eighth smaller compared to multiple emulsions prepared from xanthan alone.

Emulsions can behave like microcapsules, microspheres, or mesophasic lyotropic liquid crystals, since the interfacial film is rather thick and, in most cases, multilayered. Increasing the xanthan content in the WPI/xanthan system to 1 wt% in the external aqueous phase decreases droplet diameter from 11.8 to 3.1 µm (Figure 5.17), but no significant improvement in droplet size occurs between ratios of 4 to 0.5 and 4 to 1. It was also found that by increasing the protein-to-polysaccharide ratio, the multiple emulsion droplets were smaller and stability was increased.

The WPI/pectin droplet sizes immediately after preparation were not independent of the WPI/pectin ratio and were around 5 µm in size. Similarly no significant droplet size change was observed after aging, independent of the ratio of the protein to the gum, indicating that the droplets were well protected from coalescence.

On the other hand, in multiple emulsions stabilized with WPI alone, the droplet size decreased with time from 14.6 to 10.5 µm after 20 days, probably due to internal aqueous phase explosion. As the ratio WPI/MP decreases (e.g., to a 4 to 0.1 wt ratio), the multiple emulsion droplet size can increase from 4.7 to 7.2 µm after 20 days. Lack of exhibited hydrocolloid is attributed to the weakness of the emulsifier film adsorbed onto the external oil-water interface, as reported in the case of multiple emulsions stabilized with pure WPI (Uruakpa and Arntfield, 2005).

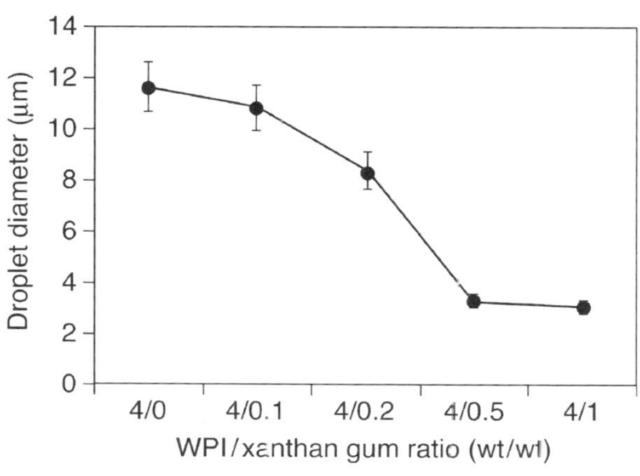

Figure 5.17 WPI/xanthan gum ratio (wt/wt) influence on the droplet diameter (µm) of multiple emulsion droplets, immediately after preparation (Benichou et al., 2002b).

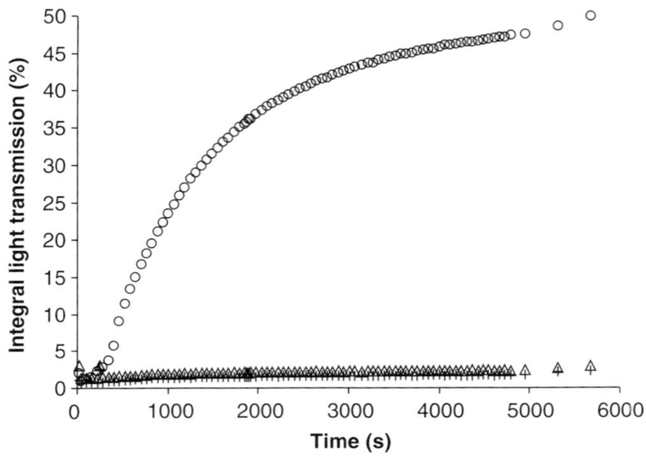

Figure 5.18 Integral light transmission as a function of time (seconds), as measured by LUMiFuge, in multiple emulsions stabilized with 70 wt% of WPI/MP at different ratios: 4/0.1 wt/wt (○), 4/0.5 wt/wt (△), and 4/1 wt/wt (+), at pH 6, after preparation. 30 wt% of the inner emulsion contains 30 wt% water (17 wt% glucose), pH 6, 10 wt% PGPR, 60 wt% R(+)-limonene (Garti and Wicker, 2005).

At a biopolymer content greater than that required for complex formation, the addition of one of the two biopolymers will only slightly affect the multiple emulsion droplet stability, and will add only a small depletion stabilizing effect to the emulsion droplets.

In WPI/pectin, complex formation occurs at a 4 to 0.1 ratio; in WPI/xanthan, the hybrid formation is at a 4 to 0.5 ratio. Thus there is no change in the multiple emulsion droplets in the WPI/pectin ratio from ratio 4/0.1, while in the WPI/xanthan there is change only at a 4/0.5 ratio.

The improved stability of the multiple emulsion is related to depletion stabilization, meaning there is less creaming. Stability against creaming, as measured by LUMiFuge, is significantly influenced by the WPI/pectin ratio (Figure 5.18), which describes a set of integrals of light transmission through the sample as a function of time. There is no change in the integral light transmission over 6000 seconds at 2500 rpm for multiple emulsions stabilized with WPI/MP ratios of 4/0.5 and 4/0.1. On the contrary, there is a rapid increase in the integral transmission in the system stabilized with a 4/0.1 ratio. In the 4 to 0.1 wt ratio system, the serum layer thickness is about 30% (Figure 5.18). At low pectin levels, the system is very unstable with respect to creaming and creams immediately after preparation. However, multiple emulsions stabilized with WPI/pectin in ratios of 4/0.5 and 4/1 are stable to creaming. Creaming stability depends on the type of conjugated WPI/pectin, as studied above, and also the viscosity of the multiple emulsion systems and specific gravity differ-

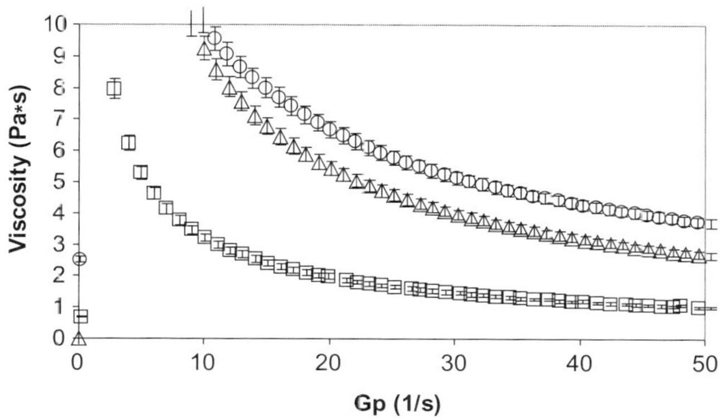

Figure 5.19 Viscosity as a function of shear rate in multiple emulsions stabilized by various ratios of WPI/MP multiple emulsion: 4/0.1 (□), 4/0 5 (△), and 4/1 (○), at pH 6, 20°C. The inner emulsion contains 30 wt% water (17 wt% glucose), pH 6, 10 wt% PGPR, and 60 wt% R(+)-limonene (Garti and Wicker, 2005).

ences between the droplets. Multiple emulsion viscosity as a function of shear rate is summarized in Figure 5.19. The viscosities of all the systems decrease with the shear rate, which is typical of non-Newtonian flow behavior of multiple emulsion systems (Jiao and Burgess, 2003). Increasing the modified pectin content leads to an increase in multiple emulsion viscosity, producing a shear rate of 0 to 50 s^{-1}. For a shear rate of 10 s^{-1}, the viscosity of multiple emulsion stabilized with a 4 to 0.1 ratio is 3.21 Pa·s. For a 4 to 0.5 ratio, the viscosity is 9.25 Pa·s, and for a 4 to 1 ratio, the highest viscosity measured is 10.16 Pa·s. This trend is seen throughout the shear rate range – increasing the viscosity of multiple emulsions stabilized the multiple emulsions by depletion stabilization.

Increasing levels of the modified pectin in the blends stabilizes the multiple emulsion. This is probably due to steric stabilization, resulting from the formation of strongly conjugated WPI/modified pectin in addition to the depletion stabilization. This is evidenced by the viscosity enhancement of the outer phase of the multiple emulsion.

Rheological studies can provide useful information on the stability and internal microstructure of the multiple emulsions. Some attention was given to this subject in recent years, and the results help clarify certain aspects of stability and release properties of the multiple emulsions (Benichou et al., 2002b). Oscillatory dynamic tests and steady-state analyses, and parameters such as shear or complex modulus (G^*), the lag phase between stress and strain (δ), the storage modulus (G'), and the loss modulus (G''), provide quantitative characterization of the balance between the viscous and elastic properties of the multiple emulsions. Oscillatory measurements generally indicate that multiple emulsions are predominantly viscous in that the loss modulus falls above the storage modulus over the entire frequency range investigated. Upon aging, the storage and loss moduli of the multiple emulsions show a significant increase.

Rheological measurements were performed on multiple emulsion droplets stabilized with WPI-xanthan gum hybrids. The phase angle δ was defined as Arctan(G''/G'), where G' is the storage modulus, G'' is the loss modulus, and $\tan(\delta) = G''/G'$. It was found that at low levels of gum (0.1 wt%), the emulsions had high viscosity, with phase angle (δ) close to 90°, indicating self-assembly of the two biopolymers onto the external oil–water interface. It was suggested that the emulsion droplets are mainly stabilized by steric interactions between the macromolecular recognition hybrids that are adsorbed onto the oil. Conversely, at high levels of gum (0.3–1 wt%), the emulsions exhibit more elasticity—regarded as a physical property derived from a depletion stabilization mechanism. Protein will preferentially adsorb onto the oil–water external interface; the uncomplexed gum will migrate to the bulk and contribute to the stabilization by a depletion mechanism (Figure 5.20). The protein-to-gum weight ratio of 4/0.5, at which maximum complexation occurred between the protein and the gum, corresponds to the intermediary viscoelasticity of the system.

At gum concentration of 0.5 wt%, the protein concentration does not affect the rheological behavior of the multiple emulsion. The emulsion conserves its elasticity properties at all ratios, with phase angle (δ) values being approximately 25° at all protein contents.

Knoth et al. (2006) and Fechner et al. (2006) investigated the stabilization of multiple emulsion in two systems. The first system was based on whey protein as the external emulsifier and carboxymethylcellulose (CMC) as the

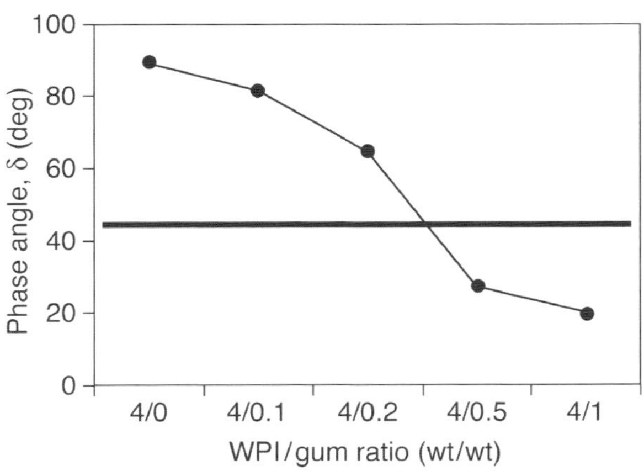

Figure 5.20 WPI/xanthan gum ratio (wt/wt) influence on the phase angle δ (degrees) of multiple emulsion droplets, immediately after preparation. The phase angle δ was defined as arctan(G''/G'), where G' is the storage modulus, G'' is the loss modulus, and $\tan(\delta) = G''/G'$ (Benichou et al., 2002b).

stabilizer, at pH 4. Addition of the CMC significantly reduced the droplet size, prevented aggregation, improved the release rate, and changed the rheological behavior of the multiple emulsion. The second multiple emulsion, stabilized by sodium caseinate-dextran, was also found to be an improved multiple emulsion. The droplet sizes were smaller with a more narrow size distribution when the conjugated sodium caseinate and dextran were used (compared to multiple emulsions stabilized solely by caseinate).

Protein-polysaccharide blends seem to provide improved multiple emulsion properties, such as reducing droplet size, reducing distribution of droplet size, preventing the instability related to the multiple emulsion, and better encapsulating the inner droplets that contribute to controlling the release rate. With the protein polysaccharide blend, the viscosity and rheological properties of the multiple emulsion change; the multiple emulsion became generally more viscous with increased elastic properties. These properties are gains due to the addition of hydrocolloids that create strong multianchoring of flexible macromolecules forming a thick multilayered coating on the droplets.

O/W/W A number of researchers have prepared a new type of multiple emulsion termed O/W/W, based on the thermodynamic incompatibility of mixed biopolymer solutions. For example, oil-in-water-in-water (O/W/W) emulsions can be prepared by mixing an O/W emulsion with a W/W emulsion.

W/W emulsion is created when the free energy of mixing two biopolymers is positive (becoming thermodynamically incompatible)—the system separates into two phases without special properties. In other cases phase separation includes certain changes in the morphology of the solution that contribute to special properties such as rheology and lower interfacial tension. This type of thermodynamic incompatibility often occurs in systems where one or both of the biopolymers are uncharged or where the biopolymers have similar electrical charges. These phase separations form a water-in-water (W/W) emulsion that consists of water droplets dispersed in continuous water (Antonov et al., 2004; Capron et al., 2001).

Kim et al. (2006) showed that various types of multiple emulsions can be prepared by mixing an O/W emulsion with a mixed biopolymer solution that separates into two phases and creates a W/W emulsion. Under certain conditions the biopolymers formed a two-phase system consisting of an (HD-heat denaturated) HD-WPI-enriched lower phase (W_1) and a pectin-enriched upper phase (W_2), which was attributed to thermodynamic incompatibility of the two biopolymers. Water-in-water emulsions (W_1/W_2 or W_2/W_1) could be formed by blending incompatible upper (U) and lower (L) phases at different ratios (Figure 5.21). Whey protein stabilized oil droplets preferentially partition into the HD–WPI rich phase of two-phase biopolymer systems. This enabled preparation of multiple emulsions by mixing a whey protein stabilized O/W emulsion with incompatible upper and lower phases. The multiple emulsions were either of the oil-in-water-in-water ($O/W_1/W_2$) type or the mixed oil-in-water–water-in-water (O/W_1–W_2/W_1) type, depending on the initial biopolymer composition

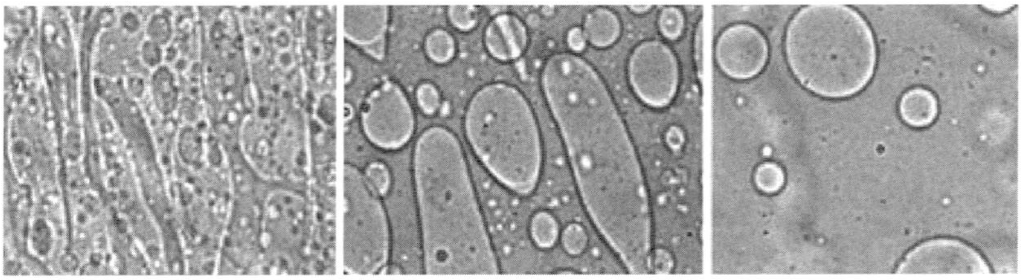

0 min 3 min 12 min

Figure 5.21 Evolution of the microstructure of a phase-separated biopolymer mixture consisting of one part upper phase (rich in pectin) and three parts lower phase (rich in whey protein) after being placed on a microscope slide. It is proposed that this system forms a W_2/W_1 emulsion, where W_1 is HD-WPI rich and W_2 is pectin rich (Kim et al., 2006).

of the system. These kinds of multiple emulsions may prove useful for the creation of food emulsions with improved physicochemical properties or for the development of novel delivery systems (Figure 5.22).

5.8 CONCLUSIONS AND FUTURE PERSPECTIVES

In the past, multiple emulsions were considered extremely promising systems for slow and controlled release of entrapped active matter from the inner phase to the outer continuous phase. Multiple emulsions have numerous potential applications in almost all areas of life. The preparation methods are simple and inexpensive. Thus multiple emulsions have been extensively studied in the last 20 years.

Much has been done in developing new and advanced methods of preparation of multiple emulsions. One of the most impressive technologies is the use of specific separation membranes.

However, intrinsic stability and release problems discouraged scientists and technologists and these promising formulations were almost abandoned. In early research, low-molecular-weight emulsifiers were used to stabilize multiple emulsions, but they did not solve these problems. Much progress has been made with the introduction of amphiphilic macromolecules as emulsifiers. This multianchoring of flexible macromolecules can improve the steric stabilization by forming a thick multilayered coating on the droplets and by depletion (and mechanical) stabilization. Coalescence of the inner phase is now minimized by reducing the droplet sizes, better coating of the interface, and use of polymeric amphiphiles. Gelation or viscosifying the inner phase are additional options. Steric stabilization, mechanical stabilization, and depletion stabilization can improve stability considerably.

Most of the amphiphilic macromolecules or surfactant polymers that have been used are not allowed in the food industry. Their use in proteins or hydro-

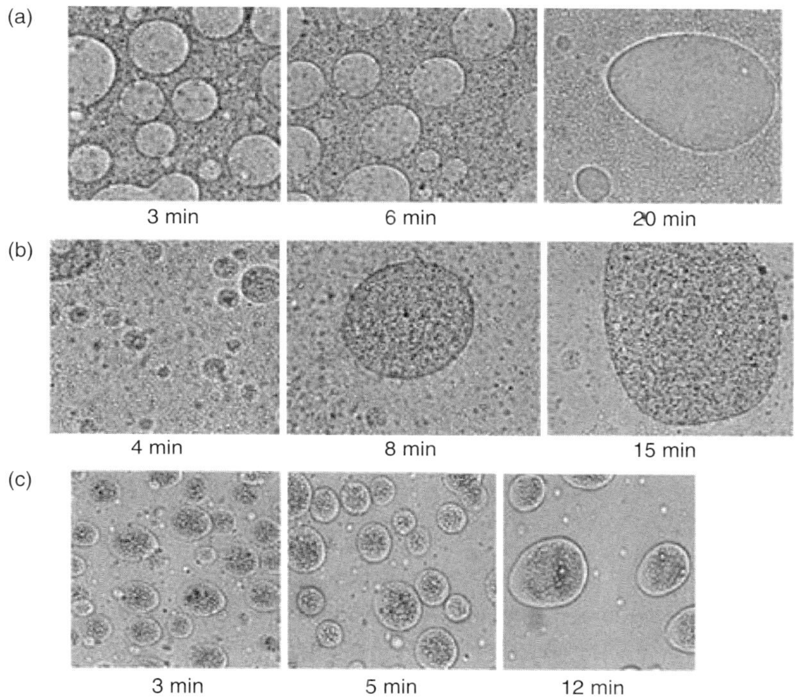

Figure 5.22 (*a*) Evolution of microstructure of phase-separated biopolymer mixtures (1U:3L) containing 0.5 wt% WPI-stabilized droplets after being placed on a microscope slide. The large circles are the water droplets, and the small circles are the oil droplets. It is proposed that this system forms a W_2/W_1-O/W_1 emulsion, where O is oil, W_1 is HD-WPI rich, and W_2 is pectin rich. (*b*) Evolution of the microstructure of phase-separated biopolymer mixtures (2U:2L) containing 0.5 wt% WPI-stabilized droplets after being placed on a microscope slide. It is proposed that this system forms an $O/W_1/W_2$ emulsion, where O is oil, W_1 is HD-WPI rich, and W_2 is pectin rich. (*c*) Evolution of microstructure of phase-separated biopolymer mixtures (3U:1L) containing 0.5 wt% WPI-stabilized droplets after being placed on a microscope slide. It is proposed that this system forms an $O/W_1/W_2$ emulsion, where O is oil, W_1 is HD-WPI rich, and W_2 is pectin rich (Kim et al., 2006).

colloids causes the breakdown in stability of multiple emulsion. Blends of hydrocolloids and protein can create a variety of hybrid complexes that significantly improve the stability of each biopolymer separately. These molecules can considerably improve the stability and slow the release rates of multiple emulsions.

Multiple emulsions are much more stable, with smaller globule size and smaller inner droplet size, more monodispersed and more viscous compare to multiple emulsions stabilized with monomeric surfactant. Today's multiple emulsions can retain the solute for longer periods of time on the shelf.

The best multiple emulsions are likely to be prepared by the separation membrane technique, with polymeric amphiphilic, viscosity-enhancing agents, emulsified microemulsions, and proper selection of the amphiphilic blends of biomacromolecules (proteins and hydrocolloids) in each of the interfaces. The main goal remains to obtain submicron multiple emulsion droplets with long-term stability, possibly by replacing the inner emulsion with a microemulsion that is thermodynamically stable (emulsified microemulsions). Better control of the release pattern from the inner phase is still far from being achieved.

REFERENCES

Adachi T, Ida J, Wakita M. 1999. Preparation of spherical and porous chitosan particles by suspension evaporation with O/W/O multiple emulsions. *Polym J* 31: 319–323.

Antonov YA, Van Puyvelde P, Moldenaers P. 2004. Interfacial tension of aqueous biopolymer mixtures close to the critical point. *Int J Biol Macromolecules* 34: 29–35.

Asakura S, Oosawa F. 1954. Interaction between two bodies immersed in a solution of macromolecules. *J Chem Phys* 22: 1255–1256.

Aveyard R, Binks BP, Clint JH. 2003. Emulsions stabilized solely by colloidal particles. *Adv Colloid Interface Sci* 100–102: 503–546.

Benichou A, Aserin A, Garti N. 2002a. Protein-polysaccharide interactions for stabilization of food emulsions. *J Dispersion Sci Technol* 23: 93–123.

Benichou A, Aserin A, Garti N. 2004. Double emulsions stabilized with hybrids of natural polymers for entrapment and slow release of active matters. *Adv Colloid Interface Sci* 108: 29–41.

Benichou A, Aserin A, Garti N. 2002b. Double emulsion stabilized by new recognition hybrids of natural polymers. *Polym Adv Technol* 13: 1019–1031.

Bergenståhl B. 1997. Physiochemical aspects of emulsifier functionality. In: *Food Emulsifiers and Their Applications*, Hasenhuettl GL, Hartel RW, eds. New York: Chapman and Hall, pp 142–147.

Bibette J, Leal-Calderon F, Schmitt V, Poulin P. 2002. Double emulsions. In: *Emulsion Science Basic Principles—An Overview*. New York: Springer, pp 117–136.

Capron I, Costeux S, Djabourov M. 2001. Water in water emulsions: Phase separation and rheology of biopolymer solutions. *Rheolog Acta* 40: 441–456.

Cornec M, Wilde PJ, Gunning PA, Mackie AR, Husband FA, Parker ML, Clark DC. 1998. Emulsion stability as affected by competitive adsorption between an oil-soluble emulsifier and milk proteins at the interface. *J Food Sci* 63: 39–43.

Dickinson E. 2003. Hydrocolloids at interfaces and the influence on the properties of dispersed systems. *Food Hydrocolloids* 17: 25–39.

Dickinson E, Evison J, Owusu RK. 1991. Preparation of fine protein stabilized water-in-oil-in-water emulsions. *Food Hydrocolloids* 5: 481–485.

Dickinson E, Evison J, Owusu RK, Williams A. 1994. Protein stabilized water-in-oil-water emulsions. In: *Gums and Stabilizers for the Food Industry*, Vol. 7, Phillips GO, Williams PA, Wedlock DJ, eds. Oxford: IRL Oxford University Press, pp 91–101.

Evison J, Dickinson E, Owusu R, Williams A. 1995. Formulation and properties of protein-stabilized water-in-oil-in-water multiple emulsions. In: *Food Macromolecules and Colloids*, Dickinson E, Lorient D, eds. Cambridge: Royal Society of Chemistry, pp 235–243.

Fechner A, Knoth A, Scherze I, Muschiolik G. 2006. Stabilization of multiple emulsion by milk protein-polysaccharide interaction conjugate. *Abstracts of Papers of the Food Colloids: Self-assembly and Material Science*, Montreux, Switzerland, The University of The Fribourg, Fribourg, Switzerland 23–26 April, B24.

Feigin RI, Napper DH. 1980. Depletion stabilization and depletion flocculation. *J Colloids Interface Sci* 75: 525–541.

Fleer GJ, Scheutjens JHMH, Vincent B. 1984. The stability of dispersions of hard spherical particles in the presence of non-adsorbing polymer. In: *Polymer Adsorption and Dispersion Stability*, Goddord E, Vincent B, eds. ACS Symp. Ser. Washington, DC: American Chemical Society, pp 245–263.

Florence AT, Whitehill D. 1985. Stability and stabilization of water-in-oil-in-water multiple emulsion. In: *Macro-and Microemulsions: Theory and Application*, Shah DO, ed. ACS Symp. Ser. Washington, DC: American Chemical Society, pp 359–380.

Fredrokumbaradzi E, Simov A. 1992. Effect of bovine serum albumim (BSA) on release of sulfacetamide sodium from multiple W/O/W emulsions. *Pharmazie* 47: 388–389.

Gallarate M, Carlotti ME, Trotta M, Bovo S. 1999. On the stability of ascorbic acid in emulsified systems for topical and cosmetic use. *Int J Pharmaceut* 188: 233–241.

Garti N, Lutz R. 2004. Recent progress in double emulsions. In: *Interface Science and Technology*, Vol. 4. *Emulsions: Structure Stability and Interactions*, Petsev DN, ed. Oxford: Elsevier, pp 557–605.

Garti N, Benichou A. 2001. Double emulsions for controlled-release applications-Progress and trends. In: *Encyclopedic Handbook of Emulsion Technology*, Sjöblom J, ed. New York: Marcel Dekker, pp 377–407.

Garti N, Reichman D. 1994. Surface-properties and emulsification activity of galactomannans. *Food Hydrocolloids* 8: 155–173.

Garti N, Madar Z, Aserin A, Sternheim B. 1997. Fenugreek galactomannans as food emulsifiers. *Food Science and Technology—Lebensmittel-Wissenschaft and Technologie* 30: 305–311.

Garti N, Aserin N, Cohen Y. 1994. Mechanistic considerations on the release of electrolytes from multiples emulsions stabilized by BSA and nonionic surfactants. *J Controlled Release* 29: 41–51.

Garti N, Aserin A, Tiunova I, Binyamin H. 1999. Double emulsions of water-in-oil-in-water stabilized by alpha-form fat microcrystals. Part 1: Selection of emulsifiers and fat microcrystalline particles. *J Am Oil Chemists Soc* 76: 383–389.

Garti N, Wicker L. 2005. Pectin methylesterase modified pectin interaction with whey protein isolate and stability of double emulsions. *Abstracts of Papers of the American Chemical Society*, American Chemical Society, Washington, DC 229. U295-U296 119-CELL Part 1.

Grinberg VY, Tolstoguzov VB. 1997. Thermodynamic incompatibility of proteins and polysaccharides in solutions. *Food Hydrocolloids* 11: 145–158.

Grossiord JL, Seiller M. 2001. W/O/W multiple emulsions: a review of the release mechanisms by break-up of the oily membrane. *STP Pharma Sci* 11: 331–339.

Ibanoglu E. 2005. Effect of hydrocolloids on the thermal denaturation of proteins. *Food Chem* 90: 621–626.

Jerdjev AM, Hunter RJ, Beattie JK. 2006. Self-depletion flocculation of tetralin oil-in-water emulsions. *Langmuir* 22: 84–87.

Jiao J, Rhodes DG, Burgess DJ. 2002. Multiple emulsion stability: Pressure balance and interfacial film strength. *J Colloid Interface Sci* 250: 444–450.

Jiao J, Burgess DJ. 2003. Rheology and stability of water-in-oil-in-water multiple emulsions containing Span 83 and Tween 80. *AAPS Pharm Sci* 5 Art. No. 7.

Kanouni M, Rosano HL, Naouli N. 2002. Preparation of a stable double emulsion ($W_1/O/W_2$): role of the interfacial films on the stability of the system. *Adv Colloid Interface Sci* 99: 229–254.

Kerstens S, Murray BS, Dickinson E. 2006. Microstructure of β-lactoglobulin-stabilized emulsions containing non-ionic surfactant and excess free protein: Influence of heating. *J Colloid Interface Sci* 296: 332–341.

Kim SY, Lee YM. 1999. Lipid nanospheres containing vitamin A or vitamin E: Evaluation of their stabilities and in vitro skin permeability. *Ind Eng Chem Res* 5: 306–313.

Kim JW, Kang HH, Suh KD, Oh SG. 2003. Stabilization of water-soluble antioxidant in water-in-oil-in-water double emulsions. *J Dispersion Sci Technol* 24: 833–839.

Kim HJ, Cho YH, Bae EK, Shin TS, Choi SW, Choi KH, Park J. 2005. Development of W/O/W multiple emulsion formulation containing Burkholderia gladioli. *J Microbiol Biotechnol* 15: 29–34.

Kim HJ, Decker EA, McClements DJ. 2006. Preparation of multiple emulsions based on thermodynamic incompatibility of heat-denatured whey protein and pectin solutions. *Food Hydrocolloids* 20: 586–595.

Koberstein-Hajda A, Dickinson E. 1996. Stability of water-in-oil-in-water emulsions containing faba bean proteins. *Food Hydrocolloids* 10: 251–254.

Knoth A, Scherze I, Kramer M, Muschiolik G. 2006. Influence of milk protein-polysaccharide interaction on the properties of W/O/W multiple emulsions. *Abstracts of Papers of the Food Colloids: Self-assembly and Material Science*, Montreux, Switzerland, 23–26 April, B23.

Kuhl TL, Berman AD, Hui SW, Israelachvili JN. 1998. Part 1. Direct measurement of depletion attraction and thin film viscosity between lipid bilayers in aqueous polyethylene glycol solutions. *Macromolecules* 31: 8250–8257.

Lin CY, Wang KH. 2003. The fuel properties of three-phase emulsions as an alternative fuel for diesel engines. *Fuel* 82: 1367–1375.

Magdassi S, Garti N. 1986. A kinetic model for release of electrolytes from W/O/W multiple emulsions. *J Controlled Release* 3: 273–277.

Mao Y, Cates ME, Lekkerkerker HNW. 1995a. Depletion force in colloidal systems. *Physica A* 222: 10–24.

Mao Y, Cates ME, Lekkerkerker HNW. 1995b. Depletion stabilization by semidilute rods. *Phys Rev Lett* 75: 4548–4551.

Matsumoto S. 1986. W/O/W-type multiple emulsion with a view to possible food applications. *J Texture Stud* 17: 141–159.

Matsumoto S, Kida Y, Yonezawa D. 1976. An attempt at preparing water-in-oil-in-water multiple-phase emulsions. *J Colloid Interface Sci* 57: 353–361.

Matsumoto S, Koh Y, Michiura A. 1985. Preparation of water/olive oil/water emulsions in an edible form on the basis of phase inversion technique. *J Dispersion Sci Technol* 6: 507–521.

Maydani Y. 1994. Multiple emulsions based on lysozyme as emulsifier. Graduate thesis. Hebrew University of Jerusalem.

Mezzenga R, Folmer BM, Hughes E. 2004. Design of double emulsions by osmotic pressure tailoring. *Langmuir* 20: 3574–3582.

Muguet V, Seiller M, Barratt G, Ozer O, Marty JP, Grossiord JL. 2001. Formulation of shear rate sensitive multiple emulsions. *J Controlled Release* 70: 37–49.

Myers D. 1998a. Emulsions. In: *Surfactant Science and Technology*. New York: VCH, pp 209–253.

Myers D. 1998b. Emulsions. In: *Surfactant Science and Technology*. New York: VCH, pp 216–219.

Nunes MC, Raymundo A, Sousa I. 2005. Rheological behavior and microstructure of pea protein/κ-carrageenan/starch gels with different setting conditions. *Food Hydrocolloids* 20: 106–113.

Okochi H, Nakano M. 2000. Preparation and evaluation of W/O/W type emulsion containing vancomycin. *Adv Drug Deliv Rev* 45: 5–26.

Omotosho LA, Law KT, Whatelely TL, Florence AT. 1986. The stabilization of W/O/W emulsions by interfacial interaction between albumin and nonionic surfactant. *Colloid Surf* 20: 133–143.

Ortega-Vinuesa JL, Martin-Rodriguez A, Hidalgo-Alvarez R. 1996. Colloidal stability of polymer colloids with different interfacial properties: mechanisms. *J Colloid Interface Sci* 184: 259–267.

Oza KP, Frank SG. 1989. Multiple emulsions stabilized by colloidal microcrystalline cellulose. *J Dispersion Sci Technol* 10: 163–185.

Ozer O, Muguet V, Roy E, Grossiord JL, Seiller M. 2000. Stability study of W/O/W viscosified multiple emulsions. *Drug Dev Ind Pharm* 26: 1185–1189.

Radford SJ, Dickinson E. 2004. Depletion flocculation of caseinate-stabilised emulsions: what is the optimum size of the non-adsorbed protein nano-particles? *Colloids Surf A* 238: 71–81.

Ribotta PD, Ausar SF, Beltramo DM, Leon AE. 2005. Interactions of hydrocolloids and sonicated-gluten proteins. *Food Hydrocolloids* 19: 93–99.

Schulz PC, Rodriguez MS, Del Blanco LF. 1998. Emulsification properties of chitosan. *Colloid Polym Sci* 276: 1159–1165.

Seifriz W. 1925. Studies in emulsion III. Double reversal of oil emulsions occasioned by the same electrolyte. *J Phys Chem* 29: 738–749.

Shan-Chen Y, Bochot A, Le Bas G, Chéron, Mahuteau J, Grossiord JL, Seiller M, Duchêne D. 2003. Effect of camphor/cyclodextrin complexation on the stability of O/W/O multiple emulsions. *Int J Pharmaceut* 261: 1–8.

Shields M, Ellis R, Saunders BR. 2001. A creaming study of weakly flocculated and depletion flocculated oil-in-water emulsions. *Colloids Surf A* 178: 265–276.

Shima M, Tanaka M, Fujii T, Egawa K, Kimura Y, Adachi S, Matsuno R. 2006. Oral administration of insulin included in fine W/O/W emulsions to rats. *Food Hydrocolloids* 20: 523–531.

Su J, Flanagan J, Hemar Y, Singh H. 2006. Synergistic effects of polyglycerol ester of polyricinoleic acid and sodium caseinate on the stabilisation of water-oil-water emulsions. *Food Hydrocolloids* 20: 261–268.

Tolstoguzov VB. 2003. Some thermodynamic considerations in food formulation. *Food Hydrocolloids* 17: 1–23.

Uruakpa FO, Arntfield SD. 2006. Surface hydrophobicity of commercial canola proteins mixed with kappa-carrageenan or guar gum. *Food Chem* 95: 255–263.

Uruakpa FO, Arntfield SD. 2005. Emulsifying characteristics of commercial canola protein-hydrocolloid systems. *Food Research Int* 38: 659–672.

Vasiljevic D, Parojcic J, Primorac M, Vuleta G. 2006. An investigation into the characteristics and drug release properties of multiple W/O/W emulsion systems containing low concentration of lipophilic polymeric emulsifier. *Int J Pharmaceut* 309: 171–177.

Vaziri A, Warburton B. 1994. Some preparative variable influencing the properties of W/O/W multiple emulsions. *J Microencapsul* 11: 649–656.

Versteeg J. 1978. Stabilized static liquid membrane compositions. US patent 4,083,798.

Walz JY, Sharma A. 1994. Effect of long range interactions on the depletion force between colloidal particles. *J Colloid Interface Sci* 168: 485–496.

Wang YF, Tao Z, Gang H. 2006. Structural evolution of polymer-stabilized double emulsions. *Langmuir* 22: 67–73.

Weiss J, Scherze I, Muschiolik G. 2005. Polysaccharide gel with multiple emulsions. *Food Hydrocolloids* 19: 605–615.

Wen LX, Papadopoulos KD. 2001. Effects of osmotic pressure on water transport in $W_1/O/W_2$ emulsions. *J Colloid Interface Sci* 235: 398–404.

Wilde PJ, Clark DC. 1993. The competitive displacement of β-lactoglobulin by Tween 20 from oil-water and air-water interface. *J Colloid Interface Sci* 155: 48–53.

Yoshida K, Sekine T, Matsuzaki F, Yanaki T, Yamaguchi M. 1999. Stability of vitamin A in oil-in-water-in-oil-type multiple emulsions. *J Am Oil Chemists Soc* 76: 195–200.

Yu SC, Bochot A, Cheron M, Seiller M, Grossiord JL, Le Bas G, Duchene D. 1999. Design and evaluation of an original o/w/o multiple emulsion. *STP Pharma Sci* 9: 273–277.

Zhang W, Miyakawa T, Uchida T, Goto S. 1992. Preparation of stable W/O/W type multiple emulsion containing water-soluble drugs and in-vitro evaluation of its drug releasing properties. *J Pharmaceut Soc Jap* 112: 73–80.

CHAPTER 6

Recent Developments in Manufacturing Particulate Products from Double-Emulsion Templates Using Membrane and Microfluidic Devices

GORAN T. VLADISAVLJEVIĆ and RICHARD A. WILLIAMS

Contents

Multiple Emulsions: Technology and Applications, Edited by Abraham Aserin

6.1 INTRODUCTION

The first investigation on using membrane emulsification can be traced back to the 1980s when Nakashima and colleagues (Nakashima and Shimizu, 1986) fabricated a particular glass membrane, called Shirasu porous glass (SPG), and successfully produced uniform-sized kerosene-in-water and water-in-kerosene emulsions (Nakashima et al., 1991a, b). Since this time the method has continued to attract attention because of its effectiveness in producing narrow particle size distributions over a wide range of mean droplet sizes at low energy consumption. Although the majority of investigations on using membrane emulsification have been undertaken by Japanese workers, other novel research is being carried out in European countries (Schröder et al., 1998; Joscelyne and Trägårdh, 1999; Williams et al., 1998; Abrahamse et al., 2002; Giorno et al., 2003) and elsewhere (Jing et al., 2005). To date, in addition to experimentation using SPG membranes, investigations of a broad range of other types of membranes, such as ceramic, metallic, polymeric and microengineered devices, have been reported (Table 6.1). The table indicates the form of the membrane, the nature of its surface, mean pore size, and a typical effective membrane surface area deployed in the example cited. Other types of microengineered materials can be utilized such as parallel arrays of microgrooves formed by etching on a single-crystal silicon substrate (Kawakatsu et al., 2000) and microfluidic devices of quartz glass (Nisisako et al., 2004), poly(dimethyl siloxane) (PDMS) (Yi et al., 2003), polyurethane (PU) (Nie et al., 2005). The silicon-based arrays of microgrooves and soft microfluidic devices are not considered in Table 6.1.

It has been stated that the development of emulsification methods for production of mono-sized droplets must be rooted in one of two possible manufacturing approaches (Williams et al., 2001a): (1) reduction of process length scales of the turbulent perturbations and enhancement of their uniformity in the mixing processes that rupture the liquids, and (2) the creation of droplets individually (drop by drop). The production of emulsions using membrane and microfluidic devices represents a typical example of the second approach. This chapter aims to introduce the latest development on the utilization of the membrane and microfluidic emulsification techniques for the preparation of double emulsions, as well as micro- and nanoparticles from double-emulsion precursors.

The polydispersity of the droplets throughout this chapter will be expressed in terms of the coefficient of variation (relative standard deviation) or the relative span factor. For a specific size distribution the coefficient of variation is defined as $CV = (\sigma/d_{av}) \times 100$, where σ is the standard deviation of the droplet diameters and d_{av} is the number–average mean droplet diameter. The relative span factor (or span) is given by $(d_{90} - d_{10})/d_{50}$, where d_{x0} is the diameter corresponding to $x0$ vol% on a relative cumulative droplet diameter distribution curve.

TABLE 6.1 Microporous membranes other than Shirasu porous glass (SPG) used by different investigators for membrane emulsification

Membrane Material	Membrane Form	Surface Affinity	Mean Pore Size	Effective Membrane Area	Authors
Coated α-alumina or zirconia	Tubular	Hydrophilic	0.02–5 μm	50–460 cm^2	Williams et al. (1998), Jing et al. (2005), Schröder (1999), Berot et al. (2003), Vladisavljević et al. (2003)
Anodic porous alumina	Flat	Hydrophilic	0.125 μm	—	Yanagishita et al. (2004)
Sol-gel porous glass (sol-gel PG)	Tubular	Hydrophilic	0.6 μm	—	Fuchigami et al. (2000)
Monolithic silica	Flat	Hydrophilic	2 μm	25 mm^2	Hosoya et al. (2005)
Glass plate with rectilinear capillaries	Flat	Hydrophilic	4 μm	25 mm^2	Hosoya et al. (2005)
Stainless steel with laser drilled pores	Tubular	Hydrophilic	100–150 μm	4–26.7 cm^2	Dowding et al. (2001), Williams et al. (2001), Vladisavljević and Williams (2006)
Stainless steel sintered	Flat	Hydrophilic	5 μm	1.25 cm^2	Xu et al. (2005)
Microporous nickel	Tubular or flat	Hydrophilic	3–40 μm	9 cm^2 [21]	Liu et al. (2003), Kosvintsev et al. (2005)
Microporous nickel with carbon coating	Tubular rotating	Hydrophobic	3 μm	—	Schadler and Windhab (2004)
Polypropylene	Hollow fiber	Hydrophobic	0.4 μm	10^3 cm^2	Vladisavljević et al. (2002)
Polyamide	Hollow fiber	Hydrophilic	10 nm	36 cm^2	Giorno et al. (2003)
Polytetrafluoroethylene (PTFE)	Flat	Hydrophilic or hydrophobic	0.5–5 μm	3–17 cm^2	Suzuki et al. (1998, 1999), Altenbach-Rehm et al. (2002a, b), Yamazaki et al. (2002, 2003)
Polycarbonate, track-etched	Flat	Hydrophilic or hydrophobic	0.6–10 μm	1.5–14 cm^2	Kobayashi et al. (2002b), Park et al. (2001), Yafei et al. (2006), Hino et al. (2000), Kawashima et al. (1991)
Cellulose acetate	Flat	Hydrophilic	0.2–3 μm	4 cm^2	Shima et al. (2004)
Micromachined silicone nitride microsieve	Flat	Hydrophilic	2–7 μm	4 mm^2	Abrahamse et al. (2002), Gijsbertsen-Abrahamse et al. (2003), Zhu and Borrow (2005)
Straight-through silicon microchannels	Flat	Hydrophilic	10–17 μm	1 cm^2	Kobayashi et al. (2003), Kobayashi et al. (2002a)

6.2 PRINCIPLES AND OPERATING METHODS FOR MEMBRANE EMULSIFICATION

Membrane emulsification (ME) methods reported so far in the literature are depicted schematically in Figure 6.1. In conventional direct ME (Figure 6.1*a*), fine droplets are created in situ at the membrane-continuous phase interface by pressing a pure dispersed phase through the membrane. To ensure a regular droplet detachment from the pore outlets, shear stress is generated at the membrane-continuous phase interface by recirculating the continuous phase along the membrane using a low shear pump (Figure 6.2*a*) (Nakashima et al., 2001a) or by agitation in a stirring vessel (Ha et al., 1999; Ma et al., 1999; Fuchigami et al., 2000; You et al., 2001; Ito and Makino, 2004; Kosvintsev et al., 2005; Wang et al., 2005) (Figure 6.2*b*). The rate of mixing should be high enough to provide the required tangential shear on the membrane surface, but not too excessive to induce further droplet breakup. The continuous phase is usually agitated by a magnetic stirrer and the dispersed phase is pressurized either inside (You et al., 2001) or outside (Wang et al., 2005) the membrane tube. Recently a stirred cell with a paddle rotating above a flat membrane was used for emulsification (Kosvintsev et al., 2005).

Another approach uses devices equipped with a moving membrane, in which the droplet detachment from the pore outlets is stimulated by rotation (Williams et al., 2001b; Schadler and Windhab, 2004; Vladisavljević and Williams, 2006) or vibration (Tice et al., 2004; Zhu and Barrow, 2005) of the membrane within a stationary continuous phase cylinder (Figure 6.2*c* and Figure 6.3). In our recent investigation (Vladisavljević and Williams, 2006) we used a stainless steel membrane with 100 μm laser drilled pores to produce O/W emulsions with relatively large (80–260 μm) droplets. The membrane tube, 1 cm in diameter, was rotated inside a stationary glass cylinder (Figure 6.3*c*), diameter of 3 cm, at a constant speed in the range 50 to 1500 rpm. Compared with cross ME, this method appears to be particularly advantageous to

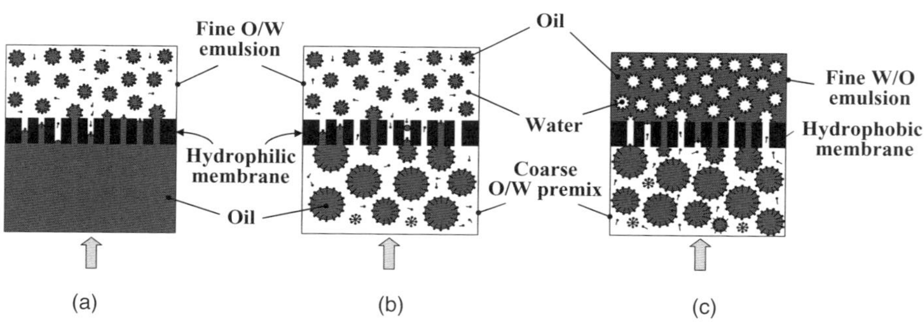

Figure 6.1 Membrane emulsification methods: (*a*) Direct ME (Nakashima et al., 1991a); (*b*) premix ME (Suzuki et al., 1996); (*c*) premix ME with phase inversion (Suzuki et al., 1999).

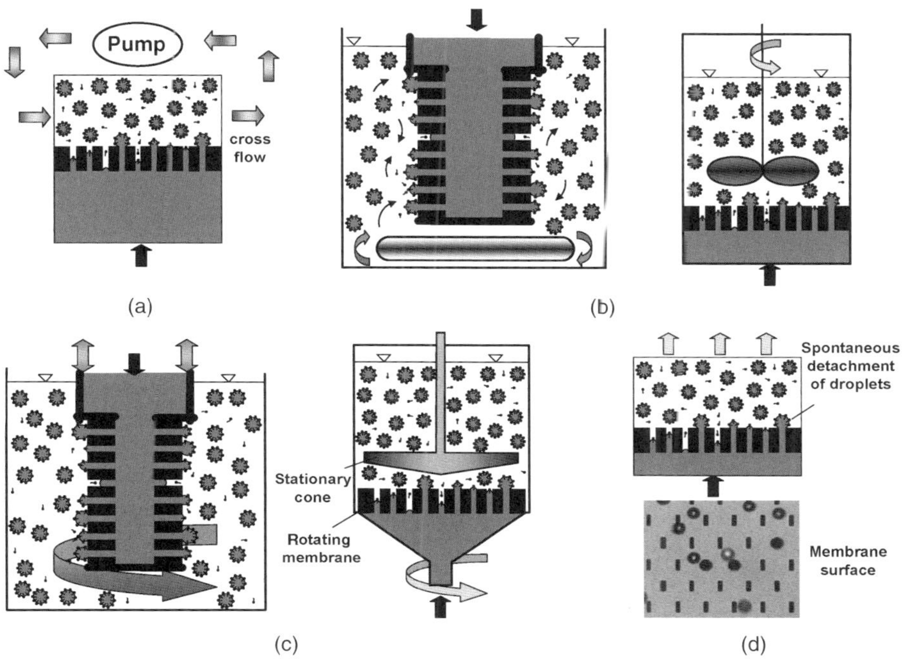

Figure 6.2 Membrane emulsification systems for controlling mechanical shear at the membrane surface employed by different investigators (direct ME): (*a*) Cross-flow system (Nakashima et al., 1991a); (*b*) stirring systems (Kosvintsev et al., 2005; You et al., 2001); (*c*) rotating/vibrating membrane systems (Vladisavljević and Williams, 2006; Kosvintsev et al., 2005); (*d*) dead-end system (Kobayashi et al., 2003).

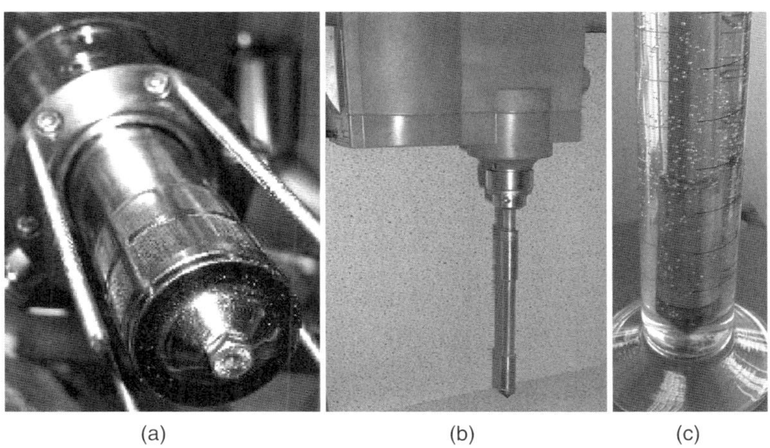

Figure 6.3 Examples of rotating membranes used for emulsification: (*a*) Rotating membrane fabricated by deposition of pure carbon layer on the microporous nickel substrate (Schadler and Windhab, 2004, 2005); (*b*) rotating stainless steel membrane with laser drilled pores (Vladisavljević and Williams, 2006); (*c*) a rotating membrane emulsification system in operation (Vladisavljević and Williams, 2006).

the production of monosized double emulsions and fragile microparticles, as the need for product recirculation is eliminated. Kosvintsev et al. (2005) modified a plate and cone rheometer so that the normally impervious plate underneath the cone was replaced by the membrane, as shown in Figure 6.2*c* right. A feed chamber was constructed to permit the injection of oil from underneath the membrane. A rotating seal was used to rotate the chamber (including the membrane) underneath the cone. Zhu and Barrow (2005) introduced membrane vibration, through piezoactuation, to help the detachment of droplets and provide extra control over droplet detachment in cross-membrane emulsification. The preliminary results showed that a vibrating membrane had a significant effect in reducing the average size of emulsion droplets but only at very low frequencies (0–100 Hz).

Even in the absence of any tangential shear, droplets can be spontaneously detached from the pore outlets at small disperse phase fluxes (Figure 6.2*d*), particularly in the presence of fast-adsorbing emulsifiers in the continuous phase and for a pronounced noncircular cross section of the pores, such as for narrow slots (41 × 11 μm) shown on the micrograph in Figure 6.4*d* (Kobayashi et al., 2003). Kobayashi et al. found that slot aspect ratios exceeding a threshold ratio of approximately 3 were needed to successfully prepare monodisperse emulsions with a *CV* below 2%.

A potential disadvantage of "direct" ME is a relatively low maximum dispersed phase flux through the membrane of typically 0.01 to 0.1 $m^3/(m^2h)$ that has to be restrained to avoid the transition from a "size-stable" to "continuous outflow" zone (Kobayashi et al., 2003) and to avoid the steric hindrance among droplets that may be formed simultaneously at the adjacent pores (Abrahamse

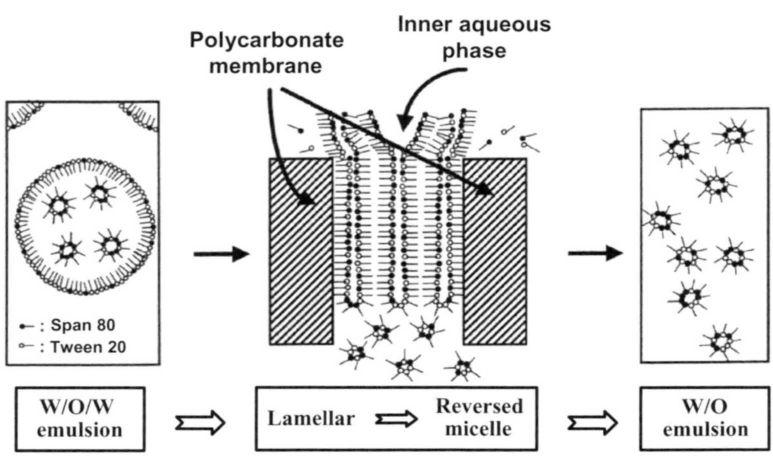

Figure 6.4 Schematic representation of phase inversion of W/O/W emulsion into a W/O emulsion by extrusion the multiple emulsion through hydrophobic polycarbonate membrane.

et al., 2002). It is observed that the formation of uniform droplets is only possible within the size-stable zone, within which the mean droplet size is almost independent on the dispersed phase flux and shear stress (Vladisavljević et al., 2004a). In the ME studies with a microsieve (Abrahamse et al., 2002) and SPG membranes (Katoh et al., 1996) the dispersed phase flux was 2.5 $m^3/(m^2h)$, but polydisperse droplets were generated. Because of the low productivity, which is due to long production times, direct ME is more suitable for the preparation of relatively diluted emulsions with disperse phase contents up to 20 vol%. Nevertheless, this process enables very narrow droplet size distributions to be produced over a wide range of mean droplet sizes. In the case of SPG membranes with a mean pore sizes between 0.1 and 20 μm. the mean droplet size of both O/W and W/O emulsions can range from less than 1 μm to over 60 μm, with the relative span factors under optimal conditions of 0.26 to 0.45 (Nakashima et al., 1993; Shimizu et al., 2002b, c; Vladisavljević and Schubert, 2002, 2003b). For ceramic membranes the droplet size range can be 0.2 to 100 μm with a range of spans (Williams, 2002).

Suzuki et al. (1996) implemented "premix" ME, in which a preliminarily emulsified coarse emulsion (rather than a single pure disperse phase) is forced through the membrane (Figure 6.1*b*) and the small droplets are formed by reducing the size of the large droplets in preexisting emulsions. This is achieved by gently mixing the two immiscible liquids using a conventional stirrer mixer, and then passing this preliminarily emulsified emulsion through the membrane. Recently Jing et al. (2005) used the direct ME under jetting flow conditions to prepare a coarse O/W emulsion. In the second stage, this coarse emulsion was dispersed through the ceramic $\alpha\text{-}Al_2O_3$ membrane into a stirring continuous phase to obtain a monodisperse emulsion.

When disperse phase of the coarse emulsion wets the membrane wall and suitable surfactants are dissolved in both liquid phases, the process results in a phase inversion; namely a coarse O/W emulsion is inverted into a fine W/O emulsion (Figure 6.1*c*), and vice versa (Suzuki et al., 1999). The main advantage of this method is that a fine emulsion can be easily prepared from a low concentration coarse emulsion at high rates. For polytetrafluoroethylene (PTFE) membrane filters with a mean pore size of 1 μm, the maximum dispersed phase volume fraction in phase-inverted emulsions was 0.9 and 0.84 for O/W and W/O emulsions, respectively (Suzuki et al., 1999). Flow-induced phase-inversion (FIPI) phenomenon was observed earlier by Akay (1998) who used a multiple expansion-contraction static mixer (MECSM) consisting of a series of short capillaries with flow dividers. Hino et al. (2000) and Kawashima et al. (1991) inverted a W/O/W emulsion made up of liquid paraffin, Span 80 (a hydrophobic surfactant), and Tween 20 (a hydrophilic surfactant) into a W/O emulsion by extrusion through polycarbonate membranes with a mean pore sizes of 3 and 8 μm. Inside the membrane pores, surfactant molecules are oriented with their hydrophobic groups toward the wall surface and with hydrophilic groups toward the solubilized water molecules as a result of a lamellar structure formed inside the pores. The structure ruptured at the pore outlets,

TABLE 6.2 Premix membrane emulsification investigations currently reported

Membrane Material	System	Mean Pore Size, d_m (μm)	Product Emulsion	Mean Droplet Size and Span	Flux ($m^3 m^{-2} h^{-1}$)	Authors
Tubular SPG	Cross flow	2.7 and 4.2	O/W	$(1.4–2.1) \times d_m$, span = 0.4–0.62	0.03–3.5	Suzuki et al. (1996)
Flat PTFE	Dead end	1.0	O/W and W/O	$(2–4.1) \times d_m$	Up to 9	Suzuki et al. (1998)
Flat PTFE	Dead end with phase inversion	1.0	O/W and W/O	$(2.8–4.0) \times d_m$	1–5.5	Suzuki et al. (1998)
Flat PTFE	Dead end, multipass (n = 1–3)	1.0	O/W	$(1.2–2.6) \times d_m$, span = 0.55–0.9	2–18	Altenbach-Rehm et al. (2002b)
Flat cellulose acetate	Dead end	0.2, 0.45, 0.8, and 3.0	W/O/W	$(1.0–3.5) \times d_m$	Not given	Shima et al. (2004)
Flat polycarbonate	Dead end, multipass (n = 1–18)	0.33, 0.38, 0.44, 0.6, and 1.0	O/W	$\leq 1.6 \times d_m$ for $n > 12$	0.2–0.6	Park et al. (2001)
Tubular SPG	Dead end multipass (n = 3)	1.1	S/O/W	$0.9 \times d_m$	1.6	Toorisaka et al. (2003)
Tubular SPG	Dead end, multipass (n = 1–5)	10.7	W/O/W	$(0.41–1.2) \times d_m$, span = 0.28–0.6	0.8–37	Vladisavljević et al. (2004b)
Tubular α-alumina	Stirring	1.5	O/W	$(1.5–1.8) \times d_m$, span = 1–1.2	0.42–0.62	Jing et al. (2005)
Flat polycarbonate	Dead end, multipass (n = 5)	—	W/O/W	0.7–2.5 μm	3.7–14.7	Yafei et al. (2006)

and this was followed by the creation of reversed micelles, as shown in Figure 6.4. Such a W/O emulsion can be redispersed into an aqueous solution of Tween 20 to form a redispersed W/O/W emulsion containing more uniform internal droplets than the original W/O/W emulsion.

Premix ME holds several advantages over "direct" ME. The optimal transmembrane fluxes with regard to droplet size uniformity are typically above $1\,m^3/(m^2h)$, which is 1 to 2 orders of magnitude higher than in direct ME; the mean droplet sizes that can be achieved using the same membrane and phase compositions are smaller than in direct ME, which can be advantageous. Also the experimental setup is generally simpler than in direct ME; for example, no moving parts such as cross-flow pump or stirrer are needed, except for the preparation of pre-emulsion. Finally, the premix ME process is easier to control and operate than direct ME, since the driving pressure and emulsifier properties are not so critical for the successful operation as in the direct ME process.

In the first premix ME study (Suzuki et al., 1996) a cross-flow system was used in which coarse emulsion was diluted by permeation into a pure continuous phase/diluted emulsion recirculating at the low-pressure side of the membrane. In the majority of subsequent investigations (Table 6.2), a dead-end system was used, in which a fine emulsion was withdrawn as a product after passing through the membrane, without any recirculation and/or dilution with the continuous phase. It enables fast preparation of emulsions with a dispersed phase content of 50 vol% or more (Vladisavljević et al., 2004b).

One of the disadvantages of premix ME is a higher polydispersity of droplets compared to direct ME. In order to combine the advantages of both techniques (i.e., high throughput of the premix process and a close particle size distribution of direct ME), a multi-pass (repeated) premix ME can be employed (Park et al., 2001; Altenbach-Rehm et al., 2002a, b; Vladisavljević et al., 2004b; Yafei et al., 2006). In this process a fine emulsion is repeatedly passed through the same membrane a number of times to achieve additional droplet size reduction and enhance droplet size uniformity. The repeated membrane homogenization was originally developed for the production of multilamellar lipid vesicles (liposomes) using track-etch polycarbonate filters (Walde and Ichikawa, 2001). In this process the coarse liposome suspension is passed under moderate pressure repetitively (usually 10 times) through the filters with progressively smaller pore sizes, which leads to a gradual break up of the large vesicles into smaller ones.

6.3 SINGLE-EMULSION PRODUCTION USING MEMBRANES

Cross-membrane emulsification (XME) using SPG and ceramic alumina membranes is one of the most commonly used ME methods for preparation of emulsions. As shown in Figure 6.5, SPG membrane contains very uniform interconnected pores with a mean pore tortuosity of about 1.3 and the wall

TABLE 6.3 Some examples of successful formulations used for manufacturing double emulsions ($W_1/O/W_2$) using membranes

Membrane			Product $W_1/O/W_2$ Emulsion		
Material and Method	Mean Pore Size	Mean Size of W_1/O Drops	ϕ	Formulation	Authors
SPG (cross-flow, direct emulsification)	1.0 μm	4.8 μm	$\phi_i = 30$ vol%	W_1: water + 5% glucose O: Soybean oil + 0.5% PC + 0.5% PGPR W_2: water + 5% glucose + 1% LPC	Mine et al. (1996)
Cellulose acetate (dead-end, premix emulsification)	0.2–3.0 μm	0.7–3.0 μm	$\phi_i = 30$ vol% $\phi_0 = 50$ vol%	W_1: Hank's solution + PTSA O: C8TG + 10% PGPR W_2: Hank's solution + 2–10% decaglycerol monolaurate	Shima et al. (2004)
SPG (stirring, direct emulsification)	10.6 and 20 μm	30.1 and 70 μm, CV = 17 and 9%	$\phi_i = 50$ vol%$_o$ = 67 vol%	W_1: water + 5.8% glucose + 1.2% epirubicin hydrochloride O: IPSO + 9% PGPR W_2: physiological saline + 1% polyoxyethylene 60 stearate	Higashi et al. (1999, 1995, 2000)
SPG (dead-end, repeated premix emulsification, n = 5–6)	10.7 μm	4.4–13.2 μm, span = 0.28–0.4	ϕ_i = 10~30 vol% ϕ_0 = 1s60 vol%	W_1: water + 5% glucose O: soybean oil + 5% PGPR W_2: water + 5% glucose + 0.5% Tween 80 + (1% sodium alginate)	Toorisaka et al. (2003)
SPG (cross-flow, direct emulsification)	5 μm	17 μm, span = 0.35–0.42	$\phi_i = 10$ vol% ϕ_0 = 1–10 vol%	W_1: water + 5% phosphate buffer (NaH_2PO_4/Na_2HPO_4) O: rapeseed oil + 10% PGPR 90 W_2: water + 2% Tween 80 + 0.5% NaCl	Vladisavljević and Schubert (2003)

SPG (direct emulsification)	4.8 μm	16.8 μm	$\phi_i = 25$ vol%	W_1: water + 1.3% $CaNa_2$-EDTA + 0.75% KCl O: cacao butter + 15% PGPR W_2: water + 0.5% HCO-60 + 0.59% NaCl + 0.95% glucose	Kukizaki et al. (2001)
SPG (premix emulsification)	1.7 μm	2.6 μm	/	W_1: water + anthocyanin O: soybean oil + 10% PGPR W_2: water + 0.4% SWA-10D + glucose	Fukui et al. (2001)
Polycarbonate SPI membrane (repeated premix emulsification)	—	0.7–2.0 μm	$\phi_i = 10$ vol% $\phi_0 = 10$ vol%	W_1: water + 0.1 M NaCl + 9×10^{-5} dextran O: dodecane + Arlacel P135 (PEG-30) W_2: water + 0.1 M NaCl + SDS (few cmc)	Yafei et al. (2006)
SPG (stirring, direct emulsification)	1.4–10.2 μm		$\phi_i \approx 30$ vol%	W_1: water + 0.67 wt% Lysozyme + 2 wt% additives (PVA, PVP, gelatin, Tween 20, Myrj 52) O: DCM/toluene mixture ((v/v): 21/79) + 0.25–7.5 wt% Arlacel 83 + 5 wt% PVA W_2: water + 1 wt% PVA + 0.067 wt% SDS + 0–2 wt% NaCl	Liu et al. (2005a, b)
SPG (cross-flow, direct emulsification)	3.16 μm			W_1: water + 1 wt% Na-alginate O: n-hexane + 0.25 wt% Span 80 + 0.25 wt% Tween 85 W_2: water + 15% $CaCl_2$	Nakashima et al. (1991a)

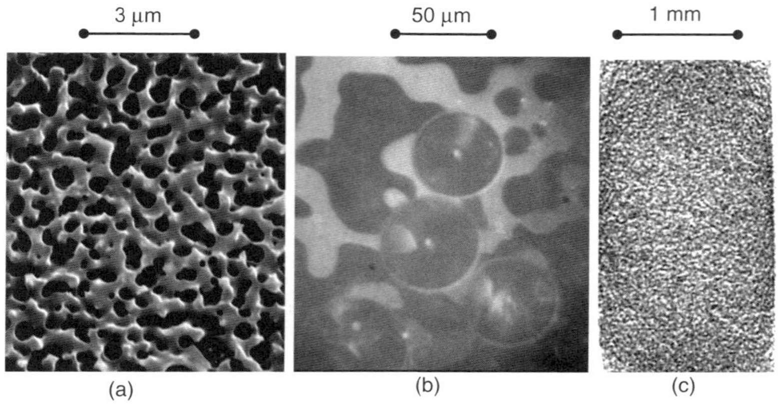

Figure 6.5 Microstructure of SPG membrane observed using different techniques: (*a*) SEM image of membrane surface; (*b*) microscopic image of droplet formation at the surface of SPG membrane; (*c*) X-ray microtomograph of an internal cross section of SPG membrane with a mean pore size of 20zyxonetwothreexyzm, showing a uniform microstructure. More than 1000 pores per square mm were identified using a high-resolution XMT scanner (Vladisavljević and Williams, 2004, unpublished).

porosity ranging from 50% to 60%. It can be fabricated with the mean pore sizes in a wide interval between 0.1 and 20 μm, and it can easily be rendered hydrophobic by coating with a silicone resin (Vladisavljević et al., 2005). Hydrophobic membranes are needed to produce W/O emulsions, whereas hydrophilic membranes are needed to produce O/W emulsions. It has been shown that the hydrophobic treatment of SPG membrane with silicone resin is not followed by pore plugging, even in the submicron range of mean pore sizes (Vladisavljević et al., 2005). However, care does have to be taken in this silinization process to prevent blockages.

The adequate choice of emulsifiers is of primary importance for the success of ME. A chosen emulsifier should rapidly adsorb to the newly formed oil-water interface to reduce the interfacial tension to an optimum level. However, the emulsifier molecules should not adsorb to the membrane surface by electrostatic interactions because it can cause the alteration of membrane polarity from hydrophilic to hydrophobic and pore blocking. The effect of dynamic interfacial tension of emulsifiers on direct ME has been investigated by several authors (Schröder et al., 1998; Van der Graaf et al., 2004; Rayner et al., 2005). As a rule, the faster the emulsifier molecules adsorb at newly formed interfaces, the smaller the mean droplet size of the resultant emulsion. The effect of emulsifier charge on droplet formation in direct ME has been investigated by Nakashima et al. (1993) for SPG membranes and by Kobayashi et al. (2003) for silicon microchannels. Both studies concluded that to avoid electrostatic attractions between the emulsifier molecules and the membrane wall, the functional groups of the chosen emulsifiers must not carry the charge opposite to that of the membrane surface. For example, an untreated SPG membrane has a negative surface potential of −15 to −35 mV within a pH range of 2 to 8, due to the dissociation of acid silanol groups (Si–OH ⇄ SiO^- + H^+). Hence, for this case, the use of cationic emulsifiers such as alkyl-substituted quaternary ammonium salts must be avoided.

The minimum transmembrane pressure for driving the dispersed phase through the pores is given by the Laplace equation: $p_{cap} = 4\gamma\cos\theta/d_p$, where γ is the equilibrium interfacial tension between the continuous and dispersed phase, d_p is the mean pore size, and θ is the contact angle between the dispersed phase and the membrane surface. The optimal $\Delta p_{tm}/p_{cap}$ ratio for the generation of mono-sized droplets in XME using SPG membranes ranges from 1.1 to 6 (Vladisavljević et al., 2004a), depending on the shear stress. The higher the wall shear stress, the higher is the Δp_{tm} value that can be applied. The wall shear stress, σ_w, is related to the mean velocity of the continuous phase inside the membrane tube, v_t, with the equation: $\sigma_w = \lambda(\rho_c v_t^2/8)$, where ρ_c is the continuous phase density and λ is the Moody friction factor. In the special case of laminar flow inside the membrane tube ($Re_t < 2300$), $\lambda = 64/Re_t$ and $\sigma_w = 8\eta_c v_t/d_i$, where d_i is the inner diameter of membrane tube and η_c is the continuous phase viscosity. The optimal wall shear stress in XME using SPG and ceramic membranes usually ranges between 10 and 40 Pa. The influence of various process parameters on membrane emulsification was reviewed by several authors (Joscelyne and Trägårdh, 2000; Charcosset et al., 2004; Lambrich and Vladisavljević, 2004; Gijsbertsen-Abrahamse et al., 2004).

6.4 SINGLE-EMULSION PRODUCTION USING MICROFLUIDIC DEVICES

Many attempts have recently been made to generate droplets in microfluidic devices (Atencia, 2005). These microfluidic devices are based on T-junction designs (Thorsen et al., 2001; Link et al., 2004; Yi et al., 2003; Tice et al., 2004; He et al., 2005; Van der Graaf et al., 2005b; Dendukuri et al., 2005), flow-focusing designs (Anna et al., 2003; Xu and Nakajima, 2004; Davidson et al., 2005; Garstecki et al., 2005; Lewis et al., 2005; Tan et al., 2006), and parallel microchannel arrays (Kawakatsu et al., 1997; Sugiura et al., 2002b; Sugiura et al., 2004b). Applications have emerged that can take advantage of precision generation of droplets. These applications range from programmable biochemical assays to drug delivery particles. The simplest microfluidic device for producing and manipulating droplets is the "T-junction" (Figure 6.6*a*): the T-shaped channel geometry forces two flows of immiscible liquids to merge in such a way that one liquid forms droplets dispersed in the other. The droplet-forming phase should not wet the channel walls at the junction; for example, hydrophobic T-junctions are needed to generate water droplets. The use of T-junctions in series with alternating surface wettabilities produces monodisperse double emulsions in a single device (Nisisako et al., 2005; Okushima

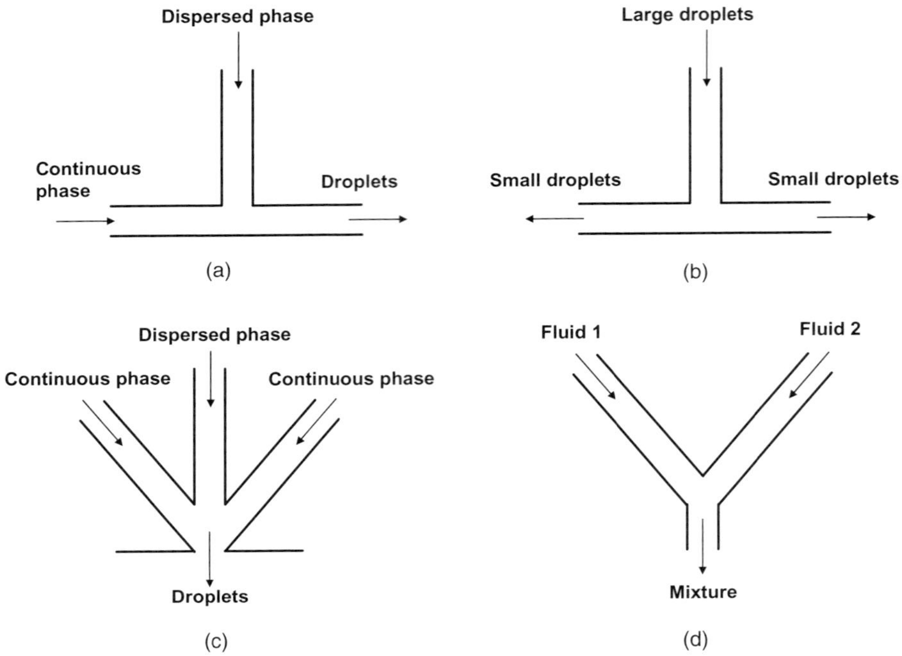

Figure 6.6 Different microfluidic junctions for droplet generation and manipulation: (*a*) T-junction at inlet (Thorsen et al., 2001); (*b*) T-junction at outlet (Link et al., 2004; Tan et al., 2004); (*c*) Sheath-flow junction (Xu and Nakajima, 2004; Nisisako et al., 2004); (*d*) Y-junction (Nisisako et al., 2004, 2005).

et al., 2004) (see Section 6.6). When reversing the flow direction, T-junctions with differently sized exit channels will passively sort droplets according to size (Tan et al., 2004) or break large droplets into smaller ones with controlled sizes (Link et al., 2004) (Figure 6.6*b*). Dispersed droplets may also be created by a microfluidic extension of Rayleigh's approach (Rayleigh, 1879), with two streams of one liquid flanking a stream of a second immiscible liquid and the combined two-phase flow is then forced through a small orifice (Garstecki et al., 2005) or enters a wide channel or "pool" (Figure 6.6*c*) (Xu and Nakajima, 2004). The pressure and viscous forces exerted by the outer fluid ultimately force the inner fluid to break into droplets, either in or just downstream of the orifice. The use of two sheath-flow junctions in series allows double emulsions or gas-in-oil-in-water dispersions to be produced, such as microbubbles stabilized with a thin oil shell and dispersed in an outer water phase for drug and gene delivery applications (Xu, 2005). The "Y-junction" is a simple microfluidic device used for mixing in situ two fluid streams of different compositions (fluids 1 and 2 in Figure 6.6*d*), for example, mixing of two differently colored monomer solutions in the production of bichromal particles for "electronic paper" (Nisisako et al., 2004). The fabrication of a planar microchannel system uses simple soft lithography, making it straightforward to adjust the geometry of the microchannel design to ensure selective generation of droplets over a range of different sizes.

In addition to soft microfluidic devices fabricated in elastomeric materials by soft lithography (Xia and Whitesides, 1998; Whitesides and Stroock, 2001), microfluidic devices can be fabricated in silicon by photolithography and wet-etching processing (Kawakatsu et al., 1997) and in stainless steel by a dicing process (Tong et al., 2001). Using photolithography and anisotropic wet-etching processing, parallel arrays of microgrooves were formed in a single-crystal silicon substrate (Figure 6.7*a* and *b*) and microchannels (MCs) were created by tightly covering the upper side of MC plate with a flat glass plate. The first MC plate was fabricated with a MC width of 6 μm and with a terrace region (Kawakatsu et al., 1997). Other MC types without terrace (Sugiura et al., 2000), with partition walls between MCs (Sugiura et al., 2001; Nakagawa et al., 2004), and with different sizes were fabricated later on. In 2002 deep ion etching (RIE) was used to fabricate new "through-type" MCs that are vertical to the plate surface (Kobayashi et al., 2002a; Kobayashi et al., 2003; Kobayashi et al., 2005a) (Figure 6.7*c*). The primary problem in straight-through MC emulsification is the difficulty in formulating monodisperse emulsions when

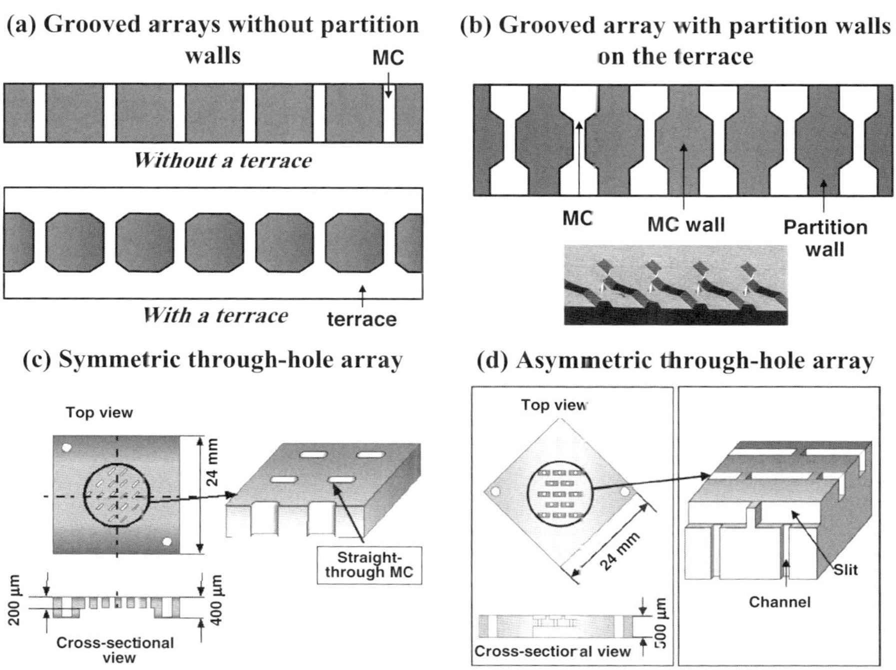

Figure 6.7 Different configurations of silicon microchannel (MC) arrays for droplet generation.

using dispersed phase with a viscosity below 1 mPa s. To overcome this problem, an array of "asymmetric" through-holes vertically fabricated on a silicon plate was recently developed (Kobayashi et al., 2005b). As shown in Figure 6.7*d*, each asymmetric through-hole is composed of a slit and a channel.

MC emulsification exploits the interfacial tension as the driving force for droplet formation. During droplet formation the dispersed phase is forced into a distorted (elongated) disk-like shape in the MCs. This disk-like shape has a higher interface area than a spherical shape, resulting in instability from the viewpoint of interface free energy. As a result the dispersed phase spontaneously breaks and transforms into spherical droplets. The droplet formation behavior drastically changes above the critical flow velocity, which can be predicted by the critical capillary number (Sugiura et al., 2002a).

Microfluidic devices can produce droplets with a very tight size distribution, but to date, only at small production scales and droplets of at least a few microns in diameter. Scale-up with membrane emulsification tends to be simpler and more versatile.

6.5 PRODUCTION OF DOUBLE EMULSIONS USING MEMBRANES AND MICROCHANNELS

A double (multiple or complex) emulsion is an emulsion in an emulsion (Garti, 1997). Two main types of double emulsions can be distinguished: water-in-oil-in-water (W/O/W) and oil-in-water-in-oil (O/W/O). A W/O/W emulsion consists of water droplets dispersed within larger oil droplets, which are themselves dispersed in an aqueous continuous phase. In an O/W/O emulsion, larger water droplets enclosing smaller oil droplets are dispersed in a continuous oil phase. W/O/W emulsions are more common than O/W/O emulsions. Other types of double emulsions such as oil-in-water-in-water (Kim et al., 2006) and ethanol-in-oil-in-water (Nakajima et al., 2003), have also been prepared and investigated.

6.5.1 Production of Water-in-Oil-in-Water Emulsions

W/O/W emulsions are usually prepared in a two-step emulsification process (Garti and Benichou, 2004) using two surfactants; a hydrophobic one designed to stabilize the internal droplets and a hydrophilic one for the external interface of the oil globules. The primary W/O emulsion is prepared under high-shear conditions to obtain small droplets, while the secondary emulsification step is carried out with less shear to avoid rupture of the oil membrane between two water phases. In conventional emulsification devices such as rotor-stator and high-pressure homogenization devices high shear stresses are needed to decrease the droplet size of the coarse emulsion. However, external flow (shear) causes internal streaming in the droplets, which increases the frequency of collision (and thus coalescence) of internal droplets with the outer water phase (Van der Graaf et al., 2005a). Besides, elongation of the droplets increases the interface available for release of internal droplets. Therefore the release rate of internal droplets is dependent on the applied shear stress, and only moderate shear can be used for the production of double emulsions if a reasonable percentage of internal phase is required. This is the reason why double emulsions manufactured using conventional emulsification processes are, in general, polydisperse.

W/O/W emulsions can be made by passing a W/O emulsion through a hydrophilic membrane or through silicon microchannel arrays (Figure 6.8). The result is droplets with much less shear, so the droplets are intact and both a high entrapment efficiency and monodispersity can be achieved (Gladman, 2003). In addition the mean size of external droplets can be precisely controlled by the membrane mean pore size (Table 6.3) or the geometry of MCs (Table 6.4). This form of manufacture is inherently scalable, since throughputs are proportional to the membrane or MC plate surface area (Gladman, 2003).

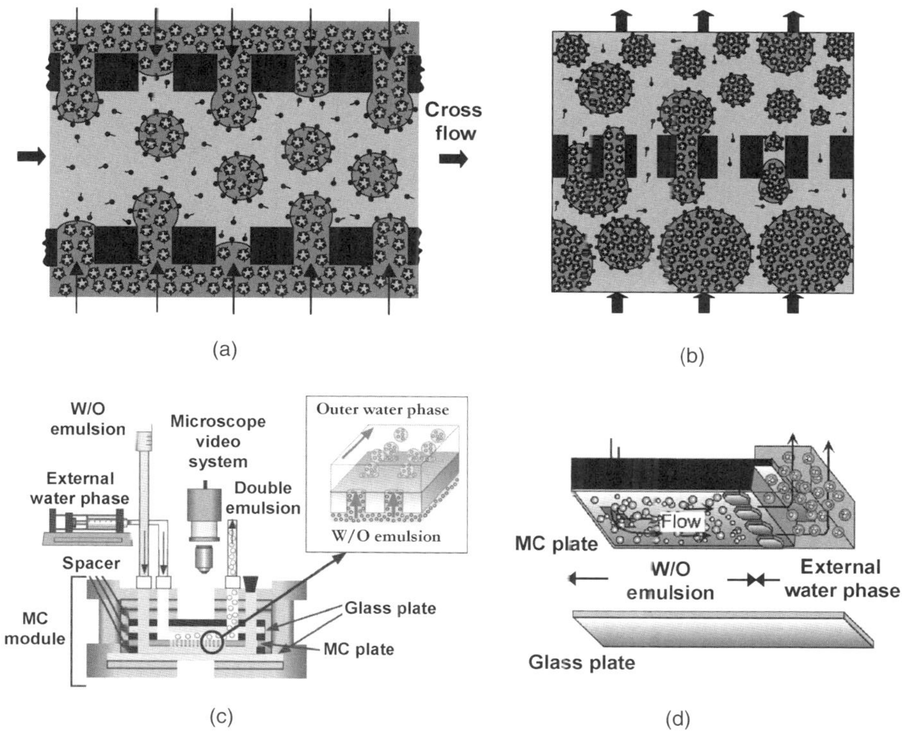

Figure 6.8 Production of double emulsions using different membrane and microfluidic experimental approaches: (*a*) XME (Vladisavljević and Schubert, 2003; Mine et al., 1996); (*b*) premix dead-end ME (Vladisavljević et al., 2004b); (*c*) Straight-through MC emulsification (Kobayashi et al., 2005c); (*d*) MC emulsification using grooved-type microchannel array (Kawakatsu et al., 2001; Sugiura et al., 2004a).

TABLE 6.4 Some examples of successful formulations for manufacturing double emulsions ($W_1/O/W_2$) using silicon microchannel arrays

MC Device			Product $W_1/O/W_2$ Emulsion		
Type	Channel Width	Mean Size of W_1/O Drops	ϕ_i	Formulation	Authors
Grooved dead-end plate, with terrace lines, without partition walls	4.2 μm	18 μm	10 vol%	W_1: water or 0.5% aqueous pectin solution O: oleic acid or triolein + 1–5% TGPR or 5% Span 20 or 5% Span 80 W_2: water + 3% Tween 20 + (0.1 M $CaCl_2$)	Kawakatsu et al. (2001)
Grooved dead-end plate, with terrace lines, without partition walls	11.7 μm	32–36 μm CV = 5.5–19%	10 vol%	W_1: water + *tris*-HCl buffer + 4 × 10^{-4} M calcein (fluorescent dye) O: Decane, ethyl oleate, MCT or triolein + 5% CR-310 (TGPR) W_2: water + *tris*-HCl buffer + 1% PGML	Sugiura et al. (2004a)
Grooved cross-flow plate, with terrace lines and partition walls	16 μm	40 μm, span = 0.45	10 vol%	W_1: water + 5% phosphate buffer (NaH_2PO_4/Na_2HPO_4) O: rapeseed oil + 10% PGPR 90 W_2: water + 2% Tween 80 + 0.5% NaCl	Vladisavljević et al. (2004a)
Symmetric straight-through MCs	42.8 × 13.3 μm	39–44 μm	10–30 vol%	W_1: milli-Q water + 5% glucose O: soybean oil + 5% TGCR or PGPR W_2: 1% Tween 80 + 5% glucose in milli-Q water	Kobayashi et al. (2005c)

Consequently products formed in this way lend themselves to scale-up and also deliver the target quality, while minimizing material wastage (e.g., associated with off-specification batches).

Mine et al. (1996) prepared double emulsions by XME using SPG membranes. First, a W/O emulsion with the mean droplet size of 0.52 μm was prepared using a microfluidizer. This W/O emulsion was then dispersed into the outer aqueous phase through SPG membrane with the mean pore size of 1 μm and a W/O/W emulsion with the mean pore size of 4.8 μm was formed. This emulsion was stable for at least six weeks when stored at 5°C. The concentration of internal water droplets should be between 30% and 50% in order to obtain a stable double emulsion, and the membrane pore size must be no less than twice the diameter of the internal droplets. W/O/W emulsions are particularly important to the pharmaceutical industry to initiate contained and targeted drug delivery (Higashi and Setoguchi, 2000). Higashi et al. (1999, 1995) prepared W/O/W emulsions for hepatic arterial chemotherapy using SPG membrane (Figure 6.9). A submicron W/O emulsion was first prepared using sonication. The water phase contained an anticancer drug (epirubicin), and the oil phase was made up of an iodized poppy seed oil (Lipiodol) with polyoxyethylene (40) hydrogenated castor oil being used as the hydrophobic surfactant. This was then followed by further emulsification through hydrophilic SPG membrane with a mean pore size of 20 μm using the W/O emulsion as the dispersed phase. Clinical trials showed that the multiple drug emulsion, when administered directly into the liver via the hepatic artery, was effective

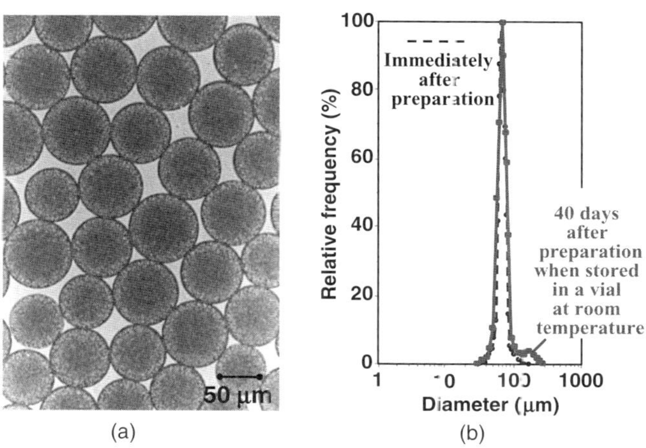

Figure 6.9 W/O/W emulsion prepared by SPG membrane emulsification for transcatheter arterial injection chemotherapy of hepatocellular carcinoma (HCC): (*a*) Microscopic view of multiple emulsion droplets. Internal water droplets containing anticancer drug are visible as black dots. (*b*) Particle size distribution of oil droplets immediately after preparation and 40 days after preparation (Nakashima et al., 2000; Higashi et al., 1995).

in contracting the liver cancer texture. It was concluded that ME enables the design of a controlled release vehicle at the much higher encapsulation efficiency in comparison with other traditional methods.

Yafei et al. (2006) produced polymer-stabilized double emulsions by high shear emulsification in the primary and membrane emulsification in the secondary step. By repeated fractionation after each emulsification, they obtained monodisperse double emulsions with the size of the external droplets ranging from submicrometer to a few micrometers. Primary W/O emulsions with an average size of water droplets of 0.1 μm were prepared by UltraTurrax homogenizer or ultrasonication. The primary emulsion was then gently mixed with a continuous aqueous phase to form a crude W/O/W emulsion. In the second emulsification step, the crude emulsion was pressured through a SPI polycarbonate membrane filter five times. The average size of external droplets was close to the pore size of a membrane filter if the applied pressure was controlled within an optimal range. Since there are commercially available polycarbonate membranes with a mean pore size between 0.01 and 20 μm, it is possible to obtain a double emulsion with different droplet size. With osmotic pressure balance between the inner and outer phases, these polymer-stabilized double emulsions remain stable for a year at room temperature without structure deterioration. Some applications also demand effective methods to break double emulsions. It was demonstrated that the release of the internal phase of polymer-stabilized double emulsions can be effectively induced using short, nonionic surfactants such as nonylphenol ethoxylates. It is a viable method to achieve controlled or targeted release of encapsulated species. A similar approach was used to prepare fine W/O/W emulsions by passing a coarse W/O/W emulsion through disposable cellulose acetate membrane filter cartridges (Shima et al., 2004). These W/O/W emulsions were utilized as carriers for hardly absorbed hydrophilic bioactive compounds that are normally deactivated or digested during the digestive process. The encapsulation efficiency of a model hydrophilic substance was more than 90% for at least one week after the double-emulsion preparation.

Vladisavljević et al. (2004b) prepared monodisperse W/O/W emulsions for controlled delivery systems by repeated SPG homogenization. A coarse W/O/W emulsion was passed five times through SPG membrane with a mean pore size of 10.7 μm using the equipment shown in Figure 6.10. The encapsulation efficiency of a marker substance ($CaNa_2$-EDTA) in freshly prepared samples after five passes was 83% to 84% at the transmembrane fluxes exceeding $20\,m^3\,m^{-2}\,h^{-2}$. Under optimal conditions the span of the particle size distribution of external droplets in the product emulsions was as low as 0.28.

Multiple emulsions can also be produced using a MC plate in much the same way as that done using a porous membrane, although the geometry of membrane pores and MCs significantly differs. Kawakatsu et al. (2001) and Sugiura et al. (2004a) investigated production of W/O/W emulsions using grooved-type MC plates. The monodisperse oil droplets with a CV of 5% to

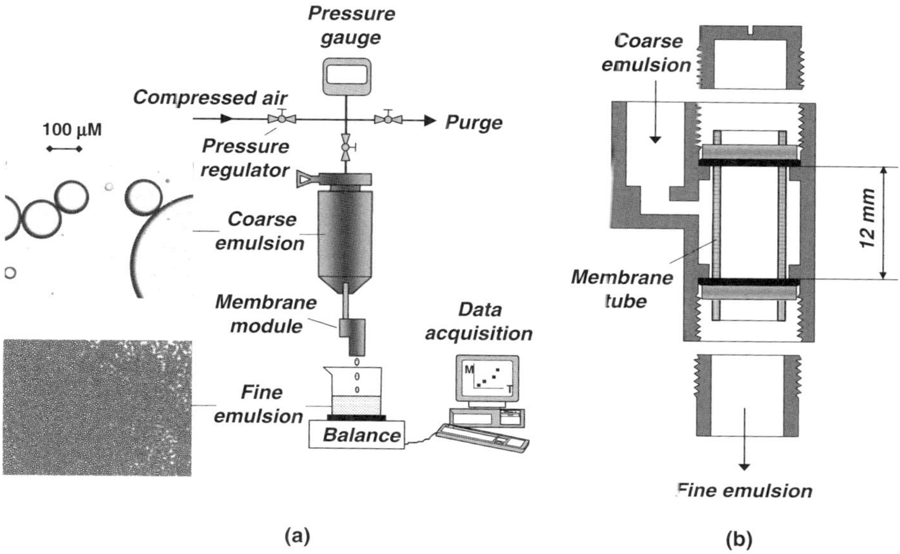

Figure 6.10 Preparation of fine W/O/W emulsions by batch SPG membrane homogenization: (*a*) Experimental setup and micrographs of feed and product emulsion droplets; (*b*) membrane homogenizer of Kiyomoto Iron Works Company (Vladisavljević et al., 2004b).

8% were prepared using 5 wt% TGPR dissolved in decane or ethyl oleate as an oil phase (Sugiura et al., 2004a). When high-viscosity oil phase containing MCT and triolein was used, larger droplets with a CV of 19% were formed. Internal droplets larger than the resultant W/O/W droplets also penetrated through the channels and were divided into several smaller droplets without disrupting the thin oil film. For decane, ethyl oleate, and MCT, internal water droplets larger than the resultant external droplets were divided into almost the same size as the external droplets and the oil phase formed a thin layer between the two aqueous phases. For MCT and triolein, internal droplets larger than the resulting W/O/W droplet were divided into several smaller droplets. The entrapment yield was measured fluorometrically using calcein solution, and estimated to be 91%. Monodisperse W/O/W emulsions were also prepared using straight-through MC emulsification (Kobayashi et al., 2005c). Figure 6.11 shows W/O/W emulsion droplets with various volume fractions of internal droplets prepared using TGCR as a hydrophobic emulsifier. The mean size of external droplets ranged from 44 μm for pure oil phase (Figure 6.11*a*) to between 41 and 39 μm for oil phase containing, respectively, 10 to 30 vol% of internal droplets (Figure 6.11*b–d*). The decrease in mean droplet size with an increase in the volume fraction of water phase was explained by an increase in the dispersed phase viscosity. The main advantage of using straight-through

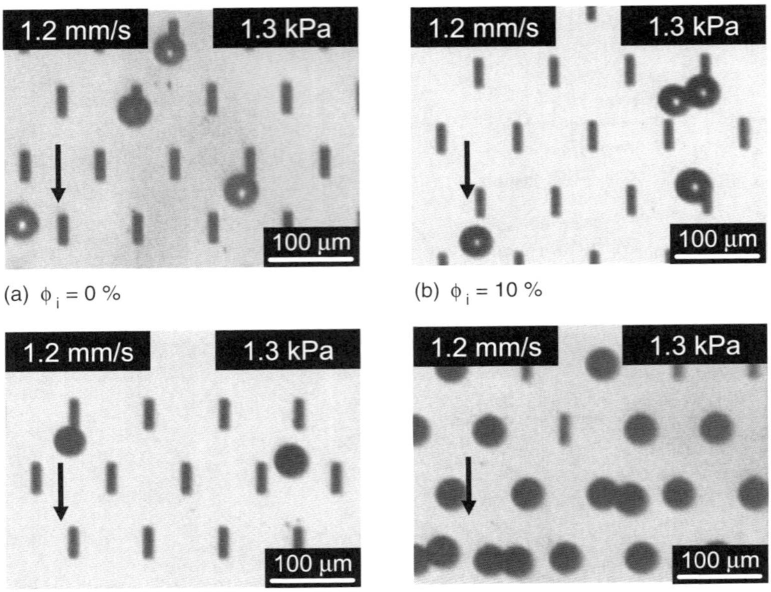

(a) $\phi_i = 0\ \%$ (b) $\phi_i = 10\ \%$

(c) $\phi_i = 20\ \%$ (d) $\phi_i = 30\ \%$

Figure 6.11 Photomicrographs showing monodisperse W/O/W emulsion droplets obtained in straight-through MC emulsification process at the pressure difference across the channels of 1.3 kPa and continuous phase cross-flow velocity of 1.2 mm/s. The concentration of internal droplets in the outer drops ranged between 0 and 30 vol% and the channel dimensions were 42.8 × 13.3 zyxonetwothreexyzm.

MC plates is a higher production rate of emulsion in comparison with grooved-type plates.

6.5.2 Production of Oil-in-Water-in-Oil Emulsions and Spreads

Okonogi et al. (1994) developed a method for producing an O/W/O-type edible emulsion, which involved dispersing an O/W emulsion into a fatty phase through a hydrophilic microporous glass membrane that was rendered hydrophobic by the pre-immersion into the fatty phase. An O/W/O-type spread can be manufactured by rapidly cooling and kneading the O/W/O emulsion consisting of plastic oils, an aqueous solution, and emulsifiers. The fatty phase content should be 25% to 75% of the final product, and the external oil phase preferably 20% and more of the total fatty phase. The authors suggested that in using membrane emulsification technology for the production of W/O-type low-fat spread and O/W/O-type spread, it is possible to reduce the amount of fatty phase in the whole compositions to 20% for the W/O-type spread and 25% for the O/W/O-type spread in comparison with other traditional methods, without using stabilizers and gelling agents as essential ingredients. To date these methods have not been widely implemented on an industrial scale.

6.5.3 Production of Ethanol-in-Oil-in-Water Emulsions

Nakajima et al. (2003) developed functional emulsions for use in food, drugs, and cosmetics that consist of ethanol as a dispersed phase and oil as a continuous phase (E/O) and also ethanol-in-oil emulsion as a disperse phase and water as a continuous phase (E/O/W). An ingredient that is insoluble or has low solubility in water and oil (polyphenols, validamycin, androstenedione, taxol, etc.) is dissolved in ethanol at a concentration of 20% to 30%, and the ethanol solution is then dispersed into oil phase using a conventional homogenizer or membrane/MC emulsification. The E/O emulsion can be further dispersed into water phase to produce E/O/W multiple emulsions with a wide range of applications as emulsion delivery systems in food, drugs, and cosmetics.

6.5.4 Production of Solid-in-Oil-in-Water Dispersions

In a solid-in-oil-in-water dispersion, oil droplets enclosing solid micro- or nanoparticles are dispersed in a continuous aqueous phase. Toorisaka et al. (2003) prepared a S/O/W dispersion for oral administration of insulin by passing a preliminary emulsified S/O/W dispersion several times (usually three) through SPG membrane with a mean pore size of 1.1 μm. The S/O suspension containing surfactant-coated insulin dispersed in soybean oil was first mixed with the aqueous solution containing a hydrophilic surfactant (sugar ester L-1695), sodium cholate, and D-glucose to prepare a coarse S/O/W dispersion. This coarse S/O/W dispersion was then homogenized by passing through the SPG membrane to prepare a fine dispersion with the mean droplet size of 1.0 μm. The surfactant-coated insulin was prepared by mixing an aqueous solution of insulin with a hexane solution containing ER-290 (sugar ester) as a hydrophobic surfactant, followed by freeze-drying. After oral administration to rats, the S/O/W dispersion showed the hypoglycemic activity for a long period, owing to the conversion of insulin into a lipophilic complex and uniform droplets of S/O suspension.

Supsakulchai et al. (2002) incorporated titanium dioxide into polymer microparticles using SPG membrane emulsification followed by solvent evaporation. As a white pigment, TiO_2 is widely used for its efficiency in scattering visible light, and imparting whiteness, brightness, and high opacity when incorporated into a plastic formulation. First, a S/O/W dispersion was prepared by extruding the S/O suspension containing fine particles of anatase-type titanium dioxide in a poly(St-*co*-AA)-toluene solution through SPG membrane into an aqueous phase containing PVA and SDS. The polymer microcapsules containing 10 to 20 wt% of TiO_2 were prepared from the S/O/W dispersion by vacuum evaporation of toluene.

6.6 PRODUCTION OF DOUBLE EMULSIONS USING MICROFLUIDIC DEVICES

Microfluidic devices can generate double emulsions in a single step, allowing precision control of the outer and inner drop sizes as well as the number of droplets encapsulated in each larger drop. Some examples of successful formulations used for manufacturing double emulsions in microfluidic devices are listed in Table 6.5. Okushima et al. (2004) reported a method for preparing monodisperse double emulsions using T-junctions in series with opposite surface wettabilities and different sizes. The smaller junction was designated for the formation of inner droplets and the larger one for outer droplets. Each junction could be modified to be hydrophobic using a silane-coupling agent. In the production of W/O/W emulsions (Figure 6.12*a*), aqueous drops of uniform size ($d_{iw} = 52\,\mu m$, $CV = 2.7\%$) were formed at the first hydrophobic junction (60 μm width × 25 μm depth) and were then carried to the second hydrophilic junction (130 μm width × 65 μm depth), while maintaining a uniform distance from each other. Each aqueous droplet was then enclosed in an organic droplet ($d_0 = 83\,\mu m$, $CV = 2.8\%$) at the downstream junction. The breakup rates at both junctions were 22 drops/s, and the entrapment efficiency was 100%. For successful encapsulation of inner drops, the operating conditions satisfying the following equation are desirable: $R_1/R_2 = N$, where R_1 is the breakup rate at the first junction, R_2 is the breakup rate at the second junction, and N is a positive integer (1, 2, . . .). By adjusting the ratio between the breakup rates at the two junctions, the number of enclosed drops can be controlled. With an increase in the oil flow rate for given flow rates of inner and outer aqueous phase, the size of the aqueous drops sheared at the first junction decreases. Since the volume flow rate of the inner aqueous phase was fixed, this reduction implies an increase in the breakup rate at the first junction, increasing the average number of inner aqueous drops per outer drop.

Using microfluidic device shown in Figure 6.12, various double emulsions can be produced by changing the wetting and geometrical properties of the junctions. O/W/O emulsions can be generated by arranging the upstream junction to be hydrophilic and the downstream junction to be hydrophobic (Figure 6.13*c* and *d*). O/W/O emulsions may find uses as precursors for the fabrication of uniform three-dimensional colloidal assemblies packed in aqueous droplets (Yi et al., 2003). Furthermore an interesting double emulsion, in which two differently colored (blue or red) aqueous drops were enclosed in a larger oil droplet, was prepared by using a cross-junction as the upstream junction (Figure 6.12*b* and 6.13*b*). Red drops and blue drops of similar size were alternately and periodically formed at the cross-junction (Figure 6.13*a*); they then flew to the downstream T-junction to be jointly encapsulated in oil droplets. This indicates that multiple components can be confined as discrete vesicles within pico/nanoliter-sized drops.

Utada et al. (2005) fabricated a microfluidic device with glass microcapillaries to prepare triple emulsions of a W/O/W/O type, such as water/silicon

TABLE 6.5 Examples of successful formulations used for manufacturing double emulsions ($W_1/O/W_2$ and $O_1/O_2/W$) in microfluidic devices other than parallel microchannel arrays

Microfluidic Device			Product Double Emulsion		
Type	Channel Width or Orifice Size	Mean Size of W_1/O Drops	ϕ	Formulation	Authors
Microfluidic device with two T-junctions in series	60 and 130 μm	83.4 μm CV = 2.8%	ϕ_i = 20 vol% ϕ_0 = 1.8 vol%	W_1: deionized water O: corn oil + 1% CR-310 (TGPR) or 0.5% lecithin W_2: water + 0.5% SDS or 0.5% MSW-7S	Okushima et al. (2004)
FFD[a], outermost liquid comes from opposite direction	50–500 μm orifice	94 μm $CV < 3\%$		W_1: distilled water O: 0.1–5% PBA-PAA + toluene/THF mixture containing 50–80 wt% toluene W_2: 80% glycerol in distilled water	Utada et al. (2005), Lorenceau et al. (2005)
Asymmetric FFD, same as above	250 μm orifice	50–300 μm, CV − 2.5%		W_1: water + 0.1 M 1,6-hexanediamine O: hexadecane + 2% (v/v) Span 80 W_2: adipoyl cloride + dichloroethane + hexadecane (1:2:30 v/v)	Takeuchi et al. (2005)
FFD, all liquids flow coaxially in the same direction	60 × 200 μm orifice	20–150 μm, $CV < 2.5\%$		O_1: silicon oil + 0.2–2 wt% Span 80 O_2: 4 wt% HCPK (photoinitiator) + TPGDA or EGDMA (monomer) W: water + 2 wt% SDS	Nie et al. (2005)

[a]FFD = Flow-focusing device.

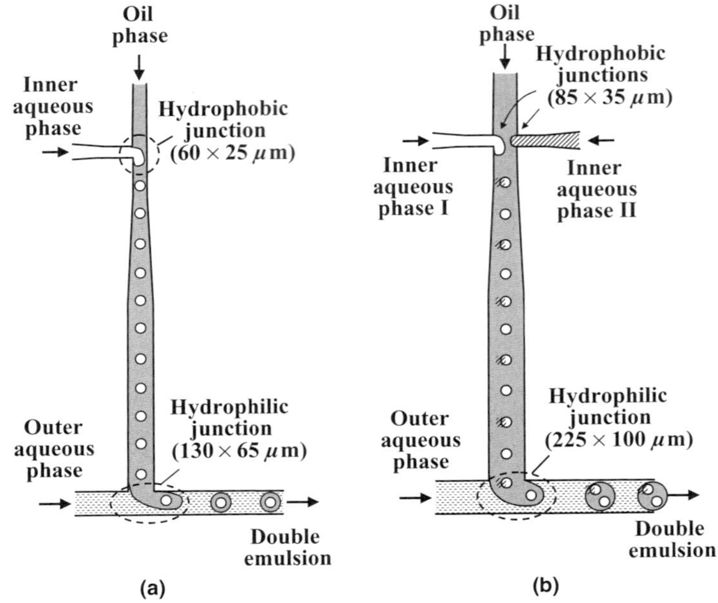

Figure 6.12 Production of double emulsions using two T-junctions in series with opposite surface wettabilities: (*a*) Production of W/O/W emulsion in which each outer drop contains a single inner droplet; (*b*) production of W/O/W emulsion in which each outer drop contains two inner droplets consisting of different aqueous phases I and II. Alternating generation of different inner droplets is achieved using an upstream cross-junction (Okushima et al., 2004).

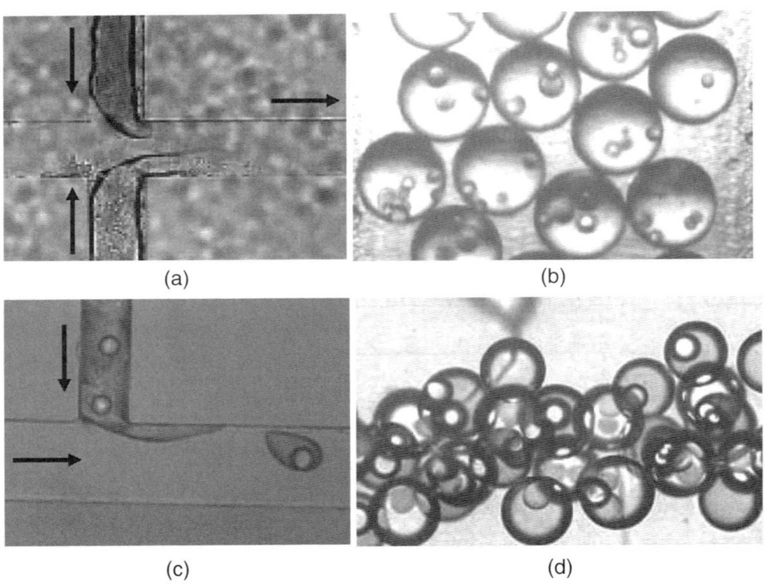

Figure 6.13 Micrographs of droplet generation using different combinations of T-junctions: (*a*) Alternating formation of differently colored droplets at a cross-junction; (*b*) W/O/W emulsion droplets containing differently colored inner droplets; (*c*) formation of O/W/O multiple emulsion drops containing a single oil droplet in each outer water drop; (*d*) uniform O/W/O emulsion droplets after generation (Okushima, 2004).

oil/water/silicon oil emulsion and water/ toluene/water/ silicon oil emulsion. The device consists of two concentric tapered tubes nested within an outer square tube, as shown in Figure 6.14*a*. The innermost fluid is pumped through the inner injection tube, the middle fluid 2 through the outer injection tube, and the middle fluid 1 through the square tube outside the injection tubes. The outermost fluid is pumped through the square tube from the opposite direction, and all fluids are forced through the exit orifice formed at the entrance of the "collection tube." This geometry results in hydrodynamic focusing and rupturing of three coaxial jets, followed by the formation of a triple emulsion. The volume of each phase can be tuned by changing the fluid flow rates. Typical diameters of the exit orifice range from 20 to 200 μm, which allows the drop

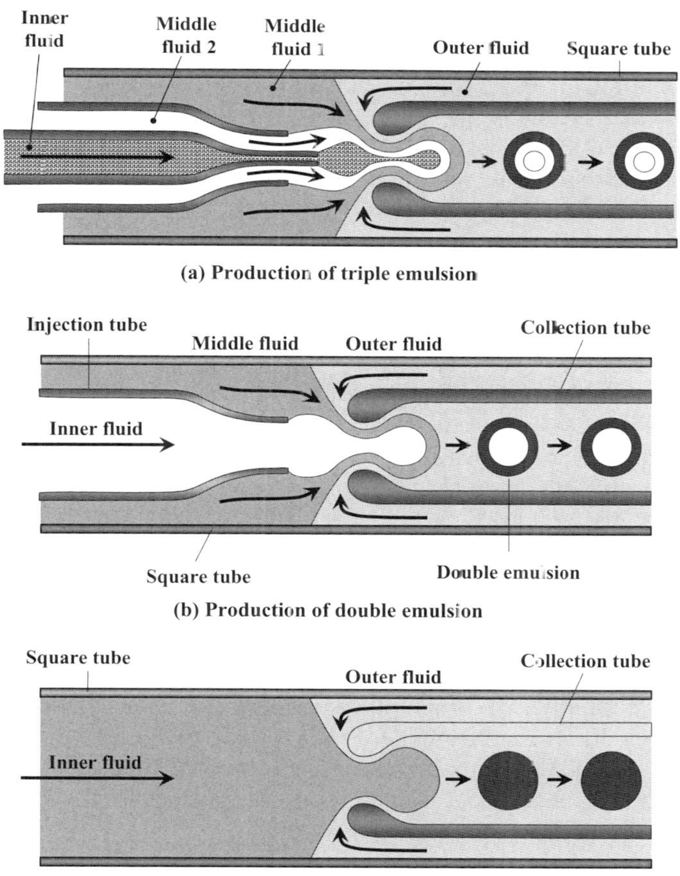

Figure 6.14 Microcapillary devices for generating monodisperse emulsions from coaxial liquid jets. This approach allows single-, double-, or triple emulsion droplets to be generated in a single step (Weitz).

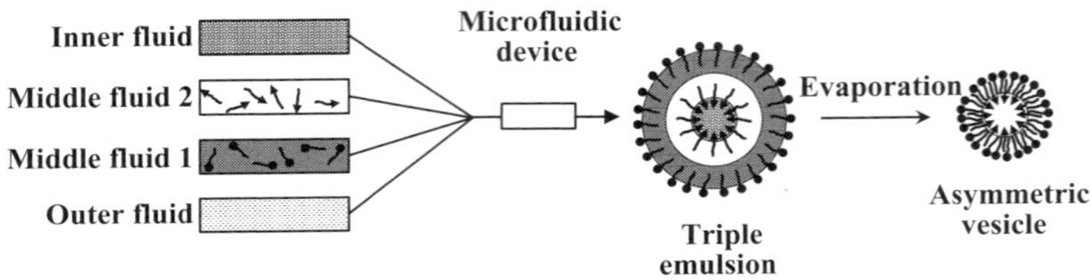

Figure 6.15 Preparation of asymmetric vesicle from a triple emulsion using microfluidic device shown in Figure 6.14*a* (Weitz).

size to be adjusted over a wide range. Triple emulsions can be used as templates to prepare capsules with two layers that consist of different chemical species, increasing enormously the versatility of encapsulating different materials. The idea is putting two different types of molecules—for example, a surfactant or a lipid—in each of the intermediate (middle) phases and evaporates them. The final product is a two-layer capsule, like an asymmetric vesicle shown in Figure 6.15. Pautot et al. (2003) fabricated asymmetric vesicles, where the inner leaflet was composed of POPC (1-palmitoyl-2-oleoyl-*sn*-glycero-3-phosphocholine) and the outer leaflet composed of POPS (1-palmitoyl-2-oleoyl-*sn*-glycero-3-phospho-L-serine), or the reverse.

The microfluidic device shown in Figure 6.14*b* generates double emulsions in a single step. The innermost fluid is pumped through the injection tube, and the middle fluid is pumped through the outer capillary in the same direction. The outermost fluid is pumped through the outer capillary from the opposite direction, and all fluids are forced through the exit orifice formed by the collection tube. The flow passes through the exit orifice and subsequently ruptures to form drops. At low outer flow rates, double-emulsion drops are formed within one orifice diameter of the entrance through the dripping mechanism; by contrast at higher flow rates, the neck is stretched into a coaxial thread that breaks further downstream through the jetting mechanism. Drops formed in the dripping regime have a lower polydispersity, generally less than 3% (Lorenceau et al., 2005). The number of small aqueous droplets contained within the larger oil drops depends on the relative frequency of drop formation for the innermost and middle fluids. If the innermost fluid breaks into droplets more rapidly than the middle fluid, large oil drops that contain many small aqueous droplets are obtained. However, if the rate of drop formation is the same, double emulsions with a single inner droplet are formed. The ratio of shell thickness to outer drop radius in these drop-in-drop emulsions can range from 3% to 40% (Utada et al., 2005).

Figure 6.14*c* shows a single emulsion device that does not have an injection tube and the outer fluid simply focuses the inner fluid; the fluid thread breaks into drops inside the collection tube. The frequency of drop production in the dripping regime can range from 100 to 7000 Hz, which enables formation of up to about 10^7 emulsion drops per hour (Utada et al., 2005).

6.7 PRODUCTION OF SOLID MICROPARTICLES VIA DOUBLE EMULSIONS USING MEMBRANE AND MICROFLUIDIC DEVICES

Vladisavljević and Williams (2005) reviewed the expanding opportunities for the production of particulate products using membranes with a broad range of possible applications to the pharmaceutical, chemical, electronic, and food industries. Williams (2001a) listed some of the engineering and technological challenges that researchers could face in manufacturing "designer particles" in large quantities. The route to success lies in applying knowledge on the fundamental principles of the process and in the formation of multiskilled teams that can address the various aspects of molecular assembly, modeling, and process design (Williams, 2001a).

Table 6.6 lists microparticles manufactured through double-emulsion routes by membrane and microfluidic devices. It can be seen that the methods afford the production of a variety of products by means of sequential secondary reactions/processes in the double-emulsion droplets, such as polymerization, gelation, evaporation, freeze-drying, and crystallization.

6.7.1 Production of Solid Lipid Microcarriers and Gel Microbeads

Solid lipid nano- and microparticles represent an alternative drug and nutrient carrier system to traditional colloidal carriers, such as emulsions, liposomes, and polymeric microparticles (Müller et al., 2000). Nakashima (2002) reported production of three kinds of solid lipid microcarriers through double-emulsion routes using SPG membranes. These carriers were particularly investigated for oral administration of irinotecan hydrochloride (CPT-11), which is a highly effective anticancer drug with some unfavorable side effects. A preparation flowsheet of the so-called solid microcarrier and water-in-oil (W/O) carrier is similar (Figure 6.16). A submicron W/O emulsion containing more than 50 mg/liter CPT-11 in aqueous solution as a disperse phase and 10 wt% TGCR in tripalmitin as a continuous phase was first prepared using a conventional rotor-stator emulsification device at 80°C to prevent oil solidification. This was followed by dispersion of the W/O emulsion in an external water phase at 80°C by a stirrer. The coarse polydisperse W/O/W emulsion was homogenized by passing through the SPG membrane with a mean pore size of about 11 μm. The W/O microcarrier was obtained by cooling down the W/O/W emulsion and subsequent filtration of the solidified droplets. Figure 6.17 shows a SEM photograph of the solidified W/O particle. The estimated entrapment efficiency of CPT-11 in the W/O carrier was more than 90% under optimal conditions (Shimizu et al., 2002a).

A solid microcarrier for encapsulating flavor-enhancing ingredients such as 5′-ribonucleotide was prepared from W/O/W emulsion using cacao butter as a melted lipid phase (Kukizaki et al., 2001). This microcarrier was used in Japan as a supplement in miso paste. The W/O emulsion with a mean particle size of 0.48 μm was produced at 40°C using a rotor-stator homogenizer. This W/O

TABLE 6.6 Examples of microparticles fabricated through multiple emulsion preparation routes using membrane or microfluidic devices

Product Type	Example	Secondary Reaction/Process after Emulsification	Authors
Solid lipid microcarriers	W/O microcarrier	Crystallization of oil phase from W/O/W emulsion	Shimizu et al. (2002a), Nakashima (2002)
	S/O microcarrier	Freeze-drying of S/O/W emulsion	Shimizu (2002a), Nakashima (2002)
	Solid microcarrier	Freeze-drying of W/O/W emulsion	Kukizaki et al. (2001)
Gel microbeads	Calcium pectinate beads	Crosslinking of bioplymer in inner water droplets of W/O/W emulsion	Kawakatsu et al. (2001)
	Calcium alginate beads		Nakashima (1991a)
Inorganic microparticles	Silica particles	Polymerization of silicic acids by interfacial reaction	Nakashima (1991a)
Polymeric microspheres	PSt-PAAm composite spheres	Suspension polymerization in W/O/W emulsion	Ma et al. (2004)
	Biodegradable PLA spheres	Solvent evaporation from oil phase in W/O/W emulsion	Liu et al. (2005a, b)
Polymeric core-shell microcapsules and hollow particles	Cured Norland Optical Adhesive (NOA)	Photopolymerization of NOA in core-shell double-emulsion precursors	Utada et al. (2005)
	P(TPGDA) particles with various shapes and morphologies	Photopolymerization of TPGDA in W/O/W emulsions	Nie et al. (2005)
Simple and multicompartment polymerosomes	PBA-PAA polymerosomes	Evaporation of organic solvent from W/O/W emulsions containing a single inner drop or many internal drops	Lorenceau et al. (2005)
Asymmetric vesicles	POPC-POPS asymmetric vesicle	Evaporation of middle fluids from triple W/O/W/O emulsion followed by molecular self-assembly	Weitz (2005)

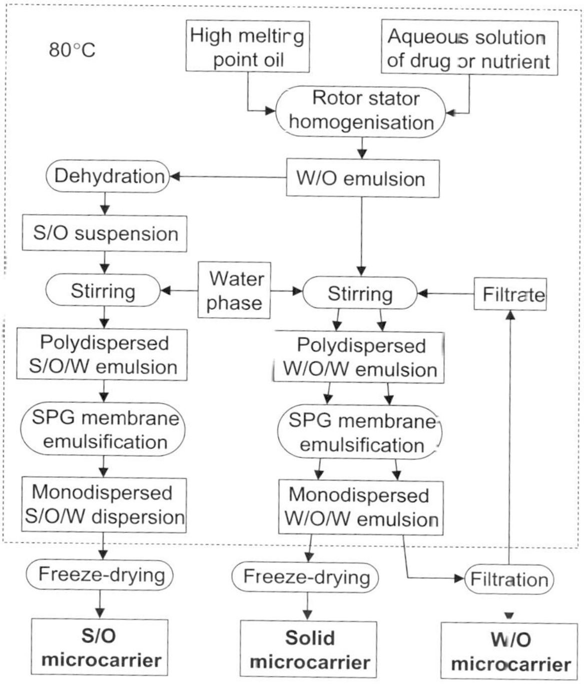

Figure 6.16 Flowsheet for the manufacture of three different types of solid lipid microcarriers for controlled nutrient or drug delivery using membrane homogenization approach. The dashed area represents the temperature-controlled operations carried out at 80°C (above the melting point of oil).

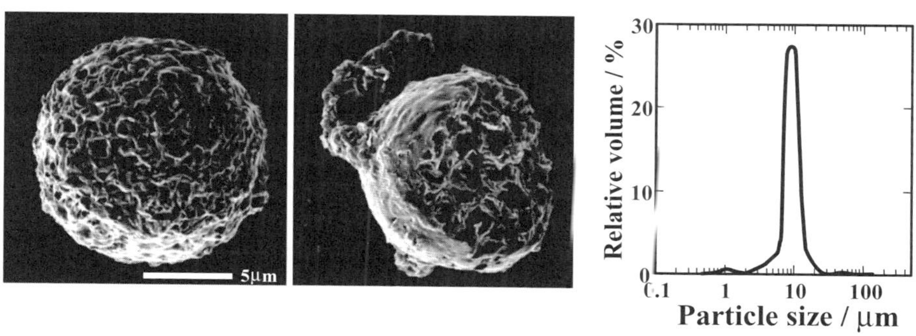

Figure 6.17 SEM photographs and size distribution of a W/O solid lipid carrier with a mean size of 8.4zyxonetwothreexyzm for oral administration of anticancer drug irinotecan hydrochloride (CPT-11). The carrier was prepared by a temperature-controlled emulsification at 30 kPa using SPG membrane with a mean pore size of 11.2zyxonetwothreexyzm, followed by cooling down and filtration of the solidified W/O/W emulsion (Shimizu et al., 2002a).

emulsion was then extruded at 40°C and 15 kPa through 4.8 μm SPG membrane into outer aqueous phase containing 0.5 wt% HCO-60, 0.59 wt% NaCl, and 0.95 wt% glucose. The osmotic pressure of both aqueous phases was 0.68 MPa. The solid carrier was produced by freeze-drying of the resultant W/O/W emulsion. The leakage of $CaNa_2$-EDTA from the microcarrier redispersed in water at 20°C was 3% over 10 days.

Kawakatsu et al. (2001) prepared calcium pectinate microcapsules from W/O/W emulsion by MC emulsification using the equipment shown in Figure 6.8*d*. The microcapsules were prepared by first homogenizing a water phase (0.5 wt% pectin solution) and oil phase (3 wt% TGCR in oleic acid) using a conventional homogenizer. The W/O emulsion was then forced through a MC plate into outer aqueous phase containing $CaCl_2$ and Tween 20. The resultant W/O/W emulsion was converted into microbeads by gelation of the internal water droplets due to the chemical reaction between pectin macromolecules and calcium ions. Similarly calcium alginate microbeads were prepared from the W/O/W emulsion by SPG emulsification using the formulation given in Table 6.3 (Nakashima et al., 1991a).

6.7.2 Production of Polymer Microspheres

Uniform polymer microspheres with a mean diameter ranging from several micrometers to 100 μm—hydrophilic and hydrophobic, smooth and rough, solid and hollow, porous and uniform, and with different morphologies (spherical, snowman-like, popcorn-like, hemispherical)—were successfully manufactured from O/W emulsions via SPG emulsification followed by suspension polymerization (Omi, 1996). These microspheres were investigated for use as packing materials for column chromatography techniques such as gel permeation chromatography (GPC), high-performance liquid chromatography (HPLC), capillary electrochromatography and size-exclusion chromatography, as pH sensors, carriers of enzymes (glucoamylase), drug delivery systems, dry and liquid toners (Hatate et al., 1997), spacers for liquid crystal displays, and so forth (Vladisavljević and Williams, 2005). Double emulsions can be used as intermediate products in this technique. For example, Ma et al. (2004) prepared polystyrene/polyacrylamide (PST-PAAm) composite microspheres from a W/O/W emulsion using SPG membrane emulsification and suspension polymerization. An acrylamide aqueous solution was emulsified by ultrasonication into the oil phase containing styrene, polystyrene, initiator, and Span 85 to obtain the W/O emulsion. Then the W/O emulsion was dispersed through SPG membrane into the outer aqueous phase to form a uniformly sized W/O/W emulsion. Polystyrene was added to increase the viscosity of the oil phase, retarding the phase separation between the inner water phase and the oil phase during the emulsification process. After the W/O/W emulsion was polymerized, the composite particles containing spherical PAAm-rich microdomains distributed inside and on the surface of a PSt matrix were obtained. Because amides can easily be hydrolyzed to carboxylic acids or degraded to amines, these PST-PAAm particles are suitable for use as immobilization media of biologically active substances such as proteins and cells.

Liu et al. (2005a, b) prepared biodegradable poly(lactide) (PLA) microcapsules by combining a SPG membrane emulsification technique and multiple emulsion-solvent evaporation method. An aqueous phase containing lysozyme or recombinant human insulin was used as the inner water phase, and a mixture of dichloromethane and toluene dissolving PLA and Arlacel 83, were used as the oil phase. These two solutions were emulsified by a homogenizer to form a W/O primary emulsion. The primary emulsion was permeated through the SPG membrane into the external water phase to form W/O/W droplets. The PLA microcapsules were obtained by evaporating the solvent. The drug encapsulation efficiency was found to depend on several factors, including PLA molecular weight, additive type, and concentration in the inner water phase, the emulsifier type and concentration in the oil phase, and the formulation of the outer water phase. SPG membrane emulsification enabled much higher drug encapsulation efficiency than the stirring method.

6.7.3 Production of Core-Shell Microparticles

Polymer capsules with liquid cores have applications in the encapsulation of drugs, cells, pesticides, perfumes, liquid inks, paints, and toners (Peyratout and Dähne, 2004). One major benefit of flow-focusing microfluidic devices is the ability to continuously fabricate reproducible core-shell particles. Utada et al. (2005) fabricated core-shell capsules by ultraviolet curing a Norland Optical Adhesive (Norland Products, Cranbury, NJ) in the middle fluid of core-shell droplets shown in Figure 6.14*b*. The adhesive was diluted by 30% with acetone to decrease its viscosity. After generating the double emulsions, the shells were cured in situ with an ultraviolet (UV) light source for approximately 10 seconds as the double emulsions traversed the collection tube. Nie et al. (2005) employed flow of three immiscible liquids in a microfluidic device shown in Figure 6.18 to obtain a coaxial jet of silicon oil and monomer in the continuous aqueous phase. Controlled breakup of the coaxial liquid thread led to the

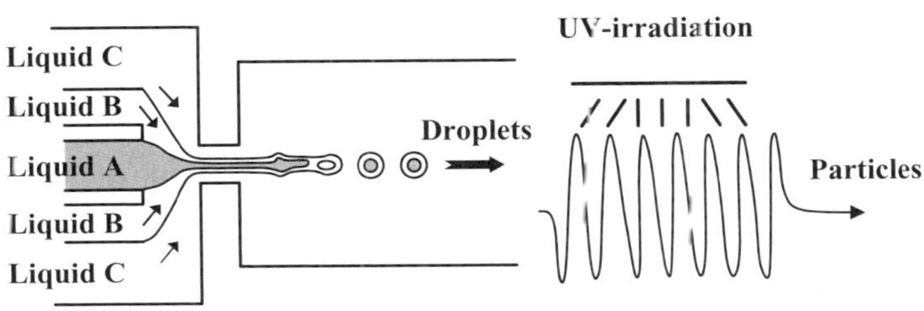

Figure 6.18 Production of core-shell particles in a continuous microfluidic reactor (Nie et al., 2005). The composition of the liquids A, B, and C in the multiple emulsion is listed in Table 6.5.

production of highly monodipserse droplets with various morphologies. The resultant droplets were continuously photopolymerized in the wavy channel to generate spherical polymer capsules with controllable diameters and shell thicknesses. The polymerization time was typically from 2 to 800 s. Conversion of monomer to polymer was close to 100%. Following polymerization the dimensions of the particles decreased by about 5% to 7%, in comparison with the corresponding droplets. The productivity of the microfluidic reactor was from 200 to 1000 s^{-1}. Particle polydispersity did not exceed 2.5%, which is close to the polydispersity of the corresponding droplets.

6.7.4 Production of Polymerosomes

The use of diblock copolymers to generate polymer vesicles is an attractive way to create new structures for encapsulation. These structures are called polymerosomes, and like more traditional liposomes, they can encapsulate nano- to picoliter volumes of liquid. The low toxicity of certain diblock copolymers makes these structures promising for drug delivery applications. The flexibility afforded by the use of diblock copolymers significantly increases the control over the properties of polymerosomes. For example, the membrane thickness can be controlled by varying the degree of polymerization of the individual diblock molecules, whereas fluidity and permeability of the membrane can be adjusted by changing the glass transition temperature of the hydrophobic block. Similarly control over the nature of the individual polymer blocks can lead to alternative mechanisms to trigger release.

The microfluidic device depicted in Figure 6.14*b* was used to generate W/O/W emulsions with a diblock copolymer dissolved in the intermediate hydrophobic fluid (Lorenceau et al., 2005). The inner aqueous phase consists of distilled water, and outer phase is a mixture of 80% (v/v) glycerol in distilled water. Glycerol was added to the outer fluid to increase its viscosity, which improves the efficiency of the flow focusing. The intermediate phase is a volatile organic solvent; the evaporation of the organic solvent causes the amphiphilic block copolymer poly(normal-butyl acrylate)–poly(acrylic acid) (PBA-PAA) to self-assemble into layers on the concentric interfaces of the double-emulsion droplets, forming a polymerosome. The PBA is the hydrophobic block in the copolymer molecule, whereas the PAA is the hydrophilic block. It is critical that the dissolved macromolecules are mostly unimers as opposed to larger aggregates because such aggregates do not efficiently stabilize the inner droplets against coalescence with the outer phase. In addition to simple polymerosomes, Lorenceau et al. (2005) form a new class of polymersome structures consisting of many internal compartments from double-emulsion drops that contain more than a single internal water droplet. The inner water droplets retain their integrity as the organic fluid evaporates, resulting in the formation of a foamlike, multi-compartmentalized vesicular structure.

6.8 CONCLUSIONS

Membrane emulsification offers great potential in manufacturing "made-to-measure" double emulsions. The process is reliable, and suitable to large-scale productions, as scale-up is mainly based on multiplication of small scale processing units (e.g., membrane modules) rather than the simple enlargement of processing vessel sizes. Double emulsions with various droplets sizes ranging from 0.7 to 60 μm, and a typical coefficient of variation of 10% to 15%, have been successfully prepared by employing different pore sizes and types of membranes. It is envisaged that direct membrane emulsification can be very attractive technique for small-scale manufacture of low-viscosity, low-concentration double emulsions, due to its high capability of droplet size and distribution controls but limited emulsifying rate. Premix membrane emulsification is a technique potentially suitable for large-scale manufacture of high-concentration emulsion products, due to its high emulsifying rate and simple operation.

Microfluidic devices can generate double- and even triple-emulsion droplets in a single step with a polydispersity in the dripping regime of generally less than 3% and an entrapment efficiency of 100% allowing precision control over the outer and inner drop sizes as well the number of droplets encapsulated in each larger drop. Silicon and polymeric microchannel arrays microfabricated vertically or horizontally to the plate surface can produce double-emulsion drops with a coefficient of variation of less than 5% at much higher production scales due to larger number of individual channels (more than 10,000 in the case of "straight-through" MC plates).

Membrane and microfluidic devices have also been adopted for the precision manufacture of solids from double-emulsion templates. To date, several different types of particles have been successfully produced by incorporating use of various membrane and microfluidic devices in processes of polymerization, gel formation, crystallization, and molecular or particle self-assembly. Membrane emulsification is more suited to the fabrication of less sophisticated particulates, such as solid lipid micro-/nanoparticles, gel microbeads, coherent polymeric microspheres, and inorganic particles such as silica microparticles. Microfluidic devices allow more sophisticated particle designs to be created, such as colloidosomes, polymerosomes, 3D colloidal assemblies, asymmetric vesicles, core-shell polymer particles, and bichromal particles.

ACKNOWLEDGMENT

Research work of G. T. Vladisavljević in the area of membrane and microchannel emulsification has been supported by the Alexander von Humboldt Foundation (Bonn), the Japan Society for the Promotion of Science (Tokyo), the Royal Society (London), and the Fulbright Agency (Washington, DC). G. T. Vladisavljević sincerely extends thanks to these organizations for their support.

REFERENCES

Abrahamse AJ, Van Lierop R, Van der Sman RGM, Van Der Padt A, Boom RM. 2002. Analysis of droplet formation and interactions during cross-flow membrane emulsification. *J Membr Sci* 204: 125–137.

Akay G. 1998. Flow-induced phase inversion in the intensive processing of concentrated emulsions. *Chem Eng Sci* 53: 203–223.

Altenbach-Rehm J, Schubert H, Suzuki K. 2002a. Premix-Membranemulgieren mittels hydrophiler und hydrophober PTFE-Membranen zur Herstellung von O/W-Emulsionen mit enger Tropfengrößenverteilung. 2002. *Chem Ing Tech* 74: 587–588.

Altenbach-Rehm J, Suzuki K, Schubert H. 2002b. Production of O/W-emulsions with narrow droplet size distribution by repeated premix membrane emulsification. In: *Proceedings of the 3rd World Congress on Emulsions*. Lyon.

Anna SL, Bontoux N, Stone HA. 2003. Formation of dispersions using "flow focusing" in microchannels. *Appl Phys Letts* 82: 364–366.

Atencia J, Beebe DJ. 2005. Controlled microfluidic interfaces. *Nature* 437: 648–655.

Berot S, Giraudet S, Riaublanc A, Anton M, Popineau Y. 2003. Key factors in membrane emulsification. *Chem Eng Res Des A* 81: 1077–1082.

Charcosset C, Limayem I, Fessi H. 2004. The membrane emulsification process—A review. *J Chem Technol Biotechnol* 79: 209–218.

Davidson MR, Harvie DJE, Cooper-White JJ. 2005. Flow focusing in microchannels. *ANZIAM J* 46: C47–C58.

Dendukuri D, Tsoi K, Hatton TA, Doyle PS. 2005. Controlled synthesis of nonspherical microparticles using microfluidics. *Langmuir* 21: 2113–2116.

Dowding PJ, Goodwin JW, Vincent B. 2001. Polymer beads with a narrow size distribution using a cross-flow membrane and a continuous tubular reactor. *Colloids Surf A* 180: 301–309.

Fuchigami T, Toki M, Nakanishi K. 2000. Membrane emulsification using sol-gel derived macroporous silica glass. *J Sol-Gel Sci Techn* 19: 337–341.

Fukui K, Matsugano K, Shimizu M, Matsuda S. 2001. Preparation of anthocyanin-encapsulated emulsion and suppression of anthocyanin leakage. In: *Proceedings of the 36th SPG Forum "Various Applications of Membrane and Particles in Healthcare Fields,"* Industrial Technology Center, Miyazaki: p 66.

Garstecki P, Stone HA, Whitesides GM. 2005. Mechanism for flow-rate controlled breakup in confined geometries: A route to monodisperse emulsions. *Phys Rev Letts* 94: Art. No. 164501, pp 1–4.

Garti N. 1997. Double emulsions: Scope, limitations and new achievements. *Colloids Surf A* 123–124: 233–246.

Garti N, Benichou A. 2004. Recent developments in double emulsions for food applications. In: *Food Emulsions*. Eds. Friberg SE, Larsson K, Sjöblom J. New York: Marcel Dekker Inc. pp 353–412.

Gijsbertsen-Abrahamse AJ, Van der Padt A, Boom RM. 2003. Influence of membrane morphology on pore activation in membrane emulsification. *J Membr Sci* 217: 141–150.

Gijsbertsen-Abrahamse AJ, Van der Padt A, Boom RM. 2004. Status of cross-flow membrane emulsification and outlook for industrial application. *J Membr Sci* 230: 149–159.

Giorno L, Li N, Drioli E. 2003. Preparation of oil-in-water emulsions using polyamide 10kDa hollow fiber membrane. *J Membr Sci* 217: 173–180.

Gladman D, Williams R. 2003. Manufacturing with membranes. *TCE* 748: 32–34.

Ha YK, Song HS, Lee HJ, Kim JH. 1999. Preparation of core particles for toner application by membrane emulsification. *Colloids Surf A* 162: 289–295.

Hatate Y, Ohta H, Uemura Y, Ijichi K, Yoshizawa H. 1997. Preparation of monodispersed polymeric microspheres for toner particles by the Shirasu porous glass membrane emulsification technique. *J Appl Polym Sci* 64: 1107–1113.

He M, Edgar JS, Jeffries GDM, Lorenz RM, Shelby JP, Chiu DT. 2005. Selective encapsulation of single cells and subcellular organelles into picoliter- and femtoliter-volume droplets. *Anal Chem* 77: 1539–1544.

Higashi S, Shimizu M, Nakashima T, Iwata K, Uchiyama F, Tateno S, Setoguchi T. 1995. Arterial-injection chemotherapy for hepatocellular-carcinoma using monodispersed poppy-seed oil microdroplets containing fine aqueous vesicles of epirubicin—Initial medical application of a membrane-emulsification technique. *Cancer* 75: 1245–1254.

Higashi S, Tabata N, Kondo K, Maeda Y, Shimizu M, Nakashima T, Setoguchi T. 1999. Size of lipid microdroplets effects results of hepatic arterial chemotherapy with an anticancer agent in water-in-oil-in-water emulsion to hepatocellular carcinoma. *J Pharmacol Exp Ther* 289: 816–819.

Higashi S, Setoguchi T. 2000. Hepatic arterial injection chemotherapy for hepatocellular carcinoma with epirubicin aqueous solution as numerous vesicles in iodinated poppy-seed oil microdroplets: clinical application of water-in-oil-in-water emulsion prepared using a membrane emulsification technique. *Adv Drug Deliv Rev* 45: 57–64.

Hino T, Kawashima Y, Shimabayashi S. 2000. Basic study for stabilization of W/O/W emulsion and its application to transcatheter arterial embolization therapy. *Adv Drug Deliv Rev* 45: 27–45.

Hosoya K, Bendo M, Tanaka N, Watabe Y, Ikegami T, Minakuchi H, Nakanishi K. 2005. An application of silica-based monolithic membrane emulsification technique for easy and efficient preparation of uniformly sized polymer particles. *Macromol Mater Eng* 290: 753–758.

Ito F, Makino K. 2004. Preparation and properties of monodispersed rifampicin-loaded poly(lactide-co-glycolide) microspheres. *Colloids Surf B* 39: 17–21.

Jing W, Wu J, Xing W, Jin W, Xu N. 2005. Emulsions prepared by two-stage ceramic membrane jet-flow emulsification. *AIChE J* 51: 1339–1345.

Joscelyne SM, Trägårdh G. 1999. Food emulsions using membrane emulsification: conditions for producing small droplets. *J Food Eng* 39: 59–64.

Joscelyne SM, Trägårdh G. 2000. Membrane emulsification—A literature review. *J Membr Sci* 169: 107–117.

Katoh R, Asano Y, Furuya A, Sotoyama K, Tomita M. 1996. Preparation of food emulsions using a membrane emulsification system. *J Membr Sci* 113: 131–135.

Kawakatsu T, Kikuchi Y, Nakajima M. 1997. Regular-sized cell creation in microchannel emulsification by visual microprocessing method. *J Am Oil Chem Soc* 74: 317–321.

Kawakatsu T, Trägårdh G, Kikuchi Y, Nakajima M, Komori H, Yonemoto T. 2000. Effect of microchannel structure on droplet size during crossflow microchannel emulsification. *J Surfactants Deterg* 3: 295–302.

Kawakatsu T, Trägårdh G, Trägårdh C. 2001. Production of W/O/W emulsions and S/O/W pectin microcapsules by microchannel emulsification. *Colloids Surf A* 189: 257–264.

Kawashima Y, Hino T, Takeuchi H, Niwa T, Horibe K. 1991. Shear-induced phase inversion and size control of water/oil/water emulsion droplets with porous membrane. *J Colloid Interface Sci* 145: 512–523.

Kim HJ, Decker EA, McClements DJ. 2006. Preparation of multiple emulsions based on thermodynamic incompatibility of heat-denatured whey protein and pectin solutions. *Food Hydrocolloid* 20: 586–595.

Kobayashi I, Nakajima M, Chun K, Kikuchi Y, Fujita H. 2002a. Silicon array of elongated through-holes for monodisperse emulsion droplets. *AIChE J* 48: 1639–1644.

Kobayashi I, Yasuno M, Iwamoto S, Shono A, Satoh K, Nakajima M. 2002b. Microscopic observation of emulsion droplet formation from a polycarbonate membrane. *Colloids Surf A* 207: 185–196.

Kobayashi I, Nakajima M, Mukataka S. 2003. Preparation characteristics of oil-in-water emulsions using differently charged surfactants in straight-through microchannel emulsification. *Colloids Surf A* 229: 33–41.

Kobayashi I, Mukataka S, Nakajima M. 2005a. Production of monodisperse oil-in-water emulsions using a large silicon straight-through microchannel plate. *Ind Eng Chem Res* 44: 5852–5856.

Kobayashi I, Mukataka S, Nakajima M. 2005b. Novel asymmetric through-hole array microfabricated on a silicon plate for formulating monodisperse emulsions. *Langmuir* 21: 7629–7632.

Kobayashi I, Lou XF, Mukataka S, Nakajima M. 2005c. Preparation of monodisperse water-in-oil-in-water emulsions using microfluidization and straight-through microchannel emulsification. *J Am Oil Chem Soc* 82: 65–71.

Kosvintsev SR, Gasparini G, Holdich RG, Cumming IW, Stillwell MT. 2005. Liquid–liquid membrane dispersion in a stirred cell with and without controlled shear. *Ing Eng Chem Res* 44: 9323–9330.

Kukizaki M, Shimizu M, Nakashima T. 2001. Preparation of W/O/W emulsion using hardened oil and supression of leakage of encapsulated material. In: *Proceedings of the 36th SPG Forum "Various Applications of Membrane and Particles in Healthcare Fields,"* Industrial Technology Center, Miyazaki: p 58.

Lambrich U, Vladisavljević GT. 2004. Emulsification using microstructured systems. *Chem Ing Tech* 76: 376–383.

Lewis PC, Graham RR, Nie Z, Xu S, Seo M, Kumacheva E. 2005. Continuous synthesis of copolymer particles in microfluidic reactors. *Macromolecules* 38: 4536–4538.

Link DR, Anna SL, Weitz DA, Stone HA. 2004. Geometrically mediated breakup of drops in microfluidic devices. *Phys Rev Letts* 92: Art. No. 054503.

Liu XD, Bao DC, Xue WM, Xiong Y, Yu WT, Yu XJ, Ma XJ, Yuan Q. 2003. Preparation of uniform calcium alginate gel beads by membrane emulsification coupled with internal gelation. *J Appl Polym Sci* 87: 848–852.

Liu R, Ma GH, Wan YH, Su ZG. 2005a. Influence of process parameters on the size distribution of PLA microcapsules prepared by combining membrane emulsification technique and double emulsion-solvent evaporation method. *Colloids Surf B* 45: 144–153.

Liu R, Ma GH, Meng FT, Su ZG. 2005b. Preparation of uniform-sized PLA microcapsules by combining Shirasu porous glass membrane emulsification technique and multiple emulsion-solvent evaporation method. *J Controll Release* 103: 31–43.

Lorenceau E, Utada AS, Link DR, Cristobal G, Joanicot M, Weitz DA. 2005. Generation of polymerosomes from double emulsions. *Langmuir* 21: 9183–9186.

Ma GH, Nagai M, Omi S. 1999. Preparation of uniform poly(lactide) microspheres by employing the Shirasu porous glass (SPG) emulsification technique. *Colloids Surf A* 153: 383–394.

Ma GH, Sone H, Omi S. 2004. Preparation of uniform-sized polystyrene-polyacrylamide composite microspheres from a W/O/W emulsion by membrane emulsification technique and subsequent suspension polymerization. *Macromolecules* 37: 2954–2964.

Mine Y, Shimizu M, Nakashima T. 1996. Preparation and stabilization of simple and multiple emulsions using a microporous glass membrane. *Colloids Surf B* 6: 261–268.

Müller RH, Mäder K, Gohla S. 2000. Solid lipid nanoparticles (SLN) for controlled drug delivery: A review of the state of the art. *Eur J Pharm Biopharm* 50: 161–177.

Nakagawa K, Iwamoto S, Nakajima M, Shono A, Satoh K. 2004. Microchannel emulsification using gelatin and surfactant-free coacervate microencapsulation. *J Colloid Interface Sci* 278: 198–205.

Nakajima M, Nabetani H, Ichikawa S, Xu QY. 2003. Functional emulsions. US Patent 6,538,019.

Nakashima T, Shimizu M, Kukizaki M. 1991a. *Membrane Emulsification Operation Manual*. Miyazaki: Department of Chemistry, Industrial Research Institute of Miyazaki Prefecture.

Nakashima T, Shimizu M, Kukizaki M. 1991b. Membrane emuylsification by microporous glass. *Key Eng Mater* 61–62: 513–516.

Nakashima T, Shimizu M, Kukizaki M. 1993. Effect of surfactant on production of monodispersed O/W emulsion in membrane emulsification. *Kag Kog Ronbunshu* 19: 991–997 (in Japanese).

Nakashima T, Shimizu M, Kukizaki M. 2000. Particle control of emulsion by membrane emulsification and its applications. *Adv Drug Deliv Rev* 45: 47–56.

Nakashima T, Shimizu M. 1986. Porous glass from calcium alumino boro-silicate glass. *Ceramics Jap* 21: 408–412 (in Japanese).

Nakashima T. 2002. History of SPG technology and its recent advances. In: *Proceedings of the 38th Int. SPG Forum "Membrane and Particle Science and Technology in Food and Medical Care"*, Industrial Technology Center, Miyazaki: p 63.

Nie Z, Xu S, Seo M, Lewis PC, Kumacheva E. 2005. Polymer particles with various shapes and morphologies produced in continuous microfluidic reactors. *J Am Chem Soc* 127: 8058–8063.

Nisisako T, Torii T, Higuchi T. 2004. Novel microreactors for functional polymer beads. *Chem Eng J* 101: 23–29.

Nisisako T, Okushima S, Torii T. 2005. Controlled formulation of monodisperse double emulsions in a multiple-phase microfluidic system. *Soft Matter* 1: 23–27.

Okonogi S, Katoh R, Asano Y, Yuguchi H, Kumazawa R, Sotoyama K, Takahashi K, Fujimoto M. 1994. Methods for producing emulsions, low-fat spread and oil-in-water-in-oil type spread. US patent 5,279,847.

Okushima S, Nisisako T, Torii T, Higuchi T. 2004. Controlled production of monodisperse double emulsions by two-step droplet breakup in microfluidic devices. *Langmuir* 20: 9905–9908.

Omi S. 1996. Preparation of monodisperse microspheres using the Shirasu porous glass emulsification technique. *Colloids Surf A* 109: 97–107.

Park SH, Yamaguchi T, Nakao S. 2001. Transport mechanism of deformable droplets in microfiltration of emulsions. *Chem Eng Sci* 56: 3539–3548.

Pautot S, Frisken BJ, Weitz DA. 2003. Engineering asymmetric vesicles. *PNAS* 100: 10718–10721.

Peyratout CS, Dähne L. 2004. Tailor-made polyelectrolyte microcapsules: From multilayers to smart containers. *Angew Chem* 43: 3762–3783.

Rayleigh L. 1879. On the instability of jets. *Proc London Math Soc* 10: 4–13.

Rayner M, Trägårdh G, Trägårdh C. 2005. The impact of mass transfer and interfacial expansion rate on droplet size in membrane emulsification processes. *Colloids Surf A* 266: 1–17.

Schadler V, Windhab EJ. 2004. Kontinuierliches Membran-Emulgieren mittels rotierender Mikromembranen mit definierten Porenabständen. *Chem Ing Tech* 76: 1392–1392.

Schadler V, Windhab EJ. 2005. Continuous membrane emulsification by using a membrane system with controlled pore distance. *Proceedings 10th Aachen Membrane Colloquium*. Aachen: Institüt für Verfahrenstechnik der RWTH Aachen, p 585.

Schröder V, Behrend O, Schubert H. 1998. Effect of dynamic interfacial tension on the emulsification process using microporous, ceramic membrane. *J Colloid Interface Sci* 202: 334–340.

Schröder V. 1999. Herstellen von Öl-in-Wasser-Emulsionen mit mikroporösen Membranen. Ph.D. thesis. Karlsruhe: University of Karlsruhe (T.H.).

Shima M, Kobayashi Y, Fujii T, Tanaka M, Kimura Y, Adachi S, Matsuno R. 2004. Preparation of fine W/O/W emulsion through membrane filtration of coarse W/O/W emulsion and disappearance of the inclusion of outer phase solution. *Food Hydrocolloids* 18: 61–70.

Shimizu M, Kukizaki M, Nakashima T, Alvarez-Nakase AM. 2002a. Leakage of inclosed irinotecan hydrochloride and remedial measures taken during production of the W/O type oral medicine. In: *Proceedings of the 38th Int. SPG Forum "Membrane and Particle Science and Technology in Food and Medical Care"*, Industrial Technology Center, Miyazaki.

Shimizu M, Nakashima T, Kukizaki M. 2002b. Particle size control of W/O emulsion by means of osmotic pressure as driving force. *Kag Kog Ronbunshu* 28: 304–309.

Shimizu M, Nakashima T, Kukizaki M. 2002c. Preparation of W/O emulsion by membrane emulsification and optimum conditions for its monodispersion. *Kag Kog Ronbunshu* 28: 310–316 (in Japanese).

Sugiura S, Nakajima M, Tong J, Nabetani H, Seki M. 2000. Preparation of monodispersed solid lipid microspheres using a microchannel emulsification technique. *J Colloid Interface Sci* 227: 95–103.

Sugiura S, Nakajima M, Itou H, Seki M. 2001. Synthesis of polymeric microspheres with narrow size distributions employing microchannel emulsification. *Macromol Rapid Commun* 22: 773–778.

Sugiura S, Nakajima M, Kumazawa N, Iwamoto S, Seki M 2002a. Characterization of spontaneous transformation-based droplet formation during microchannel emulsification. *J Phys Chem B* 106: 9405–9409.

Sugiura S, Nakajima M, Seki M. 2002b. Prediction of droplet diameter for microchannel emulsification. *Langmuir* 18: 3854–3859.

Sugiura S, Nakajima M, Yamamoto K, Iwamoto S, Oda T, Satake M, Seki M. 2004a. Preparation characteristics of water-in-oil-in-water multiple emulsions using microchannel emulsification. *J Colloid Interface Sci* 270: 221–228.

Sugiura S, Nakajima M, Seki M. 2004b. Preparation of monodispersed polymeric microspheres over 50 μm employing microchannel emulsification. *Ind Eng Chem Res* 43: 4043–4047.

Supsakulchai A, Ma GH, Nagai M, Omi S. 2002. Microencapsulation of fine titanium dioxide powders from (S/O)/W emulsion with subsequent solvent evaporation. In: *Polymer Colloids: Science and Technology of Latex Systems* (*ACS Symposium Series 801*), Daniels ES, Sudol ED, El Aasser ME, eds. Washington, DC, pp 260–275.

Suzuki K, Shuto I, Hagura Y. 1996. Characteristics of the membrane emulsification method combined with preliminary emulsification for preparing corn oil-in-water emulsions. *Food Sci Technol Int Tokyo* 2: 43–47.

Suzuki K, Fujiki I, Hagura Y. 1998. Preparation of corn oil/water and water/corn oil emulsions using PTFE membranes. *Food Sci Technol Int Tokyo* 4: 164–167.

Suzuki K, Fujiki I, Hagura Y. 1999. Preparation of high concentration of O/W and W/O emulsions by the membrane phase inversion emulsification using PTFE membranes. *Food Sci Technol Int Tokyo* 5: 234–238.

Takeuchi S, Garstecki P, Weibel DB, Whitesides GM. 2005. An axisymmetric flow-focusing microfluidic device. *Adv Mater* 17: 1067–1072.

Tan YC, Fisher JS, Lee AI, Cristini V, Lee AP. 2004. Design of microfluidic channel geometries for the control of droplet volume, chemical concentration, and sorting. *Lab Chip* 4: 292–298.

Tan YC, Cristini V, Lee AP. 2006. Monodispersed microfluidic droplet generation by shear focusing microfluidic device. *Sens Actuator B-Chem* 114: 350–356.

Thorsen T, Roberts RW, Arnold FH, Quake SR. 2001. Dynamic pattern formation in a vesicle-generating microfluidic device. *Phys Rev Letts* 86: 4163–4166.

Tice JD, Lyon AD, Ismagilov RF. 2004. Effects of viscosity on droplet formation and mixing in microfluidic channels. *Anal Chim Acta* 507: 73–77.

Tong J, Nakajima M, Nabetani H, Kikuchi Y, Maruta Y. 2001. Production of oil-in-water microspheres using a stainless steel microchannel. *J Colloid Interface Sci* 237: 239–248.

Toorisaka E, Ono H, Arimori K, Kamiya N, Goto M. 2003. Hypoglycemic effect of surfactant-coated insulin solubilized in a novel solid-in-oil-in-water (S/O/W) emulsion. *Int J Pharm* 252: 271–274.

Utada AS, Lorenceau E, Link DR, Kaplan PD, Stone HA, Weitz DA. 2005. Monodisperse double emulsions generated from a microcapillary device. *Science* 308: 537–541.

Van der Graaf S, Schroën CGPH, Van Der Sman RGM, Boom RM. 2004. Influence of dynamic interfacial tension on droplet formation during membrane emulsification. *J Colloid Interface Sci* 277: 456–463.

Van der Graaf S, Schroën CGPH, Boom RM. 2005a. Preparation of double emulsions by membrane emulsification: A review. *J Membr Sci* 251: 7–15.

Van der Graaf S, Steegmans MLJ, Van Der Sman RGM, Schroën CGPH, Boom RM. 2005b. Droplet formation in a T-shaped microchannel junction: A model system for membrane emulsification. *Colloids Surf A* 266: 106–116.

Vladisavljević GT, Schubert H. 2002. Preparation and analysis of oil-in-water emulsions with a narrow droplet size distribution using Shirasu-porous-glass (SPG) membranes. *Desalination* 144: 167–172.

Vladisavljević GT, Brösel S, Schubert H. 2002. Preparation of water-in-oil emulsions using microporous polypropylene hollow fibers: Influence of some operating parameters on droplet size distribution. *Chem Eng Process* 41: 231–238.

Vladisavljević GT, Schubert H. 2003a. Preparation of emulsions with a narrow particle size distribution using microporous α-alumina membranes. *J Dispersion Sci Technol* 24: 811–819.

Vladisavljević GT, Schubert H. 2003b. Influence of process parameters on droplet size distribution in SPG membrane emulsification and stability of prepared emulsion droplets. *J Membr Sci* 225: 15–23.

Vladisavljević GT, Lambrich U, Nakajima M. 2004a. Production of O/W emulsions using SPG membranes, ceramic α-Al_2O_3 membranes, microfluidizer and a microchannel plate: A comparative study. *Colloids Surf A* 232: 199–207.

Vladisavljević GT, Shimizu M, Nakashima T. 2004b. Preparation of monodisperse multiple emulsions at high production rates by multi-stage premix membrane emulsification. *J Membr Sci* 244: 97–106.

Vladisavljević GT, Shimizu M, Nakashima T. 2005. Permeability of hydrophilic and hydrophobic Shirasu-porous-glass (SPG) membranes to pure liquids and its microstructure. *J Membr Sci* 250: 69–77.

Vladisavljević GT, Williams RA. 2005. Recent developments in manufacturing emulsions and particulate products using membranes. *Adv Colloid Interface Sci* 113: 1–20.

Vladisavljević GT, Williams RA. 2006. Manufacture of large uniform droplets using rotating membrane emulsification. *J Colloid Interface Sci* 299: 396–402.

Walde P, Ichikawa S. 2001. Enzymes inside lipid vesicles: preparation, reactivity and applications. *Biomol Eng* 18: 143–177.

Wang LY, Ma GH, Su ZG. 2005. Preparation of uniform sized chitosan microspheres by membrane emulsification technique and application as a carrier of protein drug. *J Control Release* 106: 62–75.

Weitz DA. Soft Condensed MAtter Group, Dept. Of Physics, Harvard University. Available at <http://www.deas.harvard.edu/projects/weitzlab/alvaroweb/Triple%20and%20Multiple%20Emulsions.htm>.

Whitesides GM, Stroock AD. 2001. Flexible methods for microfluidics. *Phys Today* 54: 42–48.

Williams RA, Peng SJ, Wheeler DA, Morley NC, Taylor D, Whalley M, Houldsworth DW. 1998. Controlled production of emulsions using a crossflow membrane. *Chem Eng Res Des A* 76: 902–910.

Williams RA. 2001a. Making the perfect particle. *Ingenia n* 7: 1.

Williams RA. 2001b. Controlled dispersion using a spinning membrane reactor. UK patent application PCT/GB00/04917.

Williams RA. 2002. Emulsifying with membranes. In: *Proceedings of the 3rd World Congress on Emulsions*, Lyon.

Xia Y, Whitesides GM. 1998. Soft lithography. *Annu Rev Mater Sci* 28: 153–184.

Xu JH, Luo GS, Chen GG, Tan B. 2005. Mass transfer performance and two-phase flow characteristic in membrane dispersion mini-extractor. *J Membr Sci* 249: 75–81.

Xu Q. 2005. Generation of microbubbles in microfluidic devices. Presented before Seminar at the National Food Research Institute, Tsukuba, Japan.

Xu Q, Nakajima M. 2004. The generation of highly monodisperse droplets through the breakup of hydrodynamically focused microthread in a microfluidic device. *Appl Phys Lett* 85: 3726–3728.

Yafei W, Tao Z, Gang H. 2006. Structural evolution of polymer-stabilized double emulsions. *Langmuir* 67: 67–73.

Yanagishita T, Tomabechi Y, Nishio K, Masuda H. 2004. Preparation of monodisperse SiO_2 nanoparticles by membrane emulsification using ideally ordered anodic porous alumina. *Langmuir* 20: 554–555.

Yamazaki N, Yuyama H, Nagai M, Ma GH, Omi S. 2002. A comparison of membrane emulsification obtained using SPG (Shirasu porous glass) and PTFE [poly(tetrafluoroethylene)] membranes. *J Dispersion Sci Technol* 23: 279–292.

Yamazaki N, Naganuma K, Nagai M, Ma GH, Omi S. 2003. Preparation of W/O (water-in-oil) emulsions using a PTFE (polytetrafluoroethylene) membrane: A new emulsification device. *J Dispersion Sci Technol* 24: 249–257.

Yi GR, Thorsen T, Manoharan VN, Hwang MJ, Jeon SJ, Pine DJ, Quake SR, Yang SM. 2003. Generation of uniform colloidal assemblies in soft microfluidic devices. *Adv Mater* 15: 1300–1304.

You JO, Park SB, Park HY, Haam S, Chung CH, Kim WS. 2001. Preparation of regular sized Ca-alginate microspheres using membrane emulsification method. *J Microencapsul* 18: 521–532.

Zhu J, Barrow D. 2005. Analysis of droplet size during crossflow membrane emulsification using stationary and vibrating micromachined silicon nitride membranes. *J Membr Sci* 261: 136–144.

ABBREVIATIONS OF CHEMICALS

Arlacel 83 Sorbitan sesquioleate

Arlacel P135 PEG-30, dipolyhydroxystearate

$CaNa_2$-EDTA Calcium disodium ethylendiaminetetraacetate

C8TG Octanoic acid triacylglycerol

DCM Dichloromethane

EGDMA Ethyleneglycol dimethacrylate

HCO-60 Polyoxyethylene hardened castor oil

HCPK 1-Hydroxycyclohexyl phenyl ketone
IPSO Iodinated poppy seed oil
LPC Lysophosphatidylcholine
MCT Medium chain triglycerides
MSW-7S Decaglycerol monostearate
PAA Poly(acrylic acid)
PBA Poly(normal-butyl acrylate)
PC Phosphatidylcholine (lecithin)
PGML Pentaglycerin monolaurate
PGPR Polyglycerol polyricinoleate (E 476)
PLA Polylactide acid
POPC 1-Palmitoyl-2-oleoyl-*sn*-glycero-3-phosphocholine
POPS 1-Palmitoyl-2-oleoyl-*sn*-glycero-3-phospho-L-serine
PVA Polyvinyl alcohol
PVP Polyvinyl pyrrolidone
SDS Sodium dodecyl sulfate
Span 20 Sorbitan monolaurate
Span 80 Sorbitan monooleate
Span 85 Sorbitan trioleate
THF Tetrahydrofuran
TPGDA Tripropyleneglycol diacrylate
Tween 20 Polyoxyethylene (20) sorbitan monolaurate
Tween 80 Polyoxyethylene (20) sorbitan monooleate
Tween 85 Polyoxyethylene (20) sorbitan trioleate
TGPR Tetraglycerol polyricinoleate

CHAPTER 7

Recent Developments in O/W/O Multiple Emulsions

AXEL BENICHOU and ABRAHAM ASERIN

Contents

7.1 INTRODUCTION

Multiple emulsions, also known as emulsions of emulsions, consist of droplets of one dispersed liquid that are further dispersed in another liquid. The inner dispersed globule/droplet in the multiple-emulsion is separated (compartmentalized) from the outer liquid phase by a layer of a third phase (Garti, 1997a, b, 1998; Garti and Bisperink, 1998; Garti and Benichou, 2001, 2003).

The most common multiple emulsions reported in the literature are of W/O/W type, but in some specific applications O/W/O emulsions can also be prepared. The terms multiple emulsion, double emulsions, and triple emulsions were used in the past to describe these systems. A more suitable and more

Multiple Emulsions: Technology and Applications, Edited by Abraham Aserin

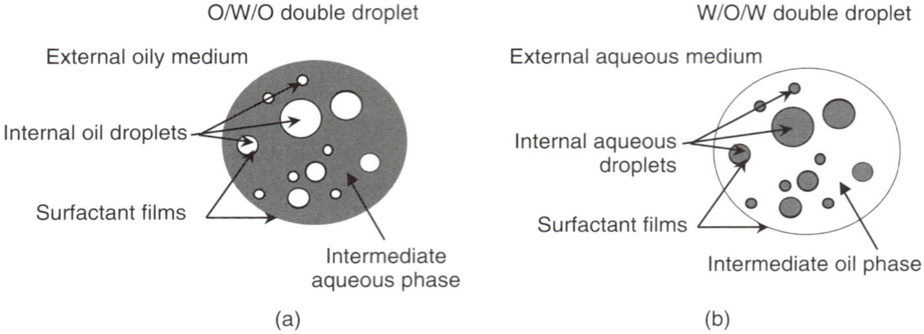

Figure 7.1 Schematic presentation of the two types of multiple emulsion droplets. (*a*) Typical O/W/O multiple emulsion droplet, (*b*) typical W/O/W multiple emulsion droplet.

accurate term for such systems should be "emulsified emulsions" as this term better describes the route to prepare the emulsions. A schematic presentation of both types of multiple emulsion droplets (W/O/W and O/W/O) is shown in Figure 7.1.

Multiple emulsions of the W/O/W type are excellent candidates for controlled and sustained release of hydrophilic drugs due to the existence of a middle oil layer that acts as a liquid membrane. Conversely, O/W/O multiple emulsions allow the entrapment of lipophilic drug owing the presence of an aqueous liquid membrane between the two oily phases. These systems are able to entrap water or oily soluble compounds in one or other of those phases. These vesicular complex structures may find applications in parenteral administration (Hino et al., 2000; Garti and Benichou, 2001; Tamilvanan, 2004), topical delivery (Laugel et al., 1998; Laugel et al., 2000), and enteral absorption of drugs and drug stabilization (Gallarate et al., 1999; Yoshida et al., 1999). Multiple emulsions have also been used as intermediates for microsphere preparation (Grossiord et al., 1996; Uchida et al., 1996; Youan et al., 1999) and in separation processes (Varadaraj, 2004; Lin and Chen, 2006). Initially one important problem with multiple emulsions was their intrinsic thermodynamic instability. However, the stabilization of these formulations has been studied and improved over years of investigation and effort.

In most cases multiple emulsions are aimed for slow and sustained release of active matter from an internal liquid reservoir into the continuous phase. In some applications the multiple emulsions can serve also as an internal reservoir to entrap matter from the outer diluted continuous phase into the inner confined space. These applications are aimed to remove toxic matter. In other applications multiple emulsions are reservoirs for improved dissolution or solubilization of insoluble materials. The materials will dissolve in part in the inner phase, in part at the internal interface, and occasionally at the external interface. Applications related to protection of sensitive and active molecules from the external phase (antioxidation) have been mentioned (Gallarate et al., 1999; Yoshida et al., 1999; Kim and Lee, 1999; Edris and Bergenstähl, 2001). Potential applications for multiple emulsions are well documented, and many of these applications have been patented (Bams and Van Megen, 1987; Gaonkar, 1994a, b; Takahashi et al., 1994; Strauel and Friour, 1994; Herb, 1996). More applications are expected to emerge in the near future with special attention to the use of multiple emulsions as intermediate in preparation of solid or semisolid microspheres (Grossiord et al., 1996; Uchida et al., 1996; Youan et al., 1999).

7.2 PREPARATION ROUTES

Multiple emulsions consist of two different interfaces that require two sets of different types of emulsifiers. In O/W/O multiple emulsions the first set of emulsifiers, for the internal interface, must be hydrophilic while the second set of emulsifiers, for the external interface, must be hydrophobic. For W/O/W multiple emulsions the order of the emulsifiers is the opposite: the inner emulsifiers are hydrophobic while the outer ones are hydrophilic. In many cases a blend of two or more emulsifiers in each set are recommended for better stabilization results. This chapter will discuss mostly O/W/O multiple emulsions but some W/O/W emulsions applications are also considered.

In early reports on the formation of multiple emulsions only one set of emulsifiers and an inversion process were used (Matsumuto and Kang, 1989). Such preparations were done in one step, but the stability was in most cases questionable. It was difficult to control the distribution of the emulsifiers within the two interfaces. There was fast migration of the emulsifiers between the phases that destabilized the emulsions. Moreover in these cases it was practically impossible to control the entrapment of an active addendum in the inner phase. In most recent emulsions formulations the emulsions are prepared in two steps. At first, a high-shear homogenization was applied on the oil that was added to the solution of the water and the hydrophilic emulsifiers, to obtain stable O/W emulsion. In the second step, the O/W emulsion is gently added with stirring (not homogenization) to the external oil and lipophilic emulsifier solution (Figure 7.2). Inner and external oil phases can be the same or different. The droplets size distribution of a typical classical multiple emulsion ranges from 10 to 50 μm.

Gaonkar claimed to have developed a method to form multiple emulsions droplets that does not require any homogenization steps (Gaonkar, 1994a). An O/W microemulsion is obtained by mixing oil, water, a second alkyl containing a polar protic solvent such as methanol, ethanol, propanol, glycerol, propylene glycol, dodecanol, and their blend, and a hydrophilic emulsifier. The O/W microemulsion is then diluted with water and does not require the addition of a lipophilic emulsifier. This dilution step results in destabilization of the microemulsion to reach the W/O/W multiple emulsion.

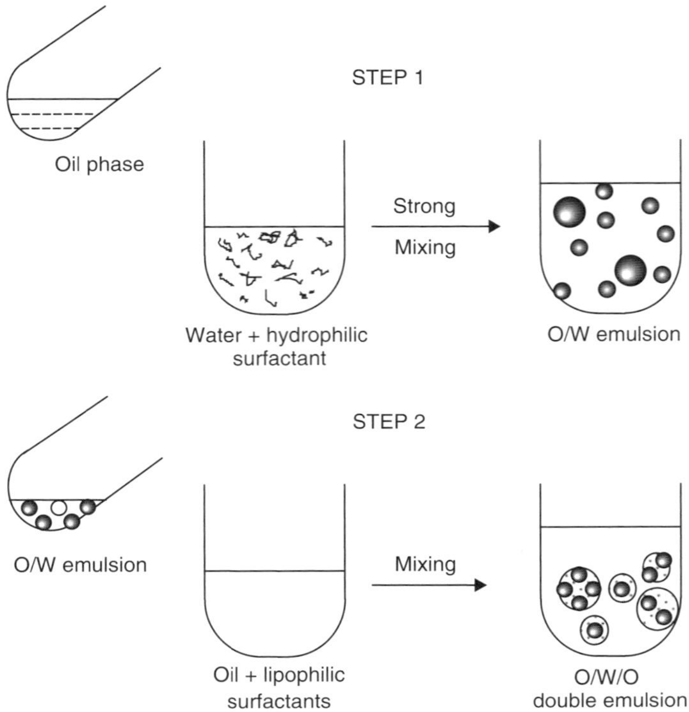

Figure 7.2 Schematic illustration of a two-step process in formation of an O/W/O multiple emulsion.

Lee et al. described a promising way to prepare multiple emulsion that is derived from the emulsion inversion path (Lee et al., 2002). An oil-in-water microemulsion based on 2-butoxyethanol/*n*-decane/water system was mixed with oleic acid, and they observed a phase inversion sequence that includes microemulsion-in-oil to oil-in-microemulsion.

Grossiord et al. discuss similar method based on "emulsified microemulsions." The idea is to disperse an oil phase within water by surfactant and to form L_2 phase (water-in-oil microemulsion) (Grossiord et al., 1998). This phase is further emulsified with water to form multiple emulsion (Figure 7.3). The problem is that there is no evidence of the formation of multiple emulsions and that the internal phase remains, after the second emulsification process, a L_2 phase of a submicronal droplets in size with intrinsic thermodynamic stability.

The same concept of "emulsified microemulsion" was earlier reported by Pilman et al. (Pilman et al., 1980) and also patented (Gaonkar, 1994a). If one can prove that the internal compartmentalization is of stable microemulsion, it might bring a breakthrough to this field, since the sizes of the external droplets could be reduced to values below one micrometer. Such formulations will

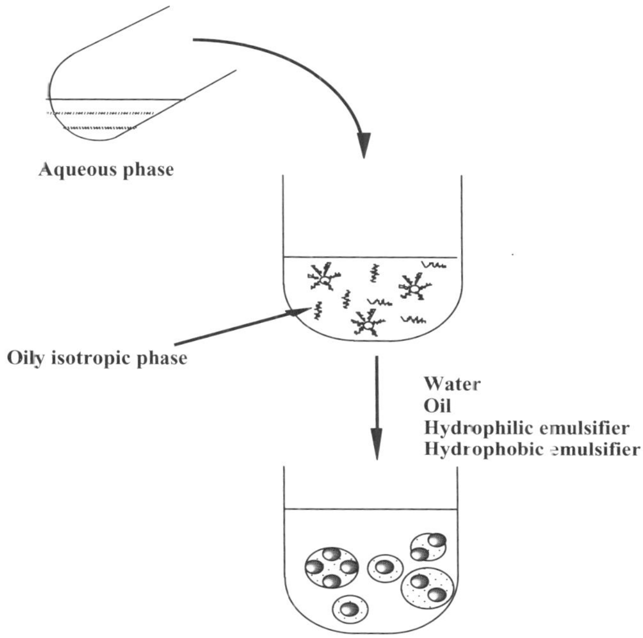

Figure 7.3 Preparation of W/O/W multiple emulsion by emulsified microemulsion technique. (Adapted from Grossiord et al., 1998.)

allow formation multiple emulsions with increased shelf life and may be even injectable.

Castro et al. reported an elegant method to entrap nifedipine, a Ca^{2+}-channel blocker that inhibits the transmembrane influx of Ca^{2+} into cardiac muscle cells and vascular smooth muscle (Castro et al., 2001). Nifedipine is highly sensitive to chemical oxidation and so needs to be protected against oxidation prior to be used. The authors of this study incorporated the drug in the inner oil phase of an O/W/O double microemulsion with the composition described in Table 7.1. Unfortunately, no more was said on the final microstructure of this system but that "multiple microemulsions" permitted to protect efficiently the drug against oxidation as shown by liquid chromatography (Figure 7.4).

In a more recent report, O/W/O multiple emulsions were prepared via one-step emulsification an aqueous phase containing polyethylene glycol and polyoxyethylene(20) sorbitan monolaurate with an oil phase comprising 1-octanol, hydroxypropyl cellulose and sorbitan monooleate (Oh et al., 2004). The formation of the O/W/O multiple emulsion droplets was attributed to the diffusion

TABLE 7.1 Composition (wt%) of O/W microemulsion and O/W/O multiple emulsion containing nifedipine

	Concentration (wt%)
O/W primary microemulsion	
Nifedipine	1.8
Isopropyl myristate	15
Brij 96V	26.25
Propylene glycol	8.75
Distilled water	48.2
O/W/O multiple microemulsion	
Primary microemulsion	20
Brij 96V	22.5
Polyethylene glycol 400	7.5
Isopropyl myristate	50

Source: Adapted from Castro et al. (2001).

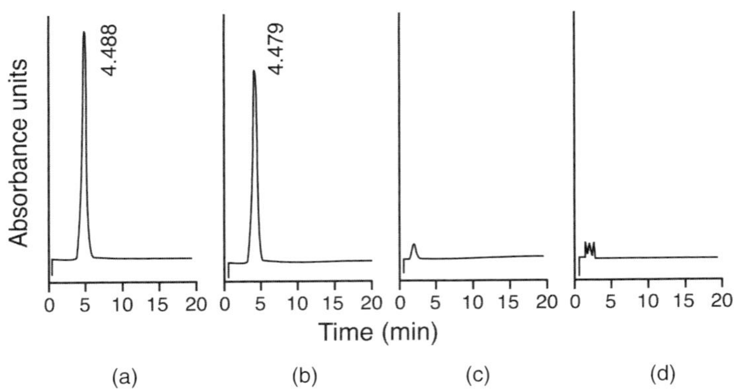

Figure 7.4 HPLC chromatograms of (*a*) nifedipine-loaded O/W/O multiple microemulsion (nifedipine concentration: $20\,g \cdot ml^{-1}$), (*b*) 20% decomposed nifedipine-loaded O/W/O multiple microemulsion, (*c*) totally decomposed nifedipine-loaded O/W/O multiple microemulsion, and (*d*) placebo O/W/O multiple microemulsion. (Adapted from Castro et al., 2001.)

of the oil phase (1-octanol) from the continuous external phase to the inner phase. A proposed mechanism of this phenomenon is shown in Figure 7.5.

Other methods to prepare multiple emulsions that were envisaged include the membrane emulsification technique (Higashi et al., 1995). In this method the primary O/W or W/O emulsion is formed by sonication and then filled into the upper chamber of a special apparatus (Figure 7.6). The external phase of the final multiple emulsion is continuously injected into the lower chamber

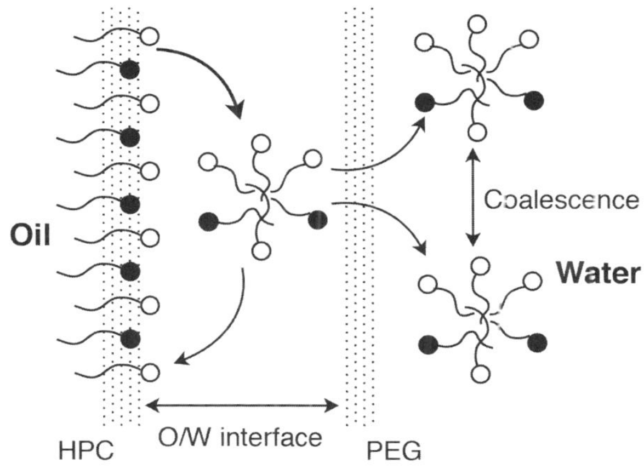

Figure 7.5 Mechanism of formation of internal oily droplets of O/W/O multiple emulsions prepared by one-step emulsification process. (Adapted from Oh et al., 2004.)

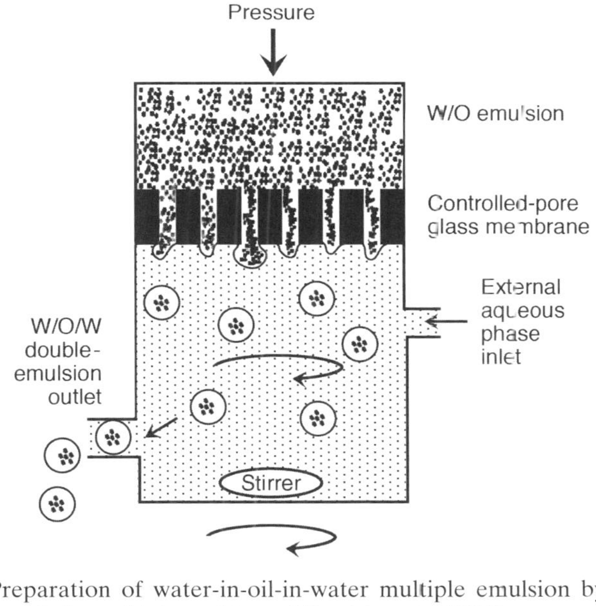

Figure 7.6 Preparation of water-in-oil-in-water multiple emulsion by a membrane emulsification technique. (Adapted from Higashi et al., 1995.)

to create a continuous flow. Nitrogen gas fed into the upper chamber initiates a permeation of the primary emulsion through the controlled-pore glass membrane into the emulsifying chamber, generating the multiple globules. The emulsion is progressively removed from the apparatus. This process should be applied in industrial scale.

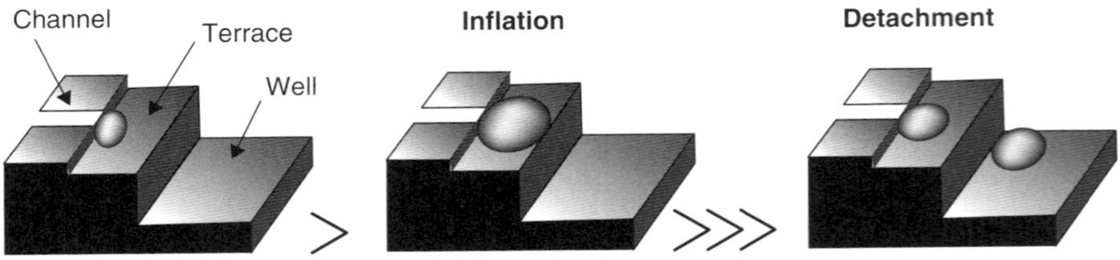

Figure 7.7 Microchannel emulsification. (Adapted from Sugiura et al., 2004.)

Monodispersed W/O/W multiple emulsion droplets were prepared by microchannel emulsification process during the second step of the emulsification (Sugiura et al., 2004). In this new technique a droplet of the to-be-dispersed phase is pushed through a channel onto a relatively flat area (the so-called terrace) upon which the droplet can expand to a disk shaped entity. Upon further inflation the disk reaches the edge of the terrace, where it can expand into the deeper well, in which spontaneous detachment of the droplet takes place (Figure 7.7). Using decane, ethyl oleate, and medium-chain triglyceride (MCT) as oil intermediate phases, monodispersed oil droplets that contained small water droplets were obtained. This technique permits very high yield of entrapment of active molecules in the aqueous core of the final multiple-emulsion droplets.

A new promising method for preparing multiple emulsions uses a modified microchannel emulsification method (Okushima et al., 2004). Figure 7.8 shows how O/W/O multiple emulsion droplets are formed. Aqueous droplets of uniform size were achieved with high reproducibly within an organic flow at the upstream T-junction. This stream is then flowed into the second T-junction, to create monodisperse organic droplets containing aqueous droplets within an external aqueous phase. The size and breakup rate of the droplets was controlled by varying the flow conditions at each junction. The number of entrapped aqueous droplets was also controlled by adjusting the relation between the breakup rates at the two junctions (Figure 7.9). O/W/O emulsions can also be produced by this method when the hydrophobic junction is prepared downstream.

7.3 STABILITY CONSIDERATIONS

Multiple emulsions are usually not empty. Soluble active materials are entrapped during the emulsification in the inner oily phase. Because of the osmotic pressure gradient, the active matter tends to diffuse and migrate from the internal phase to the external interface mostly through a controlled reverse micellar transport mechanism (Figure 7.10*a*) (Garti and Bisperink, 1998; Garti and Benichou, 2001). The dilemma that researchers were faced with was how to control the diffusion of oil molecules, as well as the emulsifier molecules

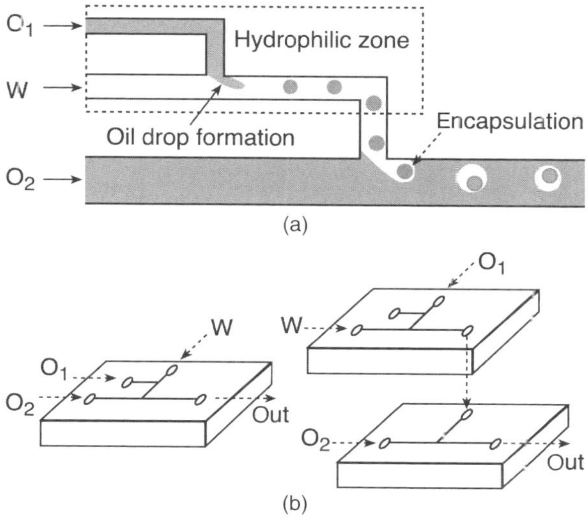

Figure 7.8 (*a*) Schematic apparatus for the formation of an O/W/O multiple emulsion by microchannel emulsification process. (*b*) Single-chip module (*left*) and double-chip module (*right*). *W*: intermediate aqueous phase, O_1: internal oil phase, O_2: external oil phase. (Adapted from Okushima et al., 2004.)

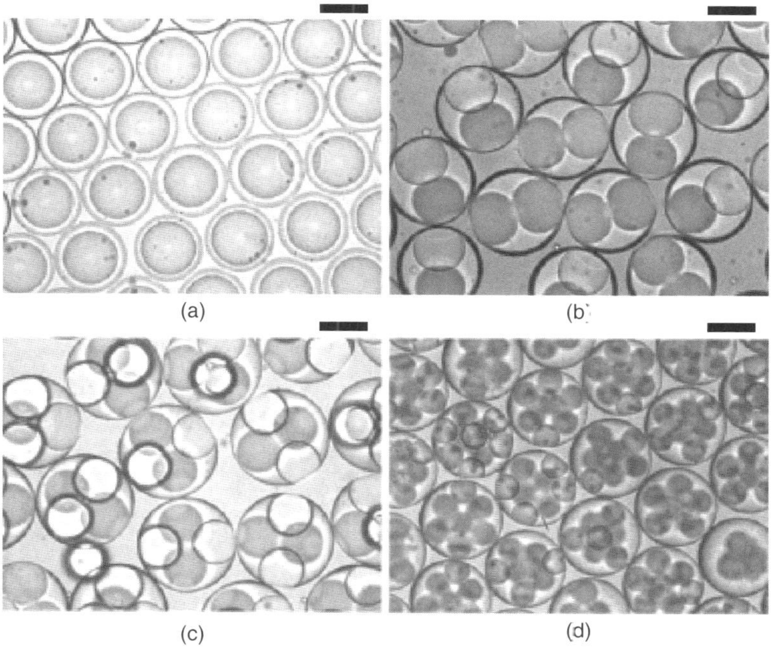

Figure 7.9 Photomicrographs of stable W/O/W multiple emulsion with a controlled number of internal droplets prepared with multiple-phase microfluidic system. The scale bar is 100 μm. (Adapted from Okushima et al., 2004.)

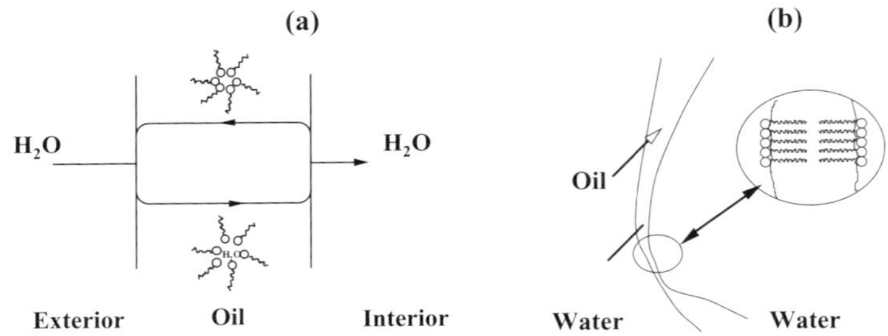

Figure 7.10 Schematic of three possible transport mechanisms in double emulsions. (*a*) Reverse micellar transport, (*b*) lamellar thinning transport of marker from the inner aqueous phase to the continuous aqueous phase.

and the active matter, from the internal phase to the outer phase (Garti, 1997a). It seemed almost impossible to retain the active material within the inner phase upon prolonged storage, mainly because the monomeric emulsifiers tend to migrate from both the inner interface and the outer interface to the intermediate layer where they aggregate to form micelles capable of solubilizing the entrapped active matter.

During years of investigation to improve stability and to control sustained and prolonged release of active materials, monomeric surfactants were progressively replaced by polymeric emulsifiers (Garti and Aserin, 1996). Polymeric amphiphilic molecules, both synthetic and naturally occurring ones, are known to improve the interfacial coverage during emulsification (Zhang et al., 1992; Vaziri and Warburton, 1994) and so ensure better encapsulation and controlled release of the addenda entrapped in the internal core of the multiple-emulsion droplets (Cornec et al., 1998).

The use of macromolecular amphiphiles and stabilizers, such as proteins and polysaccharides, was long adopted by scientists exploring stability of W/OW multiple emulsions. Gelatin (Zhang et al., 1992; Vaziri and Warburton, 1994), whey proteins (Dickinson et al., 1991; Dickinson et al., 1994; Evison et al., 1995; Cornec et al., 1998), bovine serum albumin (BSA) (Dickinson et al., 1991; Fredrokumbaradzi and Simov, 1992; Dickinson et al., 1994; Garti et al., 1994), human serum albumin (HSA), caseins and other proteins were mentioned and evaluated. The proteins were used usually in combination with other monomeric emulsifiers (Garti et al., 1994). A significant improvement in the stability of the emulsions was shown when these macromolecules were encapsulated onto the external interface and the marker transport was more controlled.

From all the scattered information that is available in the literature, we learned that the best entrapment capacities, along with best stabilities of the globules and the best retention of the addenda in the inner phase, can be achieved if the oil in the inner phase differs from that of the outer phase and the surfactants are strongly anchored to the interface. The polymeric adducts are therefore excellent candidates for such requirements.

Multiple emulsions made of low-molecular-weight emulsifiers (the so-called monomeric emulsifiers) are mostly unstable thermodynamically. This is mainly because in the second stage of the emulsification severe homogenization or shear are not recommended, and as a result large droplets are obtained. During years of research attempts have been made to find proper and more suitable combinations of emulsifiers to reduce droplets sizes and to improve the emulsion stability. Aggregation, flocculation, and coalescence (occurring in the inner phase and between the multiple-emulsion droplets) lead to rupture of droplets and separation of the phases and thus are major factors behind the instability of the emulsions.

Additional instability mechanisms and release pathways have been demonstrated and discussed in detail by various authors. These mechanisms include transport through thinned lamellae (Figure 7.10*b*), transport of adducts or complexes that are formed in the intermediate phase and act as a liquid membrane, and other variations of these mechanisms. It seems, however, that the main instability and release mechanisms are parallel or simultaneously occurring phenomena of reverse micellar transport and coalescence.

All the above-mentioned mechanisms have been well established. Yet it seems that the stability and the release patterns of these complex multiple emulsion systems depend on various parameters that simultaneously interplay and that a simplified or unique mechanism cannot explain all the in-parallel pathways that take place in the multiple emulsions (Figure 7.11).

Two types of thermodynamic instabilities are known to be responsible for the evolution of W/O/W multiple emulsions (Ficheux et al., 1998). The first type involves coalescence of the small inner droplets with the outer droplets interface, which is due to the rupture of the thin nonaqueous film that forms

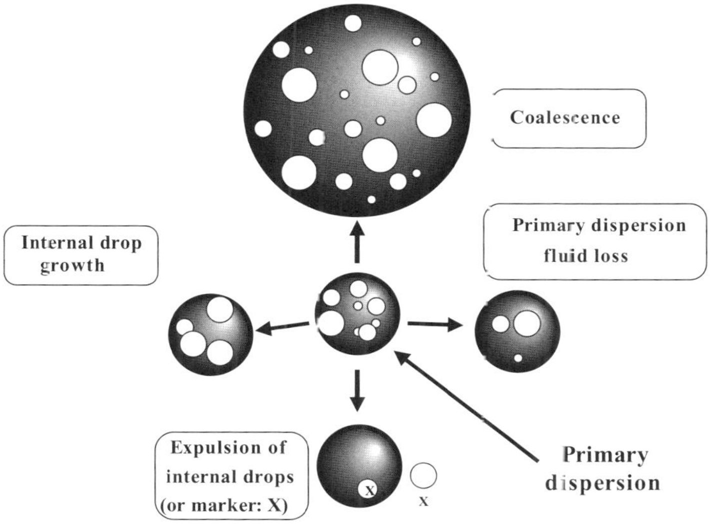

Figure 7.11 Schematic representation of instability pathways in multiple emulsions.

between the external continuous phase and the inner small water droplets. This instability irreversibly transforms a multiple globule into a simple direct emulsion droplet. Such a mechanism is suitable for delivery of water-soluble substances. The second instability involves coalescence between the small inner droplets within the oil globule. Whereas the first type of instability leads to a complete delivering of the small inner droplets toward the external phase, the second leads to an increase of the average diameter of the internal droplets and a decrease in their number.

The variations between the different suggested mechanisms are usually not dramatic, and most suggested mechanisms are basically very similar. The main causes for instability are aggregation of the internal and multiple emulsion droplets, breakage or disruption of the intermediate liquid phase on the surface of the inner droplets, seepage of the entrapped addendum from the inner to the external phase, shrinkage and swelling of the internal droplets due to osmotic gradient across the oil liquid membrane in the case of W/O/W multiple emulsions, and phase separation.

7.4 TRANSPORT PHENOMENA AND RELEASE CONSIDERATIONS

Most of release studies are done in W/O/W multiple-emulsion systems where an active water soluble molecule is present in the inner aqueous phase. Several attempts have been made to explain the transport phenomena of entrapped addenda from the inner to the outer phase of multiple-emulsion droplets. It has been demonstrated that for lipid soluble material dissolved in the oil phase, the release obeys first-order kinetics and is diffusion controlled with excellent accordance to Fick's law. Two mechanisms for the permeation through the oil intermediate phase are well accepted, the first being via the reverse micellar transport (Figure 7.10*a*) and the second via diffusion across a very thin lamellae of surfactant phase formed in areas where the oil layer is very thin (Figure 10*b*).

It is well established that the release of electrolytes from multiple emulsions in the presence of monomeric emulsifiers takes place even if the droplets are very stable to coalescence and even if the osmotic pressure of the two phases has been equilibrated (Garti and Benichou, 2001; Garti and Benichou, 2003). The model adopted for the release of matter from polymeric matrices is the Stehle and Higuchi model (Stehle and Higuchi, 1972). The model was worked out and tested by a modified release equation. The release factor B was plotted against the time t, and reciprocal initial concentrations of the solute ($1/C_0$).

$$B = \frac{3}{2}[1 - F]^{2/3} - F = \frac{3D_e t}{r_0^2 C_0},$$

where D_e is the effective diffusion coefficient, r_0 the radius of the outer phase droplets, F the release fraction, and C_0 the initial solute concentration. The linearity of the expression and the excellent correlation coefficient obtained in the experiments indicate the existence of a diffusion-controlled release mechanism.

The basic rationale for the use of O/W/O multiple emulsions for controlled delivery of lipophilic addenda is that the entrapped active molecules are then forced to diffuse through several interfaces prior to release into the external oil phase (Laugel et al., 1996). It was also reported that an external oil phase is usually preferred to an aqueous one to sustain the release of entrapped active molecules (Kundu et al., 1993). When multiple emulsion droplets are applied on a substrate (i.e., skin), the drug release profiles from O/W/O multiple emulsions always exhibit two release patterns: an initial slow release followed a rapid release phase. Laugel et al. explained that first an expected dynamical diffusion phenomenon occurs through the external oil phase and the static pattern follows because of the accumulation of the drug on the substrate surface, which acts as a local "pump" (Laugel et al., 1998). The size of the inner oil droplets continuously decrease with time until a steady-state value is reached. Simultaneously the multiple globules may undergo phase inversion and revert to simple W/O emulsion droplets (Sajjadi et al., 2002).

O/W/O multiple emulsions show a high potential for prolonged delivery of drugs. In multiple emulsions the entrapped active molecule can be transported from the internal phase to the external phase through the middle phase, which acts as a liquid membrane. Multiple emulsions of the O/W/O type were found to favor the retention of hydrocortisone in the dermis, improving the concentration of the drug within the site to be treated (Laugel et al., 1998). In this point of view, multiple emulsions can be attempted to prolong therapeutic local effect. These observations must be supported by further works with drugs with different polarities. The hydrocortisone absorption flux profiles exhibited two drug release phases, an initial slow release profile, followed by a rapid one thereafter (Figure 7.12). The slow release rates of hydrocortisone from the O/W/O emulsions was attributed to the interfacial barrier consisting of the aqueous phase. The release profile was shown to be governed by the drug partition coefficient between the inner and external oil phase and its ability to diffuse through the interfacial film as well as through the aqueous liquid membrane separating the two oil phases.

7.5 STABILIZATION BY MACROMOLECULAR AMPHIPHILES

Macromolecules adsorb onto interfaces and facilitate better coverage than monomeric emulsifiers. The adsorbed polymers are known to enhance steric stabilization mechanisms and were proved to be efficient emulsifiers in oil-in-water emulsions.

The literature distinguishes among three main mechanisms of stabilization with polymers: (1) depletion stabilization by nonadsorbing macromolecules that prevent collision among particles (droplets) and provide elasticity to the

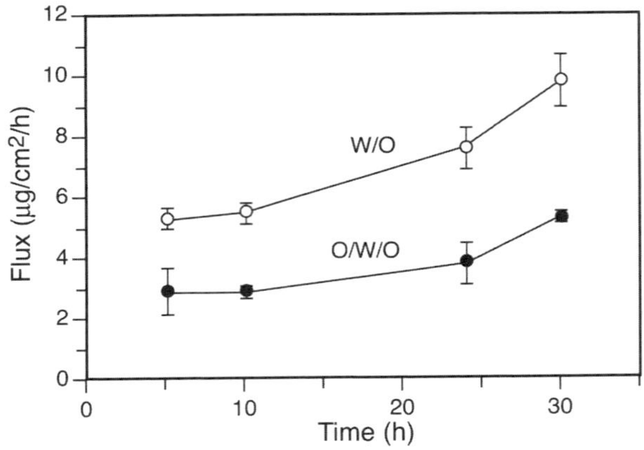

Figure 7.12 Hydrocortisone flux release from water-in-oil and oil-in-water-in-oil emulsions as a function of time across skin biopsy. (Adapted from Laugel et al., 1998.)

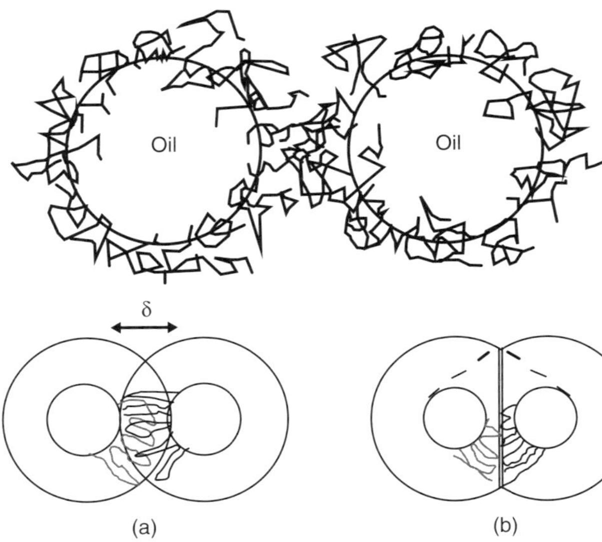

Figure 7.13 Two limiting models for steric stabilization with amphiphilic macromolecular surfactants. (*a*) Interpenetration of adsorbed layers without compression; (*b*) compression without interpenetration.

system; (2) electrostatic repulsions between two droplets carrying the same charge, and (3) steric stabilization resulting from hydrophobic interactions among adsorbed polymers. The main contribution to the stability of macromolecular-stabilized emulsions is related to droplets approaching a distance where compression due to polymer–polymer interactions occurs (without interpenetration) (Figure 7.13).

The total free energy G_i of systems stabilized with macromolecules includes three main terms: volume restriction, termed G_{vr}, mixing term, G_{mix}, and free energy of attraction, G_a. If an interaction exists between the polymeric chains adsorbed onto two neighbor particles (droplets), G_{vr} describes the reduction in the entropy of polymeric chains (configurational entropy) due to restriction of the total volume available to each chain. G_{mix} describes the buildup in polymeric amphiphile segment concentration in the interaction zone among the particles (droplets) that leads to an increase in the local osmotic pressure and in the steric free energy $\Delta G_s = \Delta G_{mix} + \Delta G_{vr}$. When the bulk solvency is decreased toward the dispersed phase, a significant minimal value of the total free energy of interaction (G_i) can be reached. Decreased solvency leads to a contraction in the thickness of the adsorbed layer, an increase in segment concentration, and consequently a minimum in the value of G_i. The segment concentration in the adsorbed layer is significantly increased, and this results in enhanced free energy of attraction (G_a). In the case of low-molecular-weight polymeric anchored chains, the effect will be more gradual than for high-molecular-weight polymers due to lower segment concentration in the interaction zone between the droplets. The different variations of energy are presented in Figure 7.14.

The use of macromolecular amphiphiles and stabilizers, such as proteins and polysaccharides, has long been adopted by scientists exploring the stability of W/O/W multiple emulsions. Gelatin (Zhang et al., 1992; Vaziri and Warburton, 1994), whey proteins (Dickinson et al., 1991, 1994; Evison et al., 1995; Cornec et al., 1998), bovine serum albumin (BSA) (Dickinson et al., 1991; Fredrokumbaradzi and Simov, 1992; Dickinson et al., 1994; Garti

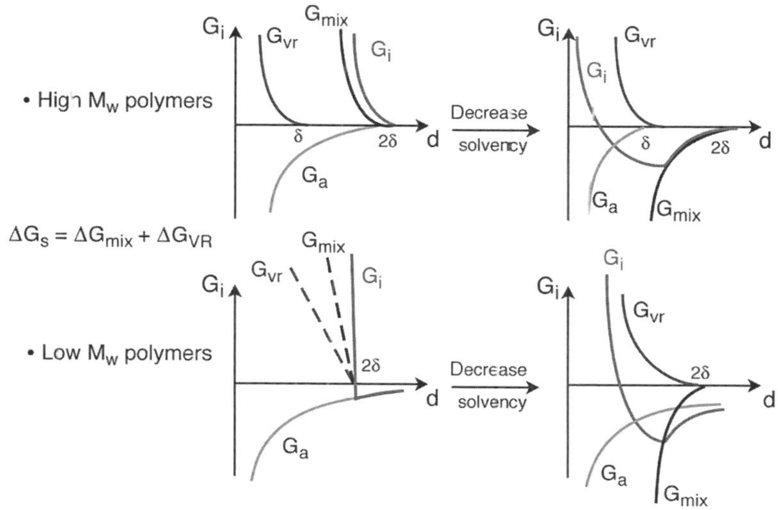

Figure 7.14 Variations on the different free energies involved in the steric stabilization mechanism.

et al., 1994), cyclodextrins (Yu et al., 1999, 2003), and chitosan (Genta et al., 1997; Schulz et al., 1998; Adachi et al., 1999) were mentioned and evaluated. The proteins were usually used in combination with other monomeric emulsifiers. A significant improvement in the stability of the emulsions was shown when these macromolecules were encapsulated onto the external interface. In most cases the macromolecule was used in low concentrations (max. 0.2 wt%) and in combination with a large excess of nonionic monomeric emulsifiers. Furthermore, from the release curves, it seems that the marker transport is more controlled. Dickinson et al. (Dickinson et al., 1991, 1994; Evison et al., 1995) concluded that proteins or other macromolecular stabilizers are unlikely to completely replace lipophilic monomeric emulsifiers in multiple emulsions. However, proteins in combination with stabilizers do have the capacity to confer some enhanced degree of stability on a multiple emulsion system, and therefore the lipophilic emulsifier concentration is substantially reduced.

In recent studies the biopolymer chitosan was used as an emulsifier in food multiple emulsions (Genta et al., 1997; Schulz et al., 1998; Adachi et al., 1999). Chitosan has surface activity and seems to stabilize W/O/W emulsions. Chitosan reacts with anionic emulsifiers such as sodium dodecylsulfate at certain ratios to form water-insoluble complex that has strong emulsification capabilities. Chitosan solution was used to form multiple emulsions of O/W/O as intermediates from which by a simple procedure of striping the water the authors formed interesting porous spherical particles of chitosan (Adachi et al., 1999).

Cyclodextrins (alpha, beta, and gamma) were shown to be potential stabilizers for O/W/O emulsions (Yu et al., 1999, 2003). The advantages of the cyclodextrins are their ability to complex with certain oil components at the oil/water interface, resulting in no need for additional surfactant. It appears that the stabilizer efficacy depends on the nature of the oil and the type of the cyclodextrin (alpha > beta > gamma). The presence of any active matter in the inner phase (e.g., benzophenone) destabilized the emulsion. The reason is the interfacial interaction between the components present at the interface allows a possible competition between the lipophilic entrapped molecules and the oil fatty acid residues to enter the cyclodextrin cavity. These formulations prepared with α-cyclodextrin may serve to entrap high-molecular weight-active molecules, since they are not able to enter the small cavity of the α-cyclodextrin molecules and thus cannot destabilize the interface.

The authors of this review envisaged (Benichou et al., 2007a, b) WPI/polysaccharide conjugates to stabilize the outer interface of W/O/W and the inner interface of O/W/O multiple emulsions and found significant improvement both in the stability and in the release of markers (glucose, vitamin B_1, lipohilic veterinary drug) compared to the use of the protein only. These new amphiphilic adducts serve as good steric stabilizers, improve stability and shelf-life, and slow the release of the markers. They therefore play a double role in the emulsions: film formation and barrier to the release of small molecules at the internal interface, and steric stabilizers of the inner oil-water interface.

7.6 STABILIZATION BY BIOPOLYMERS CONJUGATES

Recently the authors of this review envisaged the stabilization of O/W/O multiple emulsions with protein-polysaccharide conjugates at the inner oil-water interface (Benichou et al., 2007b). The basic rationale for topical use of O/W/O multiple emulsions as a means for controlled delivery of drugs to the skin is that the drug dissolved in the inner oil phase (O_1) has to diffuse through several phases, including the intermediate aqueous phase and the external oil phase (O_2) prior to delivery and diffusion through the skin (stratum corneum, epidermis, dermis) (Laugel et al., 2000). $O_1/W/O_2$ multiple emulsions for topical uses also allows differentiation of the two oil phases, O_1 and O_2, and optimization of the drug release profile, owing to the different solubility of the drug in these oils.

Multiple emulsions of the oil-in-water-in-oil type, where both the inner and the external oil phases were medium chain triglyceride (MCT), were prepared by the two-step emulsification method. The emulsifying agent at the inner oil-water interface was a combination of WPI-xanthan gum hybrid at different ratios and 3 wt% Abil EM90 (a hydrophobic polyether-polysiloxane block copolymer) as the external emulsifier. Typical multiple emulsion composition is described in Table 7.2 and release profiles of flumethrin (a veterinary drug model) at 25°C are presented in Figure 7.15.

An attempt to better control the release profile of flumethrin into the external oil phase was achieved by increasing the hydrophobicity of the external oil phase, for instance, by replacing MCT with linear alkanes or silicone oils. O/W/O multiple emulsions were prepared with different external oil

TABLE 7.2 Typical oil-in-water-in-oil double-emulsion composition (wt%) where the inner O/W interface is stabilized with 5 wt% WPI/xanthan gum mixture

	Concentration (wt%)
O/W primary emulsion	
MCT	15
Flumethrin	15
WPI/xanthan gum	5
H_2O	65
O/W/O double emulsion	
O/W primary emulsion	20
MCT	76
Abil EM90	4

Source: Adapted from Benichou et al. (2007b).

Note: Overall flumethrin overall concentration is 3 wt%.

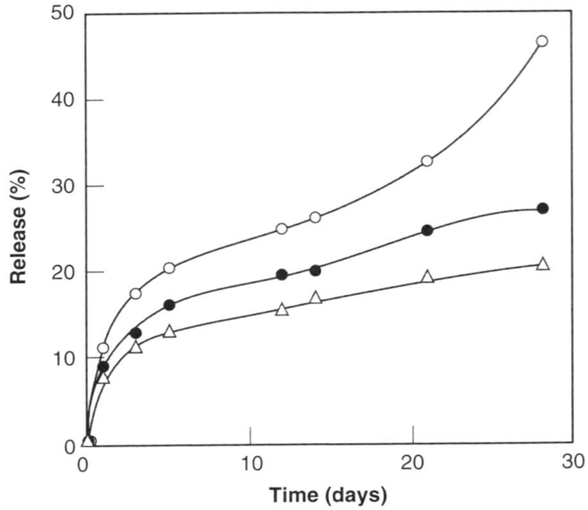

Figure 7.15 Release profiles (%) of flumethrin at 25°C from O/W/O multiple emulsions stabilized with 5 wt% WPI/xanthan gum: (○, 4/0); (●, 4/0.1); and (△, 4/0.3) as the inner emulsifier. Emulsion composition is described in Table 7.2, overall flumethrin concentration is 3 wt%. (Adapted from Benichou et al., 2007b.)

phases that contain 10 wt% 3225C commercial silicon surfactant. The flumethrin release profiles at 25°C are presented in Figure 7.16.

Emulsions prepared with linear alkanes, *n*-decane, *n*-dodecane, and *n*-tetradecane yielded stable multiple emulsions with a stability index for 28 days (ESI_{28}) in the 0.85 to 1.10 range at 25°C. Creaming (sedimentation) was quickly detected (2 hours) at all tested storage temperatures (4°C, 25°C, and 40°C). When the external oil phase (O_2) consisted of silicone oils, flumethrin releases were significantly slowed and the drug concentration in the external oil phase remained at very low levels (below 2000 ppm) after 28 days of storage at 25°C.

An attempt to reduce the inner oil phase solubility in the external oil phase was achieved by replacing it, for instance, MCT with a mixture of MCT-triacetin (3Ac) or MCT-glycerol monooleate (GMO). In these conditions it was possible to further reduce the release of the drug and the solubility of the inner oil phase into the external oil phase. Figures 7.17 and 7.18 describe the release profiles of flumethrin from multiple emulsions at 25°C where the inner oil phase (O_1) consists of a mixture of MCT and triacetin (3Ac) (Figure 7.17) or glycerol monooleate (GMO) (Figure 7.18) at different ratios.

Photomicrographs of multiple emulsion droplets after 28 days storage at 25°C are also presented (Figure 7.19). Triacetin addition in the inner oil phase (O_1) increases the hydrophilicity of the oil and can reduce the diffusion transport of O_1 to the external silicone oil phase (Laugel et al., 2000). GMO acts

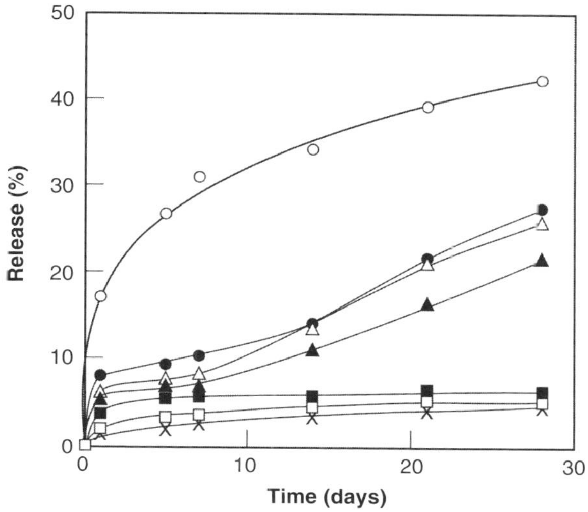

Figure 7.16 Release profiles (%) of flumethrin at 25°C from O/W/O multiple emulsions prepared with different external oil phases: MCT, ○; *n*-decane, ●; *n*-dodecane, △; *n*-tetradecane, ▲; Q7-9120 20 Cs, ■; Q7-9120 100 Cs, □; cyclomethicone 5-NF, ×. The external surfactant is 10 wt% 3225C surfactant. The primary emulsion consists of 30 wt% oil-in-water where the oil consists of flumethrin (50 wt%) dissolved in MCT and stabilized with 5 wt% WPI/xanthan gum (4/0.3.) The multiple emulsion contains 20 wt% primary emulsion so that the overall flumethrin concentration in the final multiple emulsion is 3 wt%. (Adapted from Benichou et al., 2007b.)

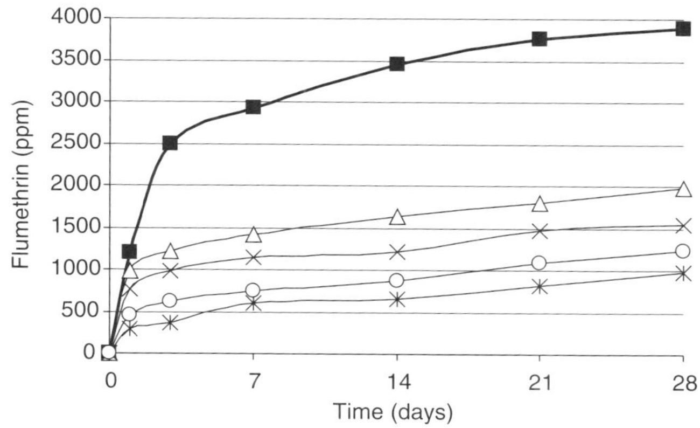

Figure 7.17 Flumethrin concentration (ppm) in the external oil phase of multiple emulsions stored at 25°C against time. The external phase is 10 wt% 3225C surfactant dissolved in Q7-9120 100 Cs silicone oil. The primary emulsion consists of 30 wt% oil phase and stabilized with 5 wt% WPI/xanthan gum (4/0.3.) The inner oil phase (O_1) is MCT, ■; MCT/3Ac, 9/1, △; MCT/3Ac, 7/3, ×; MCT/3Ac, 5/5, *; MCT/3Ac, 4/6, ○. Overall flumethrin concentration is 3 wt%. (Adapted from Benichou et al., 2007b.)

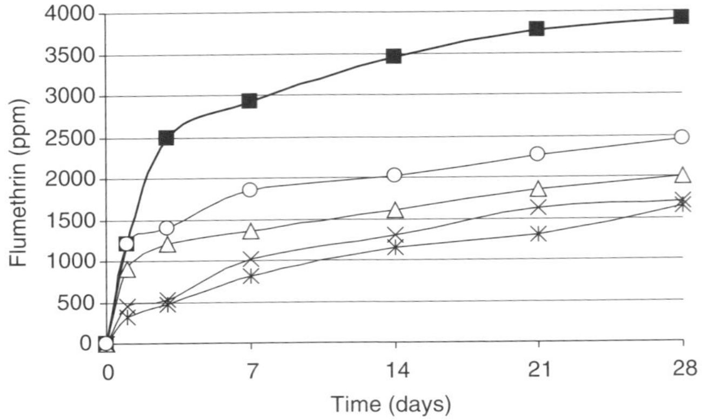

Figure 7.18 Flumethrin concentration (ppm) in the external oil phase of multiple emulsions stored at 25°C against time. The external phase is 10 wt% 3225C surfactant dissolved in Q7–9120 100 Cs silicone oil. The primary emulsion consists of 30 wt% oil phase and stabilized with 5 wt% WPI/xanthan gum (4/0.3.) The inner oil phase (O_1) is MCT, ■; 1 wt% GMO in MCT, △; 3 wt% GMO in MCT, ×; 5 wt% GMO in MCT, *; 10 wt% GMO in MCT, ○. Overall flumethrin concentration is 3 wt%. (Adapted from Benichou et al., 2007b.)

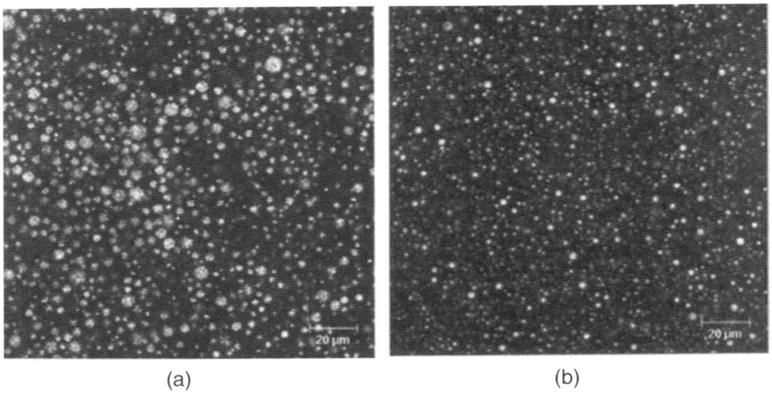

Figure 7.19 Confocal photomicrographs of multiple emulsions after 28 days of storage at 25°C. The external phase is 10 wt% 3225C surfactant dissolved in Q7–9120 100 Cs silicone oil. The primary emulsion consists of 30 wt% oil phase: (*left*) MCT/3Ac; 1/1, (*night*) 1 wt% GMO in MCT. The emulsion is stabilized with 5 wt% WPI/xanthan gum (4/0.3.) The overall flumethrin concentration is 3 wt%. The bar is 20 μm. (Adapted from Benichou et al., 2007b.)

as a co-emulsifier at the inner oil/water interface and improves the primary oil-in-water stability. The polymeric film around the oil inner droplets is then better sealed against the drug release.

These oil-in-water-in-oil multiple emulsions may find potential applications in sustained release of hydrophobic drugs for topical uses. The use of WPI-xanthan gum hybrid at the inner oil-water interface significantly improved the stability of the primary oil-in-water emulsion droplets, which resulted in a very high yield of drug entrapment (above 95%). Submicron oil-in-water primary emulsions were obtained by performing a high-pressure homogenization process during the first step of the multiple-emulsion preparation.

The absence of monomeric surfactant in the intermediate aqueous phase of the multiple emulsion prevented uncontrolled release of the drug through micellar diffusion controlled transport. Flumethrin release rates from the inner oil phase (O_1) imply that an interfacial barrier prolonged the release of the entrapped drug at a rate governed by its ability to partition into and to diffuse through the interfacial inner polymeric film as well as through the intermediate aqueous phase.

The differentiation of the two oil phases, O_1 and O_2, in O/W/O multiple emulsions will reduce the release of the entrapped active molecules while the solubility of the inner phase in the external oil phase remains constant. Drug solubility in the external oil phase did not constitute a limiting factor in the release process. Modulated release of triterpenic compounds from an O/W/O multiple emulsion formulated with dimethicones studied with infrared spectrophotometric and differential calorimetric approaches is one of these examples. Laugel et al. explored the advantages in the release of triterpenic compounds from O/W/O emulsions (Laugel et al., 2000). They found two principal advantages: the use of low molecular silicones decreased the oily touch of the final preparation, and this excipient influenced the skin's distribution of the active matter after the topical application, which is due to the large range of viscosity. The effects of different dimethicones incorporated within multiple emulsions were studied, through in vitro penetration results. The residual film on the skin was also evaluated. Correlations were established between the silicone structure and the distribution of drugs in different skin levels or between the silicone structure and the percutaneous penetration. The incorporation of silicones within O/W/O multiple emulsions seems to be an efficient means of modulating the penetration and the distribution of drugs in the skin.

In another study the stability of retinol (vitamin A alcohol) was compared in three different emulsions: oil-in-water (O/W), water-in-oil (W/O), and oil-in-water-in-oil (O/W/O) (Yoshida et al., 1999). The stability in the O/W/O emulsion was the highest among the three types of emulsions. The remaining percentages, at 50°C after four weeks, were of 56.9, 45.7, and 32.3, in the O/W/O, W/O, and O/W emulsions, respectively. However, it was also reported that with increasing peroxide value of O/W and W/O emulsifiers, the remaining percentage of vitamin A palmitate and retinol in the emulsions increased significantly, indicating that peroxides in the formulas accelerate the decomposition of vitamin A. Organophilic clay mineral tan oil gelling agent and a W/O emulsifier also affected the stability of retinol. The stability of retinol in the O/W/O emulsion increased with increasing inner oil phase ratio, whereas in O/W it was unaffected by the oil fraction. The encapsulation percentage of retinol in the O/W/O emulsion, the ratio of retinol in the inner oil phase to the total amount in the emulsion, increased with increasing the oil fraction. The remaining percentage of retinol in the O/W/O emulsion was in excellent agreement with encapsulation percentage, indicating that retinol in the inner oil phase is more stable than that in the outer oil phase. Addition of antioxidants (tert-butylhydroxytoluene, sodium ascorbate, and EDTA) to the O/W/O emulsion improved the stability of retinol up to 77.1% at 50°C after four weeks. Yoshida et al. concluded that the O/W/O emulsion is a useful formula to stabilize vitamin A.

Orange oil-in-water emulsions were encapsulated in another oil phase to form a multiple emulsion having orange oil inside its inner compartment (Edris and Bergenstähl, 2001). Although the yield was only 44.5%, it is a promising area for future research on preventing air oxidation of the oil. Spray drying of the multiple emulsion can provide a secondary coating and secure maximum protection of orange oil and this afford a free-flowing flavor powder. Spray drying of the orange oil multiple emulsion had very little destructing effect on its structure, as revealed by the light microscope. This method may have potential application in different food or pharmaceutical products where such maximum protection is required. A secondary coating was applied to flavor oil already encapsulated in a multiple emulsion. The spray-drying technique includes spraying a flavor emulsion into a stream of hot air. The water phase is then evaporated rapidly, leaving the flavor material locked-in the carrier.

O/W/O multiple emulsions have been in some cases freeze-dried for flavor encapsulation (Cho and Park, 2003). The effects of various process parameters on the emulsion stability and flavor retention have shown that high-pressure homogenization, at 68 MPa, produces more stable emulsions with small and uniform droplets. Incorporating gum Arabic in the intermediate aqueous phase created a highly viscous emulsion, which resulted in a stable O/W primary emulsion. A stable O/W/O multiple emulsion was obtained by using a blend of Span 80 and PGPR as an external emulsifier. After freeze-drying, well-formed microcapsules having high flavor retention (71%) were formed.

Ascorbic acid (vitamin C) has been incorporated in both W/O/W and O/W/O multiple-emulsions systems (Silva et al., 1997; Gallarate et al., 1999; Farahmand et al., 2006). In the first system, vitamin C was dissolved in the inner aqueous phase of a W/O/W multiple emulsion. In this case, improved protection against oxidation was reported when compared with O/W microemulsion, O/W and W/O emulsions, both at 45°C and 20°C. Incorporating vitamin C in the intermediate aqueous phase of an O/W/O multiple emulsion improved the occlusive properties, the acceptability and durability on skin, after topical application (Farahmand et al., 2006). The stability of these formulations was attributed to absence of ionization and subsequent oxidation of the external oil phase where the vitamin is insoluble (Gallarate et al., 1999). Furthermore oxygen was preferably dissolved in the oil phase than in the intermediate aqueous phase, and thus prevented auto-oxidation of the vitamin. Another factor that reduced oxidation of the vitamin was the presence of reverse micelles in the oil phase that could solubilize the vitamin molecules and add a stabilization effect.

7.7 STABILIZATION BY SOLID PARTICLES

Most work has been reported on emulsions with two-liquid phases plus a solid phase, which is slightly different from an O/W/O emulsion defined as a liquid system. These emulsions are sometimes referred to as Pickering emulsions. In these emulsions the solid particles adsorb at the oil-water interface and stabilize emulsions against coalescence. For example, some food emulsifiers such as monoglycerides undergo crystallization during their use (cooling), forming particles at the interface. The wetting conditions of the two liquids on the solid particles are the key factor in the stabilization mechanism. The type of emulsion produced depends on which phase preferentially wets the solid particles. Protection against coalescence is based on the wetting energy needed to displace the solid particles from the interface into the dispersed droplets. The energy ΔE to force a sphere to enter the most wetting phase is could be expressed by

$$\Delta E = \pi r^2 \gamma_{O/W} (1 - \cos\theta)^2,$$

where r is the spherical particle radius, is the contact angle between the interface and the wetting phase, and $\gamma_{O/W}$ is the interfacial tension between the oil and water phases. In O/W/O emulsions fat crystals are wetted by liquid oil (or plastic fat) and surround (adsorb onto) the O/W internal emulsion droplets.

During the stabilization process, particles (e.g., fat crystals) must "collect" at the emulsion droplet interface and provide a physical barrier to coalescence. The key factors that determine the influence of fat crystals on emulsion stabilization are (1) the wettability of the crystals at the interface, (2) interfacial film rheology, (3) particle microstructure (polymorphism and morphology), and (4) location of fat crystals (in the O/W dispersed emulsion or in the continuous W/O emulsion phase) (Johansson et al., 1995; Johansson and Bergenstähl, 1995; Rousseau, 2000).

The wetting behavior of particles at the interface is described by contact angles, which are related to the surface tension of each of the three interfaces by Young's equation

$$\gamma_{O/W} \cos\theta = \gamma_{O/S} - \gamma_{W/S},$$

where is the contact angle measured through the water phase and $\gamma_{O/W}$, $\gamma_{O/S}$, and $\gamma_{W/S}$ are the surface tensions of the oil-water, oil-solid and water-solid interfaces, respectively. $\gamma_{O/W}$ cos is also known as the adhesion tension. Modification of the contact angle (and therefore emulsion stability) can by achieved by a modification of the aqueous, oil or solid phase so as to alter $\gamma_{O/W}$, $\gamma_{O/S}$, or $\gamma_{W/S}$.

The concept of stabilizing emulsions by solid particles (mechanical stabilization) was described (Oza and Frank, 1986) for colloidal microcrystalline cellulose (CMCC) that is adsorbed in a solid form onto oil droplets at the interface of a W/O emulsion with improved stability. Khopade and Jain repeated use of a similar process and managed to stabilize W/O/W emulsions by using MCC (microcrystalline colloidal cellulose) particles at both interfaces (Khopade and Jain, 1998). The droplets were small, and the yield of the multiple emulsion was fairly good. The increasing concentration of MCC in either internal or external phase increased droplet sizes. These systems showed promise in tuberculosis therapy.

In another study (Jahaniaval et al., 2003) an O/W/O multiple emulsion system was prepared and stabilized using a novel method of mixing two oil-in-water (O/W) emulsions together. The first emulsion consisted liquid canola oil, water sodium caseinate, and lecithin and the second emulsion contained canola oil, palm-cotton stearin (50:50), lecithin, water, and sodium caseinate. The two emulsions were mixed and cooled at –5°C while mixing at low shear rates (2000–3000 rpm). Microstructural analysis showed that a stable O/W/O emulsion was formed with plastic fat as the continuous phase and the first O/W emulsion as the dispersed phase. O/W/O multiple emulsions were prepared by mixing two O/W emulsions, one consisting of liquid oil in water stabilized with sodium caseinate and the second being a mixture of liquid and solid fat (palm stearin:cotton stearin, 60:40) was also stabilized with sodium caseinate. The emulsions were prepared separately at 50°C and then mixed together at 45°C prior to supercooling. Phase transition during cooling yield the final the O/W/O multiple emulsion. Stable multiple emulsions with high inner oil phase content were obtained by using saturated triacyl alcohol as the liquid oil phase because of their capacity to be retained in the crystal structure of the fat during cooling.

Garti et al. tried micronized particles of the α and β′-polymorphs of tristearin fat together with polyglycerol-polyricinoleate (PGPR) as the internal emulsifiers in W/O/W multiple emulsions (Garti et al., 1999). Solid fat particles did not sufficiently stabilize the water-in-oil emulsion, and similarly the PGPR (at the concentrations used in the formulation) did not provide good stability. It was, however, shown that a blend of the two components composed of solid submicronal fat particles of α- and β′-polymorphs (which are more hydrophilic than the β-form and thus wet better the oil/water interface) precipitates onto the water droplets and covers them. The fat particles bridge between the water droplets and sinter them only if a lipophilic surfactant (PGPR) was co-adsorbed onto the water-oil interface (the W/O emulsion) (Figure 7.20). It was concluded that the fat particles adsorbed onto the hydrophobic emulsifier film

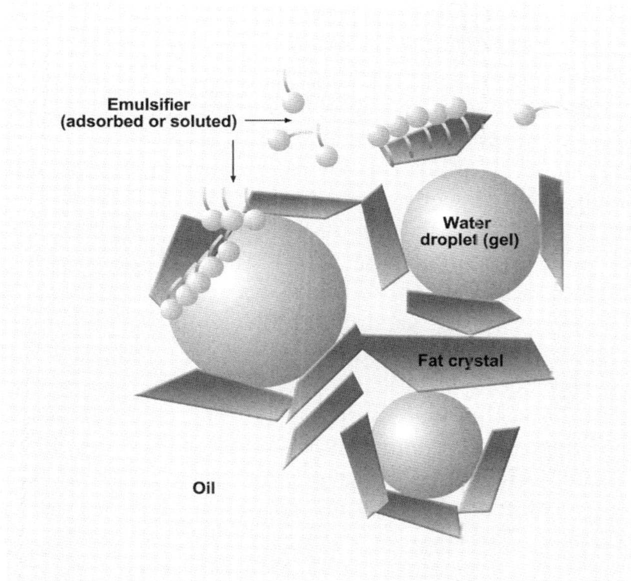

Figure 7.20 Schematic of the colloidal margarine structure showing the role of the emulsifier and fat crystals in stabilizing W/O emulsions droplets. (Adapted from Garti et al., 1999.)

and both, the solid particles and the emulsifier, wetted the water and spread at the interface.

Organophilic montmorillonite is an interesting clay that gained some interest in emulsion technology. Sekine et al. made stable O/W/O emulsions with components consisting of hydrophilic nonionic surfactant (hydrogenated ethoxylated castor oil, HCO-60), organophilic montmorillonite, and commercial nonionic surfactant (DIS-14) (Sekine et al., 1999). The montmorillonite was added in the second step at the outer W/O/W interface. The droplets sizes decreased with the increase of the HCO-60 (0.1–3 wt%) concentration. The viscosity of the multiple emulsion increased as the concentration of the montmorillonite and DIS-14 increased, demonstrating that the excess amount of inner oil phase is adsorbed by the outer oil phase. The results indicate that the weight fraction of the inner oil phase should not exceed 0.3 wt% for stable O/W/O emulsion since the viscosity of the multiple emulsion is so high that the formulation becomes semisolid.

7.8 STABILIZATION BY INCREASED VISCOSITY

Restricting the mobility of the active matter in the different compartments of the multiple emulsion will slow down coalescence and creaming, as well as decrease the transport rates of the drug or the marker from the inner phase through the intermediate liquid membrane to the external phase. Attempts were made (1) to increase the viscosity of the internal oil phase by adding gums/hydrocolloids to the inner water phase of W/O/W multiple emulsions or in the intermediate aqueous phase in O/W/O multiple emulsions, (2) to increase the viscosity of the oil phase (fatty acids salts), and (3) to thicken the gelling of the external water by gums for cosmetic or similar applications in which semisolid emulsions are directly applied (Altrock and Ritums, 1982; Gaonkar, 1994b; Terrisse et al., 1994; Okonogi, 1994; Muguet et al., 2001; Cho and Park, 2003; Tadros et al., 2006; Khan et al., 2006). The thickener was found to affect also the external continuous phase, since the entrapment is not quantitative and the yields of entrapment are limited and emulsifier-dependent. Some of the examples are topical skin care products, creams, and body lotions (Tables 7.3 and 7.4) (Susuki and Lim, 1994; Vaziri and Warburton, 1995).

Multiple emulsions that were solidified after preparation can suffer from destabilization effects. This phenomenon is scarcely considered, but in practice, it occurs very often. The solidification occurs because of temperature changes (temperatures can fluctuate from subzero of ca. −20°C to ca. 40°C) during transport or storage. Clausse et al. (Clausse, 1998; Clausse et al., 1999) studied the phenomena in W/O/W emulsions by microcalorimetric (DSC) techniques. It was concluded that out of thermodynamic equilibrium, multiple emulsions may suffer from water transfer during the solidification. This phenomenon occurs even if partial solidification takes place. In addition a change in the size distribution of emulsion droplets was observed. The mean diameter of the

TABLE 7.3 Example of W/O/W multiple emulsion stabilized with a calcium alginate gel layer in the intermediate aqueous phase

	Concentration (wt%)
Oil phase	
Hexaglycerol mixed ester	1
CSL (calcium stearyl-2-lactylate)	0.75
Soybean oil	23.25
Internal aqueous phase	
27% Solution sodium alginate low viscosity	25
Outer aqueous phase	
Polysorbate 20	0.36
Water	9.14
Vinegar	12
Sucrose	6
NaCl	2.5
Aqueous phase + 1% xanthan gum	20

Source: Adapted from Vaziri and Warburton (1995).

TABLE 7.4 Composition of low-fat spread using W/O/W multiple emulsion formulation

	Concentration (wt%)
Outer aqueous phase	
Salt	1
Gelatin	3
Maltodextrin DE = 7[a]	10
Water	85
Sodium caseinate	1
W/O emulsion	
Water	57.9
Rapeseed oil	40
PGPR	2
Flavor	0.1

Source: Adapted from Susuki and Lim (1994).
[a]DE: dextrose equivalent.

droplets in the W/O emulsion can shift toward O/W emulsion, and the multiple emulsion can invert. Therefore it is not always obvious that increasing viscosity, gelation, or partial solidification improves emulsions stability.

An attempt to improve the long-term stability of multiple emulsions and to prevent from the release of active molecules from the inner phase was achieved by incorporating various viscosying agents in the different phases. Processes for producing low fat O/W/O multiple emulsions were developed and patented (Altrock and Ritums, 1982; Gaonkar, 1994b; Okonogi, 1994). A fatty semisolid external oil phase was used to increase the viscosity of the multiple emulsion, and thus to increase stability and the desired plasticity of low-fat spreads. This way low-fat spreads with less than 65 wt% fat content could be produced and not display exudation upon storage.

Multiple emulsions with improved stability were also prepared by incorporating thickening agents in the different phases of these systems. W/O/W multiple emulsions capable of breaking and releasing their inner aqueous phase under shear rates compatible with agroalimentary, pharmaceutical, and cosmetic applications were formulated. Multiple emulsions were gelified with a synthetic polymer (Carbopol 974P®) (Terrisse et al., 1994). Others were thickened with chemically modified cellulose (hydroxypropylcellulose) (Terrisse et al., 1994). This was shown to impart specific behavior on fragmentation and release the profile of the entrapped addendum as a function of the shear rate.

A thickening coemulsifier can be added to the external water phase of W/O/W multiple emulsions. The optimized emulsifier-coemulsifier pair at the interface will lead to improved packing, elasticity, and temperature stability of the final multiple emulsion. Furthermore more controlled release kinetics of release can be achieved (Terrisse et al., 1994).

To improve the stability of W/O/W multiple emulsions containing arachis and olive oil, Vaziri et al. examined the stabilizing effect of cherry gum, in combination with acacia and gelatin (Vaziri and Warburton, 1995). The exceptional film-forming properties of this gum enhanced the stability of the multiple globules, as compared to controls. This result was attributed to formation of a liquid crystal bearing interfacial films around the external oil globules. The O/W external interface in this case was more compact and better sealed against release of the entrapped molecules, and thus these properties inhibited the transfer of phases and uncontrolled release. More thickening agents (acacia, tragacanth, sodium alginate, methyl cellulose, carboxymethylcellulose, and hydroxypropylmethylcellulose were tested and reported to improve both stability and sustained release of entrapped addendum from multiple emulsions (Khan et al., 2006).

7.9 RHEOLOGICAL CONSIDERATIONS

The understanding of the rheological behavior of multiple emulsions is important in the formulation, handling, mixing, processing, storage, and pipeline transformation of such systems. Furthermore rheological studies can provide useful information on the stability and internal microstructure of the multiple emulsions. Some attention was given to this subject in recent years, and the results help clarify certain aspects of the stability and release properties of multiple emulsions (Muguet et al., 1999).

The mechanical properties of the oil membrane in W/O/W emulsions were characterized by an aspiration technique (Geiger et al., 1999). The deformability was determined of an individual globule during total or partial flow into a cylindrical glass tube, which was calibrated under well-controlled conditions of aspiration. An analysis of the behavior of the multiple emulsion by a migration of the lipophilic surfactant to the interface between the oily and the external aqueous phases was done. It was shown that the elastic shear modulus and the interfacial tension of the oily membrane increased with the lipophilic surfactant concentration.

Grossiord et al. applied linear shear flow on the W/O/W multiple emulsions that contained an active matter, and from the rheological patterns they learned of the droplets' bursting with the release of entrapped substances as well as the composition of the system (Grossiord and Seiller 1998; Muguet et al., 1999; Geiger et al., 1999). The Grossiord and Seiller (1998) described set of two types of experiments: oscillatory dynamic tests and a steady-state analyses. They measured the stress and strain of the emulsions by applying sinusoidal shear. These parameters (shear or complex modulus G^*, the lag phase between stress and strain δ, the storage modulus G', and the loss modulus G'') provide a quantitative characterization of the balance between the viscous and elastic properties of multiple emulsions. At lag phase $\delta = 0°$ and when the lag phase equals $90°$, the system is viscoelastic. The shear sweep and the temperature sweep charac-

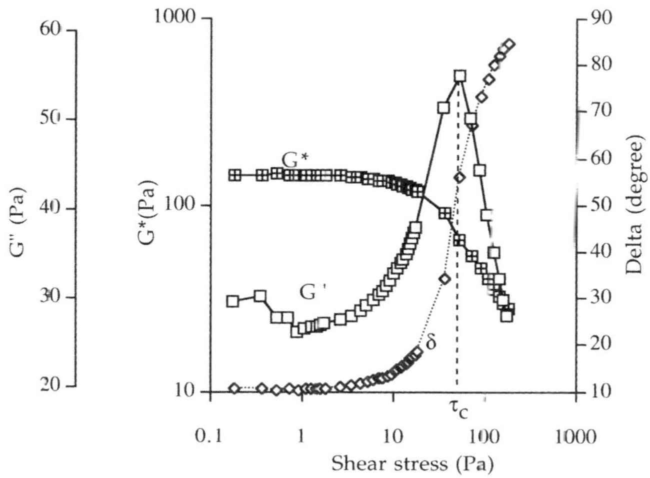

Figure 7.21 Change in G^*, G'', and d for increasing stress at fixed frequency in a typical W/O/W multiple emulsion. (Adapted from Grossiord and Seiller, 1998.)

terize the multiple emulsion at rest. Figure 7.21 describes a transition between elastic and viscous behavior, which occurs at a critical stress values. The change in these parameters indicates a pronounced structural breakdown.

Stroeve and Varanasi (1984) examined the breakup of multiple-emulsion globules in a simple shear flow and reported that the multiple emulsion exhibits behavior that is similar to that of simple emulsions. The studies showed also that the mechanisms taking place during the breakup were complex and did not always lead to a total release of the entrapped electrolyte. Some phenomena such as a partial leakage of the internal aqueous compartment or the expulsion of the aqueous microglobules covered by a residual lipophilic film were able to restrict the release.

De Cindio et al. prepared food multiple emulsions and studied their rheological behavior by steady shear and oscillatory measurements (De Cindio et al., 1991; De Cindio and Cacace, 1995). They concluded that the W/O/W appeared to have similar rheological properties to those of a simple O/W emulsions with the same fraction of dispersed phase but lower oil content. It was also demonstrated that the plots of both storage moduli G' and G'' versus oscillation frequency W are similar in all eight prepared emulsions, with the loss tangent being about 1, and that both elastic and viscous contributions to viscoelastic behavior of multiple emulsions are of similar magnitude. The influence of mixture of emulsifiers on the multiple emulsion stability was studied by an oscillatory ring-surface rheometer from which the interfacial elasticity at the oil-aqueous interface can be evaluated.

In our lab we have recently described the rheological behavior of W/O/W multiple emulsions droplets stabilized with protein-polysaccharides hybrids

(Benichou et al., 2007a). The phase angle δ was defined as arctan(G''/G'), where G' is the storage modulus, G'' is the loss modulus, and tan(δ) = G''/G'. It was found that at low levels of polysaccharide in the biopolymer complex (0.1 wt%), the emulsions has high viscosity, with phase angle δ close to 90°, indicating self-assembly of the two biopolymers onto the external oil-water interface. The emulsions were claimed to be stabilized mainly by steric interactions between the hybrids adsorbed onto the oil. Alternatively, at a high level of gum (0.3 to 1 wt%), the emulsions exhibit more elasticity, which is regarded as a physical property derived from the depletion stabilization mechanism. The protein was preferably adsorbed onto the oil/water external interface and the uncomplexed gum rather migrated to the bulk and contributed to the emulsion stabilization by a depletion mechanism (Figure 7.22). It was also determined that a protein-to-gum ratio of 4:0.5, at which complexation between the protein and the gum takes place, corresponds to intermediary viscoelasticity of the system. Microscopic observations and zeta potential measurements performed on the samples after rheological studies revealed that the emulsion droplets are still multiple droplets and that interactions between the biopolymeric molecules remained unchanged.

At a gum concentration of 0.5 wt% the protein concentration does not affect the rheological behavior of the multiple emulsion that conserves its elasticity properties at all ratios with phase angle (δ) values around 25° at all proteins contents. At the same time the yield of preparation of multiple emul-

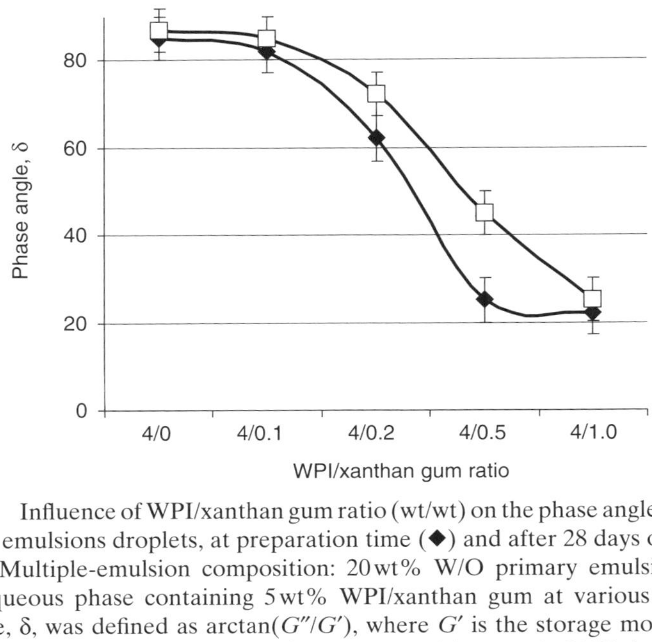

Figure 7.22 Influence of WPI/xanthan gum ratio (wt/wt) on the phase angle δ (degrees) of multiple emulsions droplets, at preparation time (◆) and after 28 days of storage at 25°C (□). Multiple-emulsion composition: 20 wt% W/O primary emulsion, 80 wt% external aqueous phase containing 5 wt% WPI/xanthan gum at various ratios. The phase angle, δ, was defined as arctan(G''/G'), where G' is the storage modulus, G'' is the loss modulus, and tan(δ) = G''/G'. (Adapted Benichou et al., 2007a.)

sions prepared with increasing amounts of protein and at 0.5 wt% xanthan gum from 64% for multiple emulsions stabilized with 5 wt% WPI to 96% with addition of 0.5 wt% xanthan gum.

Pal studied the rheology of O/W/O multiple emulsions (Pal, 1996). The simple O/W emulsions were found to be Newtonian up to a dispersed phase concentration of 45% by volume and non-Newtonian above this volume fraction. All the double O/W/O emulsions are highly non-Newtonian. The degree of shear thinning was shown to increase with the increase in primary O/W emulsion concentration. The oscillatory measurements indicate that the multiple emulsions are predominantly viscous in that the loss modulus falls above the storage modulus over the entire frequency range investigated. Upon aging, the storage and loss moduli of the multiple emulsions show significant increase. However, the increase in viscosity with aging is only marginal.

The rheological behavior of W/O/W emulsion studied under a conic plate viscometer has shown a negative thixotropic flow pattern, mostly under low shear rate (Kawashima et al., 1991). As the shear rate or the shear time was raised, an increase the shear stress was observed, which induced phase inversion to a W/O of a semisolid-type emulsion. The hydrodynamic parameters (dissipated energy, kinetic energy, and impulse applied to the emulsion by the rotating cone) causing the phase inversion were determined, and a mechanism for such inversion was suggested. Induced shear, causing phase inversion, should be considered in future applications of these systems as possible way to release drugs.

7.10 POTENTIAL APPLICATIONS OF O/W/O MULTIPLE EMULSIONS

Replacing one or more of the liquid phases in multiple emulsions with solid or semisolid particles will dramatically decrease release rates of the entrapped active molecules. Such multiple emulsions could be stored, before use, for prolonged periods of time without transporting the active matter to the outer interface. Upon use, the multiple emulsion will be heated or sheared, and the solid internal matrix will be ruptured so that the active matter is released. The major problem in practicing such technology is the difficulties arising in dispersing (and keeping stable) the micro- or nanoparticles in the continuous water phase. Microspheres and nanoparticles using solid encapsulation techniques were tested to replace the primary dispersed emulsion. Some of the experiments carried out showed that the release can be slowed down with these systems (Blanco-Prieto et al., 1994, 1996, 1997; Schugens et al., 1994; Nihant et al., 1995; Couvreur et al., 1997; Maa and Hsu, 1997; Leo et al., 1998; Zambaux et al., 1998). Multiple emulsions as intermediates for the preparation of solid microspheres or microcapsules, O/W/O multiple emulsions for improved solubilization and chemical protection of water-insoluble active matter were formulated. Multiple emulsions for selective adsorption of certain compounds for extraction and purification purposes were also described.

Drugs, cosmetic ingredients, and food additives are microencapsulated for variety of reasons, which include reducing local side effects, controlled release, site-specific (drug) delivery, and drug targeting. A tremendous amount of research work is done in a search of suitable methods to achieve a good encapsulation of water-soluble active matter. The physical characteristics of the microspheres produced largely determine their suitability for use for different objectives. Microspheres are prepared from both natural and synthetic polymers.

Among microencapsulation techniques, the multiple emulsion–solvent–evaporation method is one of the most useful methods for entrapping water-soluble compounds (Blanco-Prieto et al., 1994, 1996, 1997; Schugens et al., 1994; Nihant et al., 1995; Couvreur et al., 1997; Maa and Hsu, 1997; Leo et al., 1998; Zambaux et al., 1998). Figure 7.23 shows schematically the preparation technique. It is known that the preparation of microspheres using O/W emulsions is not an efficient preparation for the entrapment of water-soluble drugs because the compound rapidly dissolves into the aqueous continuous phase and is lost. It has been widely accepted that the problem of inefficient encapsulation of water-soluble drugs can be overcome by using the multiple emulsion solvent-evaporation technique. But water-insoluble drugs are usually satisfactorily encapsulated by O/W emulsion technique (Blanco-Prieto et al., 1997).

The W/O/W emulsions are generally used for encapsulating proteins or peptides. These highly water-soluble molecules are quantitatively introduced in the internal aqueous phase of the multiple emulsions and result in microcapsules with increased loading in comparison to particles produced by single

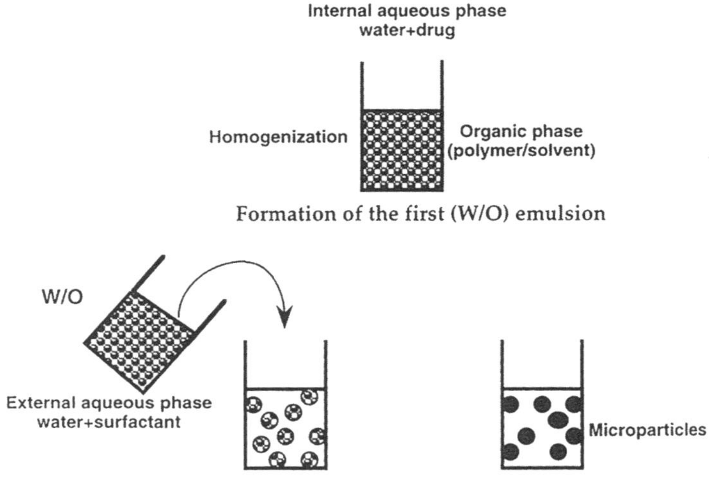

Figure 7.23 Schematic of microencapsulation technique by the multiple-emulsion solvent evaporation method.

emulsion-solvent evaporation method. The particular location of the proteins induces a stabilizing effect on the two emulsions, which in turn contributes to a successful stabilization of the multiple emulsion and loading.

The double-emulsion–solvent-evaporation technique is commonly used to prepare biodegradable hydrophobic microspheres containing hydrophilic pharmaceuticals, proteins, and polypeptides for sustained release applications (Schugens et al., 1994; Blanco-Prieto et al., 1997; Couvreur et al., 1997; Maa and Hsu, 1997; Leo et al., 1998; Zambaux et al., 1998). In most cases the microspheres are in the range size of 10 to 100 μm. However, Blanco-Prieto et al. managed to reduce the microcapsules sizes to less than 5 μm (Blanco-Prieto et al., 1997).

Couvreur et al. reviewed the preparation and characterization of many of the different types of the solvent-evaporation microspheres and mostly discuss small poly(lactic-co-glycolic acid) microspheres (mean size lower than 10 μm) containing small peptides (Couvreur et al., 1997). Three main evaporation strategies have been utilized in order to increase the encapsulation capacity: an interrupted process, a continuous process, and a rotary evaporation procedure.

Much work was devoted in recent years to prepare microparticles of narrow size distribution with different biodegradable polymers. Sizes of common microcapsules are 40 to 50 μm (Blanco-Prieto et al., 1997; Couvreur et al., 1997; Leo et al., 1998; Zambaux et al., 1998).

Liquid-to-liquid emulsification is a critical step in the multiple emulsion microencapsulation process (W/O/W or O/W/O). It was found that the size of these droplets decreases with increasing homogenization intensity and duration. The emulsion droplet size depends, as expected, on viscosity, total volume size, and the volume ratio of the continuous phase to the dispersed phase in the rotor/stator design being investigated. All these physical parameters influence the structure of the microspheres obtained by this technique.

Chitosan porous spherical particles were prepared from O/W/O multiple emulsions stabilized with a chitosan aqueous solution. The particulation was obtained by a simple evaporation technique (Adachi et al., 1999).

In the case of O/W/O multiple emulsions, the internal oil phase is used as the targeted material for microencapsulation (Hwang et al., 2005). Emulsion polymerization occurs in the water phase to prepare the encapsulating polymeric particles' substance in the internal oil phase. The encapsulating inorganic particles' substance can also be fabricated by the sol–gel reaction, hydrolysis, and the condensation reaction in the aqueous intermediate phase of the O/W/O emulsion (Yang et al., 2001, 2003). However, since encapsulating polymers are often decomposed under acidic conditions, the targeted materials are exposed to the external medium and become quickly unstable. Therefore in the recent literature inorganic particles like silica are being used as vehicles and carriers for functional materials such as drugs (Kortesuo et al., 1999; Barbe et al., 2004), peptides (Tourne-Peteilh et al., 2003), and vitamins (Murphy et al., 1992).

Oh et al. has developed a new method of producing microparticles by combining sol–gel technology with an oil-in-water-in-oil (O/W/O) multiple emulsion (Hwang et al., 2005). Microspheres 15 to 40 μm in diameter were prepared with very dense surfaces. These microspheres efficiently slowed down the release of retinol (vitamin A) into the external ethanol phase permitting higher loading of the vitamin. The method holds promise for application in the controlled-release of encapsulated active molecules since (1) silica particles of various sizes are controlled by the their original aqueous globule sizes, which serve as microreactors, and (2) the internal microstructure of silica particles can be readily adapted via sol–gel chemistry technology.

Spherical silica particles containing retinol have been fabricated using O/W/O multiple emulsion and the sol–gel method (Lee et al., 2001). O/W/O multiple emulsions were stabilized with hydroxypropyl cellulose (HPC) and surfactants such as Tween 20 and Span 80. In addition a polymeric stabilizer present in the intermediate aqueous phase was shown to improve the encapsulation efficiency. In the presence of polyvinyl alcohol the yield of encapsulation efficiency of retinol was 7%. With Pluronic P123 (a block copolymer of ethylene oxide propylene oxide) a yield of encapsulation of 31% could be reached. Figure 7.24 shows the retinol released profile from silica particles prepared in the O/W/O multiple emulsions and stabilized with different surfactant and different polymeric stabilizers in the intermediate aqueous phase.

The sol–gel reaction during the formation of silica particles in the multiple emulsion system started in the external oil phase containing the precursor alkoxide type (tetraethyl orthosilicate, TEOS), as shown in Figure 7.25. Under stirring, the TEOS molecules can penetrate the surfactant layer surrounding the aqueous phase, and then hydrolysis can start. As hydrolysis proceeds, the Si–OH based molecules diffuse and dissolve in the aqueous phase. A gel network is formed by condensation, yielding the insoluble hydrated silica encapsulating the retinol molecules. The water content in the multiple emulsion was demonstrated to impart the final shape and size distribution of the particles.

In another recent paper (Sawae et al., 2005), polyethylene glycol microspheres were prepared using an oil-in-water-in-oil double-emulsification method to encapsulate lipase complexes. The PEG microspheres exhibited heat-resistant properties that are advantageous when compared with the lipase complex itself dissolved in organic solvent. It was found that PEG microspheres retained high enantioselectivity in the esterification of phenyl ethyl alcohol and lauric acid in isooctane. The immobilized lipase complexes in the PEG microspheres can be easily recovered, and their activity remained preserved up to 10 reuses.

Hollow Cu_2O submicron spheres have been successfully prepared in an O/W/O multiple emulsion, employing the microwave irradiation technique (Liu et al., 2004). Heptane and lauryl alcohol polyoxyethylene were used as the oil

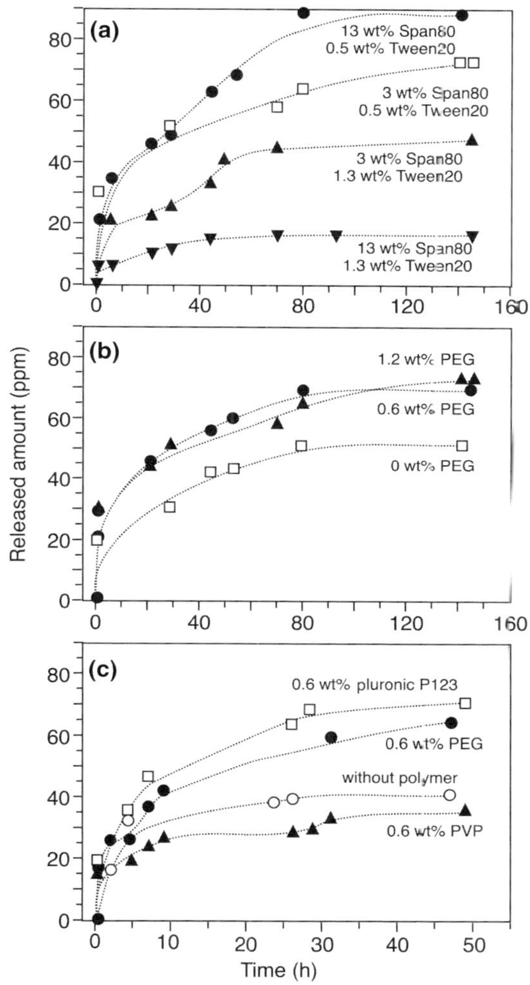

Figure 7.24 Release profiles of retinol from silica particles prepared in O/W/O multiple emulsion. (*a*) effect of surfactant concentration; (*b*) effect of PEG concentration; (*c*) effect of polymer. (Adapted from Hwang et al., 2005.)

phase and emulsifier, respectively. This multiple emulsion system plays a space-limiting role and prevents the hollow spheres from coalescing. Submicron Cu_2O hollow spheres composed of small Cu_2O nanoparticles of 22 nm in size were prepared by a multiple emulsion (O/W/O) method. The liquid droplet of the multiple emulsions serves as a template and has a space-limiting role besides preventing the reaction product from aggregation.

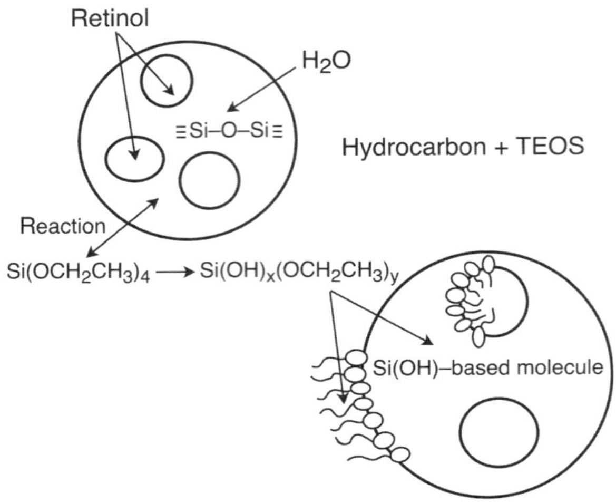

Figure 7.25 Schematic description of the reaction mechanism for the encapsulation of retinol (vitamin A) in silica particles by the O/W/O emulsification method. (Adapted from Lee et al., 2001.)

7.11 CONCLUDING REMARKS

Multiple emulsions have been widely studied over the last 20 years. Most reports deal with W/O/W multiple-emulsions types, but some new applications have involved O/W/O multiple emulsions. The internal phases (aqueous or oily) are an excellent reservoir for active molecules that needs protection and can be released in a controlled manner. However, the sizes of the droplets and the intrinsic thermodynamic instability of these systems are a significant drawback in applying this technology.

Use of the conventional low-molecular-weight emulsifier did not solve these problems. Nevertheless, much progress was made with the introduction of amphiphilic macromolecules as emulsifiers. These multianchoring flexible macromolecules were found to improve the steric stabilization of the multiple globules because they form thick multilayered coating on the droplets.

A variety of hybrids, complexes, and adducts have been studied as cosolvents between the surfactant and co-emulsifiers. These molecules improved significantly the stability and slowed the release rates. Physical methods of separation, filtration, and extraction also have been proved to have a positive effect on the release patterns of drugs or active entrapped molecules.

New techniques were developed to prepare, characterize, and study the release kinetics in these complex systems. Progress was made in characterization of the parameters and mechanisms that are involved in the coalescence, aggregation, and rupture of the multiple emulsion droplets, and a good control of the rheological parameters was achieved by better understanding of their effect on the static and shear-induced stability.

During the last decade O/W/O multiple emulsions have been researched as a potential prolonged release vehicle for oily soluble entrapped addenda. These systems also hold promise for infinite applications in micro- and nanoencapsulation of organic and inorganic active molecules. New forms are exhibiting the long-term stability necessary for release control.

Multiple-emulsions technology is already starting to be widely applied in foods, cosmetics, and pharmaceuticals.

REFERENCES

Adachi T, Ida J, Wakita M. 1999. Preparation of spherical and porous chitosan particles by suspension evaporation with O/W/O multiple emulsions. *Polym J* 31: 319–323.

Altrock WL, Ritums JA. 1982. Process for producing a low fat oil-in-water-in-oil emulsion. US patent 4,366,180.

Bams GW, Van Megen WH. 1987. Edible water-in-oil-in-water emulsion. US patent 4,650,690.

Barbe C, Bartlett J, Kong L, Finnie K, Lin HQ, Larkin M, Calleja S, Bush A, Calleja G. 2004. Silica particles: A novel drug delivery system. *Adv Mater* 16: 1959–1966.

Benichou A, Aserin A, Garti N. 2007a. W/O/W double emulsions stabilized with WPI-polysaccharide complexes. *Colloids Surf A* 294: 20–32.

Benichou A, Aserin A, Garti N. 2007b. O/W/O double emulsions stabilized with WPI-polysaccharide conjugates. *Colloids Surfaces A* 297: 211–220.

Blanco-Prieto MJ, Delie F, Fattal A, Tartar A, Puisieux F, Gulik A, Couvreur P. 1994. Characterization of V3 BRU peptide loaded small PLGA microspheres prepared by a $W_1/O/W_2$ emulsion solvent evaporation method. *Int J Pharm* 111: 137–145.

Blanco-Prieto MJ, Fattal E, Gulik A, Dedieu B, Couvreur P. 1997. Characterization and morphological analysis of a cholecystokinin derivative peptide-loaded poly(lactide-co-glycolide)microspheres prepared by a W/O/W emulsion solvent evaporation method. *J Controlled Release* 43: 81–87.

Blanco-Prieto MJ, Leo E, Delie F, Gulik A, Couvreur P, Fattal E. 1996. Study of the influence of several stabilizing agents on the entrapment and in vitro release of PBC from poly(lactide-co-glycolide) microspheres prepared by a W/O/W solvent evaporation method. *Pharm Res* 13: 1137–1139.

Castro D, Moreno MA, Lastres JL. 2001. First-derivative spectrophotometric and LC determination of nifedipine in Brij 96 based oil/water/oil multiple microemulsions on stability studies. *J Pharm Biomed An* 26: 563–572.

Cho YH, Park J. 2003. Evaluation of process parameters in the O/W/O multiple emulsion method for flavor encapsulation. *J Food Sci* 68: 534–538.

Clausse D. 1998. Thermal behavior of emulsions studied by differential scanning calorimetry. *J Therm Anal Calorim* 51: 191–201.

Clausse D, Pezron I, Komunjer L. 1999. Stability of W/O and W/O/W emulsions as a result of partial solidification. *Colloids Surf A* 152: 23–29.

Cornec M, Wilde PJ, Gunning PA, Mackie AR, Husband FA, Parker ML, Clark DC. 1998. Emulsion stability as affected by competitive adsorption between an oil-soluble emulsifier and milk proteins at the interface. *J Food Sci* 63: 39–43.

Couvreur P, Blanco-Prieto MJ, Puisieux F, Roques B, Fattal E. 1997. Multiple emulsions technology for the design of microspheres containing peptides and oligopeptides. *Adv Drug Deliv Rev* 28: 85–96.

De Cindio B, Cacace D. 1995. Formulation and rheological characterization of reduced-calorie food emulsions. *Int J Food Sci Technol* 30: 505–514.

De Cindio B, Grasso G, Cacace D. 1991. Water-in-oil-in-water emulsions for food applications. *Food Hydrocolloids* 4: 339–353.

Dickinson E, Evison J, Owusu RK. 1991. Preparation of fine protein-stabilized water-in-oil-in-water emulsions. *Food Hydrocolloids* 5: 481–485.

Dickinson E, Evison J, Owusu RK, Williams A. 1994. Protein stabilized water-in-oil-water emulsions. In: *Gums and Stabilisers for the Food Industry*, Vol. 7, Phillips GO, Williams PA, Wedlock DJ, eds. Oxford: IRL Oxford University Press, pp 91–101.

Edris A, Bergenstähl B. 2001. Encapsulation of orange oil in a spray dried double emulsion. *Nahrung Food* 45: 133–137.

Evison J, Dickinson E, Owusu Apenten RK, Williams A. 1995. Formulation and properties of protein-stabilized W/O/W multiple emulsion. In: *Food Macromolecules and Colloids*, Dickinson E, Lorient D, eds. Cambridge: Royal Society of Chemistry, pp 235–243.

Farahmand S, Tajerzadeh H, Farboud ES. 2006. Formulation and evaluation of a vitamin C multiple emulsion. *Pharm Dev Technol* 11: 255–261.

Ficheux MF, Bonakdar L, Leal-Calderon F, Bibette J. 1998. Some stability criteria for double emulsions. *Langmuir* 14: 2702–2706.

Fredrokumbaradzi E, Simov A. 1992. Effect of bovine serum albumim (BSA) on release of sulfacetamide sodium from multiple W/O/W emulsions. *Pharmazie* 47: 388–389.

Gallarate M, Carlotti ME, Trotta M, Bovo S. 1999. On the stability of ascorbic acid in emulsified systems for topical and cosmetic use. *Int J Pharm* 188: 233–241.

Gaonkar AG. 1994a. Method for preparing a multiple emulsion. US patent 5,322,704.

Gaonkar AG. 1994b. Stable multiple emulsion comprising interfacial gelatinous layer, flavor-encapsulating multiple emulsions and low/no-fat food products comprising the same. US patent 5,332,595.

Garti N. 1997a. Double emulsions—scope, limitations and new achievements. *Colloids Surf A* 123: 233–246.

Garti N. 1997b. Progress in stabilization and transport phenomena of double emulsions in food applications. *Food Sci Technol* 30: 222–235.

Garti N. 1998. Influence of the formulation on the characteristics and stability of multiple emulsions. In: *Multiple Emulsions: Structure, Properties and Applications*, Seiller M, Grossiord JL, eds. Paris: Editions de Santé, pp 81–116.

Garti N, Aserin A. 1996. Double emulsions stabilized by macromolecular surfactants. In: Chattopadhyay AK, Mittal KL, eds. *Surfactants in Solution*. Surf Sci Ser, Vol. 64. New York: Marcel Dekker, pp 297–332.

Garti N, Aserin A, Cohen Y. 1994. Mechanistic considerations on the release of electrolytes from multiples emulsions stabilized by BSA and nonionic surfactants. *J Controlled Release* 29: 41–51.

Garti N, Aserin A, Tiunova I. 1999. Double emulsions of water-in-oil-in-water stabilized by alpha-form fat microcrystals. Part I: Selection of emulsifiers and fat microcrystalline particles. *J Am Oil Chem Soc* 76: 383–389.

Garti N, Benichou A. 2001. Double emulsions for controlled-release applications-Progress and trends. In: *Encyclopedic Handbook of Emulsion Technology*, Sjöblom J, eds. New York: Marcel Dekker, pp 377–407.

Garti N, Benichou A. 2003. Recent Developments in Double Emulsions for Food Applications. In: *Food Emulsions*, 4th ed., Friberg SE, Larsson K, Sjöblom J, eds. Food sci technol ser, Vol. 132. New York: Marcel Dekker, pp 281–340.

Garti N, Bisperink C. 1998. Double emulsions: Progress and applications. *Curr Opin Colloid Interface Sci* 3: 657–667.

Geiger S, Jager-Lezer N, Tokgoz S, Seiller M, Grossiord JL. 1999. Characterization of the mechanical properties of a water/oil/water multiple emulsion oily membrane by a micropipette aspiration technique. *Colloids Surf A* 157: 325–332.

Genta I, Perugini P, Conti B, Pavanetto F. 1997. A multiple emulsion method to entrap a lipophilic compound into chitosan microspheres. *Int J Pharm* 152: 237–246.

Grossiord JL, Seiller M. 1998. Rheology of W/O/W multiple emulsions: Formulation, characterization and breakup mechanisms. In: *Multiple Emulsions: Structure, Properties and Applications*, Seiller M, Grossiord JL, eds. Paris: Editions de Santé, pp 169–192.

Grossiord JL, Seiller M, Pezron L, Clausse D. 1996. Multiple emulsions: Vesicular systems with industrial applications. *Oleagineux Corps Gras Lipides* 3: 158–162.

Grossiord JL, Seiller M, Silva-Cunha A. 1998. Obtaining multiple emulsions. In: *Multiple Emulsions: Structure, Properties and Applications*, Seiller M, Grossiord JL, eds. Paris: Editions de Santé, pp 57–80.

Herb CA. 1996. Rinse-off water-in-oil-in-water compositions. European patent application. 717,978A2.

Higashi S, Shimizu M, Nakashima T, Iwata K, Uchiyama F, Tateno S, Tamura S, Setoguchi T. 1995. Arterial-injection chemotherapy for hepatocellular carcinoma using monodispersed poppy-seed oil microdroplets containing fine aqueous vesicles of epirubicin. *Cancer* 75: 1245–1254.

Hino T, Kwashima Y, Shimabayashi S. 2000. Basic study for stabilization of W/O/W emulsion and its application to transcatheter arterial embolization therapy. *Adv Drug Deliv Rev* 45: 27–45.

Hwang YJ, Oh C, Oh SG. 2005. Controlled release of retinol from silica particles prepared in O/W/O emulsions: The effect of surfactants and polymers. *J Controlled Release* 106: 339–349.

Jahaniaval F, Kakuda Y, Abraham V. 2003. Characterization of a double emulsion system (oil-in-water-in-oil emulsion) with low solid fats: Microstructure. *J Am Oil Chem Soc* 80: 25–31.

Johansson D, Bergenståhl B. 1995. Wetting of fat crystals by triglycerides oil and water. II. Adhesion to the oil/water interface. *J Am Oil Chem Soc* 72: 933–938.

Johansson D, Bergenståhl B, Lundgren E. 1995. Wetting of fat crystals by triglycerides oil and water. I. The effect of additives. *J Am Oil Chem Soc* 72: 921–932.

Kawashima Y, Hino T, Takeushi H, Niwa T, Horibe K. 1991. Rheological study of W/O/W emulsion by a cone-and-plate viscometer: Negative thixotropy and shear-induced phase inversion. *Int J Pharm* 72: 65–77.

Khan AY, Talegaonkar S, Iqbal Z, Jalees F, Khar RK. 2006. Multiple emulsions: An overview. *Curr Drug Deliv* 3: 429–443.

Khopade AJ, Jain NK. 1998. A stable multiple emulsion system bearing isoniazid: Preparation and characterization. *Drug Dev Ind Pharm* 24: 289–293.

Kim SY, Lee YM. 1999. Lipid nanospheres containing vitamin A or vitamin E: Evaluation of their stabilities and in vitro skin permeability. *J Ind Eng Chem* 5: 306–313.

Kortesuo P, Ahola M, Karlsson S, Kangasniemi I, Kiesvaara J, Yli-Urpo A. 1999. Sol-gel-processed sintered silica xerogel as a carrier in controlled drug delivery. *J Biomed Mater Res* 44: 162–167.

Kundu SC, Kameron AD, Meltzer NM, Quick TW. 1993. Development and validation of method for determination of in vitro release of retinoic acid from creams. *Drug Dev Ind Pharm* 19: 425–438.

Laugel C, Baillet A, Ferrier D. 1996. Similar qualitative and quantitative O/W/O and W/O emulsions: Improvement of cosmetic properties. *STP Pharm Sci* 6: 376–380.

Laugel C, Baillet A, Youenang Piemi MP, Marty JP, Ferrier D. 1998. Oil-water-oil multiple emulsions for prolonged delivery of hydrocortisone after topical application: Comparison with simple emulsions. *Int J Pharm* 160: 109–117.

Laugel C, Rafidison P, Potard G, Aguadisch L, Baillet A. 2000. Modulated release of triterpenic compounds from a O/W/O multiple emulsion formulated with dimethicones: Infrared spectrophotometric and differential calorimetric approaches. *J Controlled Release* 63: 7–17.

Lee JM, Lim KH, Smith DH. 2002. Formation of two-phase multiple emulsions by inclusion of continuous phase into dispersed phase. *Langmuir* 18: 7334–7340.

Lee MH, Oh SG, Moon SK, Bae SY. 2001. Preparation of silica particles encapsulating retinol using O/W/O multiple emulsions. *J Colloid Interface Sci* 240: 83–89.

Leo E, Pecuet S, Rojas J, Couvreur P. 1998. Changing the pH of the external aqueous phase may modulate protein entrapment and delivery from poly(lactide-co-glycolide) microspheres prepared by a W/O/W evaporation method. *J Microencapsul* 15: 421–430.

Lin CY, Chen LW. 2006. Emulsification characteristics of three- and two-phase emulsions prepared by the ultrasonic emulsification method. *Fuel Process Technol* 87: 309–317.

Liu H, Ni Y, Wang F, Yin G, Hong J, Ma G, Xu Z. 2004. Fabrication of submicron Cu_2O hollow spheres in an O/W/O multiple emulsions. *Colloids Surf A* 235: 79–82.

Maa YF, Hsu CC. 1997. Effect of primary emulsions on microsphere size and protein-loading in the double emulsion process. *J Microencapsul* 14: 225–241.

Matsumuto S, Kang WW. 1989. Formation and applications of multiple emulsions. *J Dispersion Sci Technol* 10: 455–482.

Muguet V, Seiller M, Barratt G, Clausse D, Marty JP, Grossiord JL. 1999. W/O/W multiple emulsions submitted to a linear shear flow: Correlation between fragmentation and release. *J Colloid Interface Sci* 218: 335–337.

Muguet V, Seiller M, Barratt G, Ozer O, Marty JP, Grossiord JL. 2001. Formulation of shear rate sensitive multiple emulsions. *J Controlled Release* 70: 37–49.

Murphy PA, Smith B, Hauck C, O'Connor K. 1992. Stabilization of vitamin A in a synthetic rice premix. *J Food Sci* 57: 437–439.

Nihant N, Schugens C, Grandfils C, Jerome R, Teyssie P. 1995. Polylactide microparticles prepared by double emulsion-evaporation. 2. Effect of thepoly(lactide-co-glycolide) composition on the stability of the primary and secondary emulsions. *J Colloid Interface Sci* 173: 55–65.

Oh C, Park JH, Shin SI, Oh SG. 2004. O/W/O multiple emulsions via one-step emulsification process. *J Dispersion Sci Technol* 25: 53–62.

Okonogi S. 1994. Methods for producing emulsions, low-fat spread and oil-in-water-in-oil type spread. US patent 5,279,847.

Okushima S, Nisisako T, Torii T, Higushi T. 2004. Controlled production of monodisperse double emulsions by two-step droplet breakup in microfluidic devices. *Langmuir* 20: 9905–9908.

Oza KP, Frank SG. 1986. Microcrystalline cellulose stabilized emulsions. *J Dispersion Sci Technol* 7: 543–561.

Pal R. 1996. Multiple O/W/O emulsion rheology. *Langmuir* 12: 2220–2225.

Pilman E, Larsson K, Torenberg E. 1980. Inverse micellar phases in ternary systems of polar lipids/fats/water and protein emulsification of such phases to W/O/W—microemulsion—emulsions. *J Dispersion Sci Technol* 1: 267–281.

Rousseau D. 2000. Fat crystals and emulsion stability—A review. *Food Res Int* 33: 3–14.

Sajjadi S, Zerfa M, Brooks BW. 2002. Dynamic behavior of drops in oil/water/oil dispersions. *Chem Eng Sci* 57: 663–675.

Sawae H, Sakaguchi A, Nakashio F, Goto M. 2005. Important factors affecting functions of PEG microspheres containing lipase complexes. *J Chem Eng Japan* 38: 54–59.

Schugens CH, Laruelle N, Nihant N, Grandfils C, Jerome R, Teyssie P. 1994. Effect of the emulsion stability on the morphology and porosity of semicrystalline poly-L-lactide microparticles prepared by W/O/W double emulsion evaporation. *J Controlled Release* 32: 161–176.

Schulz PC, Rodriguez MS, Del Blanco LF. 1998. Emulsification properties of chitosan. *Colloid Polym Sci* 276: 1159–1165.

Sekine T, Yoshida K, Matsuzaki F. 1999. A novel method for preparing oil-in-water-in-oil type multiple emulsions using organophilic montmorillonite clay mineral. *J Surfact Detergents* 2: 309–315.

Silva MR, Contente DML, Filho PAR. 1997. Ascorbic acid liberation from O/W/O multiple emulsions. *Cosmet Toilet* 112: 85–87.

Stehle RG, Higuchi WI. 1972. In vitro model for transport of solutes in three phase system. I. Theoretical principles. II. Experimental considerations. *J Pharmaceut Sci* 61: 1922–1930.

Strauel P, Friour G. 1994. Process for preparing photographic emulsions having a low fog level. US patent 420,106.

Stroeve P. Varanassi PP. 1984. An experimental study on double emulsion drop breakup in uniform shear flow. *J Colloid Interface Sci* 99: 360–373.

Sugiura S, Nakajima M, Yamamoto K, Iwamoto S, Oda T, Satake M, Seki M. 2004. Preparation characteristics of water-in-oil-in-water multiple emulsions using microchannel emulsification. *J Colloid Interface Sci* 270: 221–228.

Susuki S, Lim JK. 1994. Microencapsulation with carrageenan, locust bean gum mixture in a multiphase emulsification technique for sustained drug-release. *J Microencapsul* 11: 197–203.

Tadros T, Levecke B, Booten K. 2006. Application of polymeric surfactants in cosmetics and personal care. *Chemica Oggi* (suppl): 55–58.

Takahashi Y, Yoshida T, Takashi T. 1994. Process for producing a W/O/W type multiple emulsion for medicines, cosmetics, etc. US patent 4,985,173.

Tamilvanan S. 2004. Oil-in-water lipid emulsions: Implications for parenteral and ocular delivering systems. *Prog Lipid Res* 43: 489–533.

Terrisse I, Seiller M, Grossiord JL, Magnet A, Le Hen-Ferrenbach C. 1994. Application of rehological analysis to W/O/W multiple emulsions: Effect of the incorporation of a coemulsifier. *Colloids Surf A* 91: 121–128.

Tourne-Peteilh C, Lerner DA, Charnay C, Nicole L, Begu S, Devoisselle JM. 2003. The potential of ordered mesoporous silica for the storage of drugs: The example of a pentapeptide encapsulated in a MSU-Tween 80. *Chem Phys Chem* 3: 281–286.

Uchida T, Yoshida K, Goto S. 1996. Preparation and characterization of polylactic acid microspheres containing water-soluble dyes using a novel W/O/W emulsion solvent evaporation method. *J Microencapsul* 13: 219–228.

Varadaraj R. 2004. Oil-in-water-in-oil emulsion. US application patent 14,821.

Vaziri A, Warburton B. 1994. Some preparative variables influencing the properties of W/O/W multiple emulsions. *J Microencapsul* 11: 649–656.

Vaziri A, Warburton B. 1995. Improved stability of W/O/W multiple emulsion by addition of hydrophilic colloid components in the aqueous phase. *J Microencapsul* 12: 1–5.

Yang MS, Cui FD, You BG, Fan YL, Wang L, Yue P, Yang H. 2003. Preparation of sustained-release nitrendipine microspheres with Eudragit RS and Aerosil using quasi-emulsion solvent diffusion method. *Int J Pharm* 259: 103–113.

Yang YY, Chung TS, Ng NP. 2001. Morphology, drug distribution, and in vitro release profiles of biodegradable polymeric microspheres containing protein fabricated by double-emulsion solvent extraction/evaporation method. *Biomaterials* 22: 231–241.

Yoshida K, Sekine T, Matsuzaki F, Yanaki T, Yamaguchi F. 1999. Stability of vitamin A in oil-in-water-in-oil-type multiple emulsions. *J Am Oil Chem Soc* 76: 195–200.

Youan BBC, Gillard J, Rollmann B. 1999. Protein-loaded poly(epsilon-caprolactone) microparticles III. Entrapment of superoxide dismutase by the (water-in-oil)-in-water solvent evaporation method. *STP Pharma Sci* 9: 175–181.

Yu SC, Bochot A, Chéron M, Seiller M, Grossiord JL, Le Bas G, Duchene D. 1999. Design and evaluation of an original O/W/O multiple emulsion containing natural cyclodextrins as the emulsifier. *STP Pharm Sci* 9: 273–277.

Yu SC, Bochot A, Le Bas G, Chéron M, Mahuteau J, Grossiord JL, Seiller M, Duchene D. 2003. Effect of camphor/cyclodextrin complexation on the stability of O/W/O multiple emulsions. *Int J Pharm* 261: 1–8.

Zambaux MF, Bonneaux F, Gref R, Maincent E, Dellacherie M, Alonso J, Labrude P, Vigneron C. 1998. Influence of experimental parameters on the characteristics of poly(lactic acid) nanoparticles prepared by double emulsion method. *J Controlled Release* 50: 31–40.

Zhang W, Miyakawa T, Uchida T, Goto S. 1992. Preparation of stable W/O/W type multiple emulsion containing water-soluble drugs and in-vitro evaluation of its drug releasing properties. *J Pharm Soc Jap* 112: 73–80.

CHAPTER 8

Potentialities of W/O/W Multiple Emulsions in Drug Delivery and Detoxification

JEAN-LOUIS GROSSIORD and MONCEF STAMBOULI

Contents

8.1 INTRODUCTION

Multiple emulsion systems have various applications. They can be used for isolating an entrapped substance before release (smell or taste disguising), for protecting that substance from the eventual alterations induced by the external environment (oxidation, light, enzymatic degradation), and also for obtaining sustained and controlled release of the substance through a large variety of mechanisms. Yet another application is the great potential of multiple

Multiple Emulsions: Technology and Applications, Edited by Abraham Aserin

emulsions for detoxification. W/O/W multiple emulsions have been studied in pharmaceuticals, cosmetics, the food industry, and the environment, as attested by the many papers that have been published over the last two decades (Davis, 1981; Dickinson et al., 1991; Tadros, 1992; Garti, 1997; Grossiord and Seiller, 1998; Grossiord and Seiller, 2001). The large interest in multiple emulsions can be explained mainly by the fact that the stability problems that restricted their utilization have been partly overcome. Moreover systematic studies have showed that the release of molecules, initially encapsulated in the internal phase, can result from various mechanisms that can be controlled by many formulation and manufacture parameters. More recently some studies have demonstrated the effectiveness of W/O/W multiple emulsions in detoxification (Morimoto et al., 1979; Morimoto et al., 1982; Assouly et al., 2003; Hamoudeh et al., 2006; Stambouli, 2006; Frasca et al., 2006; Grosber et al., 2006).

For drug delivery applications, it is clear that the release from W/O/W multiple emulsion occurs either by transport through the oily membrane or by its breakdown. In the first case, and depending on the affinity of the molecule for the oily phase, the transport is due to molecular diffusion (Fick diffusion) or to diffusion facilitated by certain surfactants that take on the role of carriers. Transport through the oily membrane will not be considered here. This chapter is devoted to an examination of the breakup mechanisms, as are produced by swelling or under shear.

For detoxification applications the choice of an appropriate "extractant–trapping agent" couple is crucial:

- The extractant acts as a carrier to transport the pollutant or the toxic component as a lipophilic complex through the oily membrane.
- The trapping agent plays two roles simultaneously: it maintains the pollutant or the toxic component in the internal phase by breaking up the complex, and regenerates the extractant.

This chapter will present the main significant results that were obtained in our two laboratories over the last 15 years (Potier et al., 1992; Grossiord et al., 1993; Raynal et al., 1993, 1994; Jager-Lezer et al., 1997; Geiger et al., 1998, 1999; Grossiord and Seiller, 1998; Muguet et al., 1999, 2000, 2001; Olivieri et al., 2001, 2003; Djedour et al., 2006; Agnely et al., 2006).

8.2 DRUG DELIVERY MECHANISMS

8.2.1 Release by Breakup after Swelling

Basic Principle The swelling/breakdown process occurs only if there is a concentration gradient between the internal and the external aqueous phases. The resulting osmotic water flow is observed to travel from the external phase to the internal one, thereby reducing the concentration gradient. This water

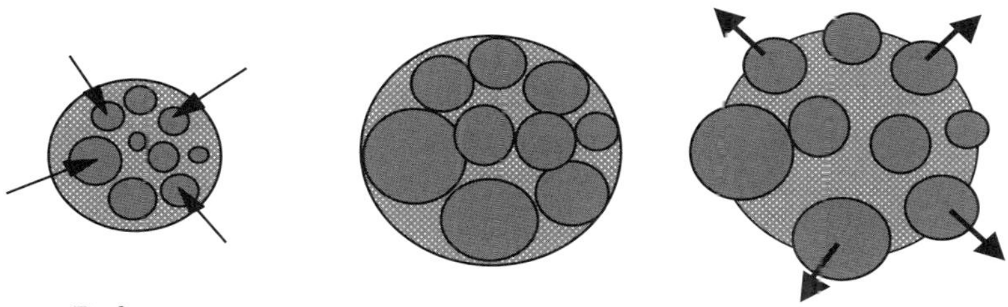

Figure 8.1 Schematic view of a swelling/breakdown kinetics. (From Grossiord and Seiller, 2001. STP Pharma.)

flow causes the drops to swell until a critical size is reached. Beyond the critical size, a breakup of the oily membrane occurs (Figure 8.1). The development of such osmotic swelling has been considered by various authors (Matsumoto et al., 1980; Tomita et al., 1982) not for inducing a controlled release but for estimating the stability of multiple emulsions.

In order to prevent the swelling/breakdown kinetics from the beginning, immediately after making the multiple emulsion, which is as soon as a concentration gradient is established, very concentrated multiple emulsions are prepared (typically the volume fraction is higher than 0.75). At such a high concentration the dispersed drops are close packed, so there is not enough room for the drops to swell. When the multiple emulsions are very concentrated, they remain stable with respect to the swelling/breakdown mechanism. When the multiple emulsions are diluted, the swelling/breakdown mechanism can occur.

Experimental Results It is possible to follow the swelling/breakdown kinetics, by recording the variation of the viscosity over time (Grossiord et al., 1993). For example, as Figure 8.2 shows, the evolution of the viscosity is a function of time in two different dilution conditions:

1. In an iso-osmotic dilution (defined as the same concentration in both the internal and external phases), no evolution of the viscosity is observed. As expected, there is no water flow when there is no concentration gradient.
2. In a hypo-osmotic dilution (higher concentration in the internal phase), an increase up to a maximum value is observed, followed by a decrease, until equilibrium is reached. The first part of the curve corresponds to the swelling step, whereas the last part corresponds to the breakdown.

In effect the concentration gradient is the driving force of the swelling/breakdown process (Raynal et al., 1994; Matsumoto et al., 1980).

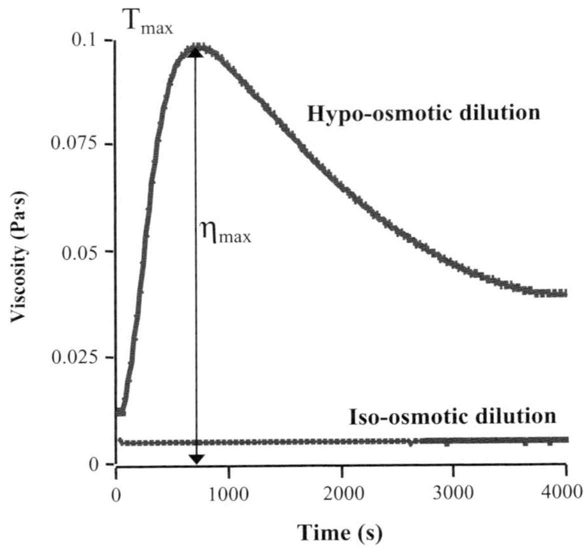

Figure 8.2 Graphs of viscosity versus time for iso-osmotic and hypo-osmotic dilutions. (From Grossiord and Seiller, 2001. STP Pharma.)

The swelling/breakdown kinetics depends on various parameters: the concentration gradient, the nature, and the concentration of the two surfactants, and the properties of the oil. The results below emphasize the effects of the two most critical parameters, namely the concentration gradient and the lipophilic surfactant concentration.

Influence of the Concentration Gradient Figure 8.3 represents the evolution of the viscosity versus time, for three different values of the concentration gradient: 50, 100, and 200 mM (Grossiord and Seiller, 1998). As expected, the maximum swelling was obtained for the highest concentration gradient.

Influence of the Lipophilic Surfactant Concentration Figures 8.4 and 8.5 illustrate the changes over time in viscosity and rate of release for seven different concentrations of lipophilic surfactant, ranging from 1% to 10%. Notice that the maximum swelling increases with the surfactant concentration and the rate of release decreases with the surfactant concentration. The same results were found using lipophilic surfactants of different properties, both polymeric and not.

The higher swelling, when the release is lower can be explained as follows: the excess of the lipophilic surfactant, initially located in the oily membrane, migrates at the external interface and fills up the free spaces. As a result breakdown can be delayed and even prevented (Jager-Lezer et al., 1997; Geiger et al., 1998, 1999).

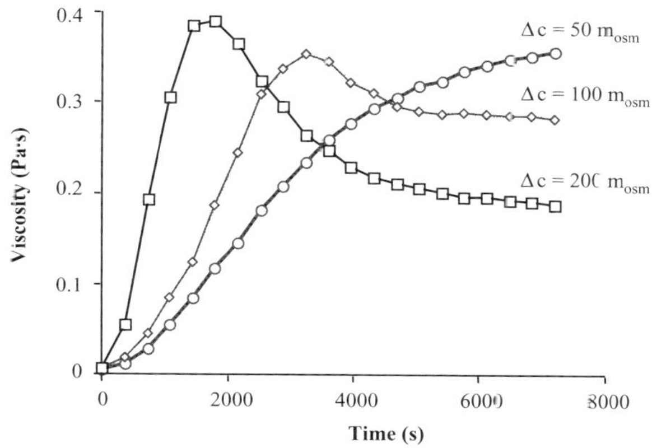

Figure 8.3 Change of viscosity versus time for different concentration gradients. (From Grossiord and Seiller, 2001. STP Pharma.)

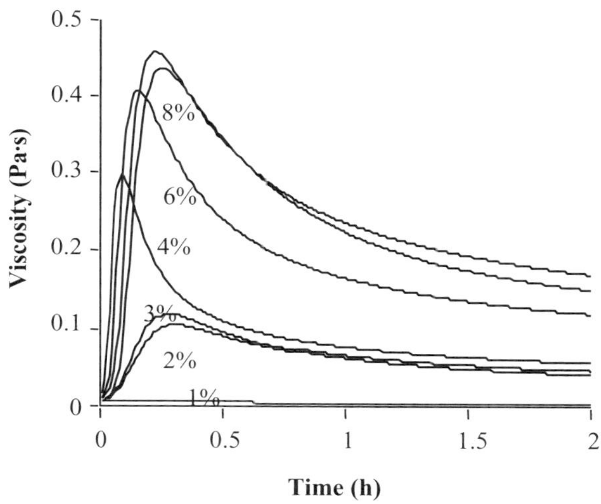

Figure 8.4 Change of viscosity versus time for different lipophilic surfactant concentrations. (From Grossiord and Seiller, 2001. STP Pharma.)

8.2.2 Release by Breakup under Shear

Theory The release under shear of an active molecule that is initially encapsulated in the aqueous phase of a W/O/W multiple emulsion is a very promising phenomenon for applications in cosmetics or pharmaceuticals. Taylor (Taylor, 1932, 1934) was first to study the deformation of molecules under shear and their bursting in a simple, dilute emulsion. He considered that breakup occurred when shear stress exceeds cohesion stress. He defined this breakup by way of a capillary number, *Ca*:

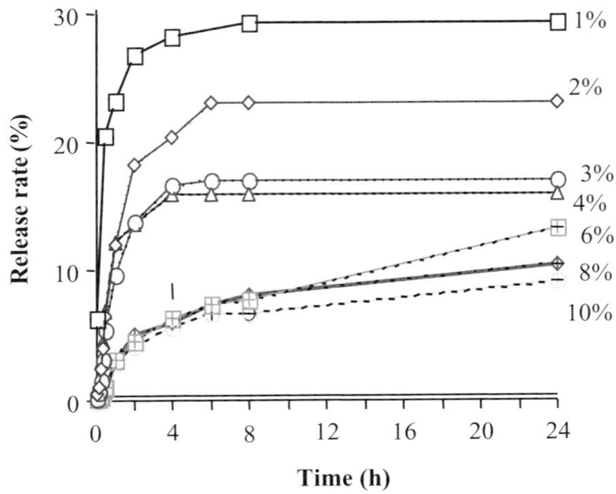

Figure 8.5 Change of release rate versus time for different lipophilic surfactant concentrations. (From Grossiord and Seiller, 2001. STP Pharma.)

$$Ca = \frac{\tau r}{\sigma} = \frac{\eta_c \dot{\gamma} r}{\sigma},$$

where τ is the shear stress, r the globule radius at rest, σ the interfacial tension between oil and water, η_c the continuous phase viscosity, and $\dot{\gamma}$ the shear rate. Bursting occurs when the capillary number exceeds a critical value $(Ca)_{cr}$ close to unity. This relation shows that the thickening of the continuous phase induces the globules to burst.

Experimental Results Our group has confirmed that the thickening of the external aqueous phase of W/O/W multiple emulsions increases or even creates the release under shear (Muguet et al., 1999, 2000, 2001). To this end, the multiple emulsions were sheared at different increasing rates beyond the breakup shear rate. After shearing, the granulometric distributions were recorded and compared to the profile at rest, as previously. An example of the size distributions of one such formulation, at rest and after the samples were sheared at 100, 200, 400, 800, and 1600 s^{-1}, is shown in the graphs of the Figure 8.6.

As was expected, the distributions were observed to shift toward the lower diameter as the shear was increased. It was possible to compare quantitatively the experimental diameters d_{43} (diameter moment/volume) of the globules after shear with the theoretical diameters determined from the expression of the capillary number given previously. The correlation between experiment and theory was satisfactory. So we could conclude that the Taylor theoretical model does apply to the breakup of the multiple emulsions globules, at least at the first approximation.

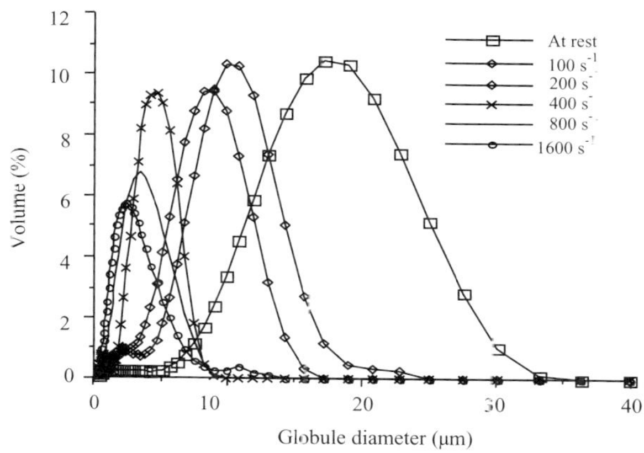

Figure 8.6 Size distributions at rest and after application of different shear rates. (From Grossicrd and Seiller, 2001. STP Pharma.)

Correlation between Fragmentation and Release (Muguet et al., 1999, 2000, 2001; Olivieri et al., 2001) In order to study the correlation between release under shear, NaCl and $MgSO_4$ electrolytes were encapsulated in the internal aqueous phase electrolyte molecules. While it would have been more satisfactory to use real drugs, we were only interested in the controlled release under shear of an encapsulated molecule. The electrolytes used were very hydrophilic probes, so we were sure that being hydrophilic, they could not cross the oily membrane.

The conductometric analyses were carried out in correlation with the granulometric measurements, to determine the release of the encapsulated molecules. The study was carried out for different multiple emulsions prepared with the same lipophilic and hydrophilic surfactants, but with different dispersed mass fractions (20%, 30%, 40%) and thickened by the same hydrophilic polymer at various concentrations (0.5%, 1%, 2%). The release from an unthickened and very concentrated multiple emulsion (80%) was also measured. Figure 8.7 shows the evolution of the release rate as a function of the capillary number *Ca*, in order to take into account not only the shear rate but also the initial diameter and the viscosity values.

It can be observed that there is a good correlation between release and fragmentation. The rates of release were increasing with the capillary number, up to 90%, when the value of *Ca* was five times its critical value. It is worth noting that the different curves corresponding to the different multiple emulsions are very close. However, had the release rates been plotted as a function of the shear rate, the different graphs would not have been superimposed at all. It may be possible to define a single curve from the evolution of the release rate versus the capillary number and thus to describe the behavior of all the formulations (whatever the hydrophilic polymer concentration and the mass fraction). The figure shows that the mechanisms responsible for the release under shear are probably the same and are defined by a normalized value of

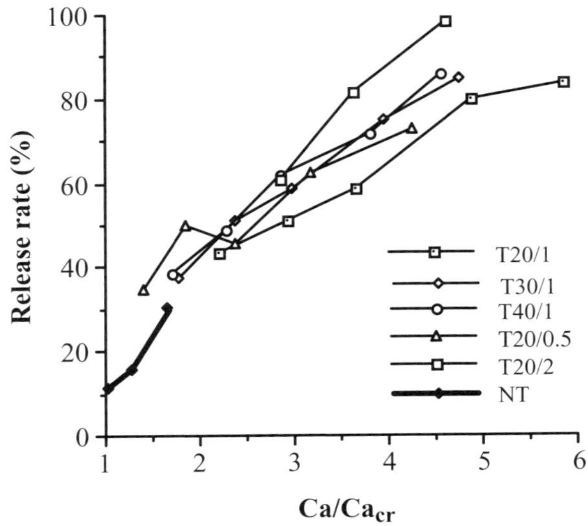

Figure 8.7 Evolution of the release rate as a function of the capillary number for different preparations. (From Grossiord and Seiller, 2001. STP Pharma.)

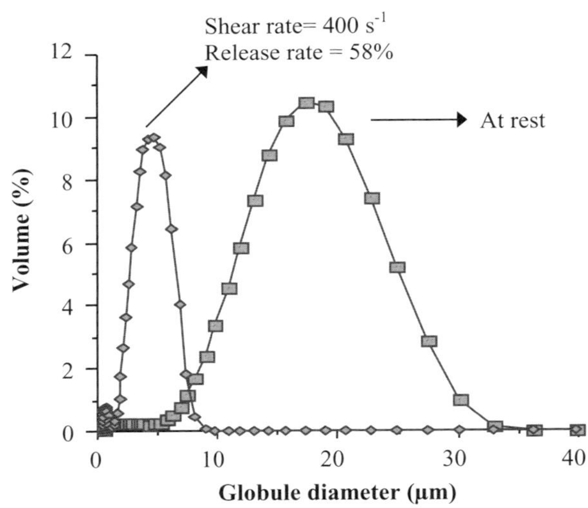

Figure 8.8 Size distributions of a preparation at rest and after application of a shear rate of 400 s^{-1}. (From Grossiord and Seiller, 2001. STP Pharma.)

the capillary number. Nevertheless, this curve may not be universal and may depend on the type of surfactants used for the formulation of the emulsions.

It is important to note that the fragmentation does not necessarily result in the release of encapsulated molecules. The two graphs in Figure 8.8 represent granulometric distributions of the same multiple emulsion at rest and after it is sheared at a rate of $400\,s^{-1}$. Note that the two distributions are almost completely separated. This means that the whole initial population of drops was fragmented into smaller drops. If this complete fragmentation had been accompanied by complete release, the rate of release would have been expected to be 100% at $400\,s^{-1}$, but it was in fact only 58%.

Two mechanisms could account for this partial release. The first is that almost all the aqueous microglobules do not necessarily leak at the very moment when fragmentation occurs. It is even probable that some of the globules remain inside the fragments that result from the breaking. This could be due to an excess of hydrophilic surfactant, or even of lipophilic surfactant. The second one is the phenomenon shown experimentally by Srinivasan and Stroeve (Srinivasan and Stroeve, 1986). They observed microglobules expulsed under shear to be covered by a thin film of oily phase that prevented the dispersion of the electrolyte in the external phase. Thus it seems that partial leakage and/or the persistence of a lipid film around the aqueous microglobules occurs during fragmentation of the multiple globules.

Development of Shear-Sensitive Thermogelling Emulsions (Olivieri et al., 2001, 2003; Djedour et al., 2006; Agnely et al., 2006) The thickening of the aqueous external phase at ambient temperature is liable to lead to globules bursting during fabrication. In order to solve this apparent paradox in topics applications, new thermoresponsive hydrogels have been introduced in the external aqueous phase. This way, the corresponding multiple emulsion will gel only near skin temperature, which can increase the active ingredient's delivery when topically applied.

So multiple emulsions present interesting advantages: (1) an easy fabrication process at room temperature with a higher entrapment yield and (2) a higher fraction released at skin temperature compared with the results obtained at room temperature. Two types of thermothickening polymers were studied:

- *EMP Hydrogel.* This is a poly(ethylene oxide)-b-poly(propylene oxide)-b-poly(ethylene oxide)-g-poly(acrylic acid) copolymer. EMP Hydrogel™ is a copolymer of poloxamer 407 and poly(acrylic acid) (average molecular weight $3 \times 10^6\,g{\cdot}mol^{-1}$), which was obtained via dispersion/emulsion polymerization of acrylic acid along simultaneous grafting of poly(acrylic acid) onto poloxamer backbone from Lev Bromberg (Bromberg, 1998a). The association mechanism that is envisioned for this kind of thermoassociating polymer was reported by Bromberg (1998b). The thermoviscosifying behavior of 4 wt% EMP Hydrogel is illustrated in Figure 8.9. As is seen, a 10^4-fold increase in Hydrogel's viscosity (equilibrium viscosity) occurred over a temperature range of 22°C to 37.5°C (Bromberg, 1998c). Note that the sol–gel transition was progressive and not instantaneous, so it was not possible to define the sol–gel transition temperature, *Tgel*. However, this transition was perfectly reversible.

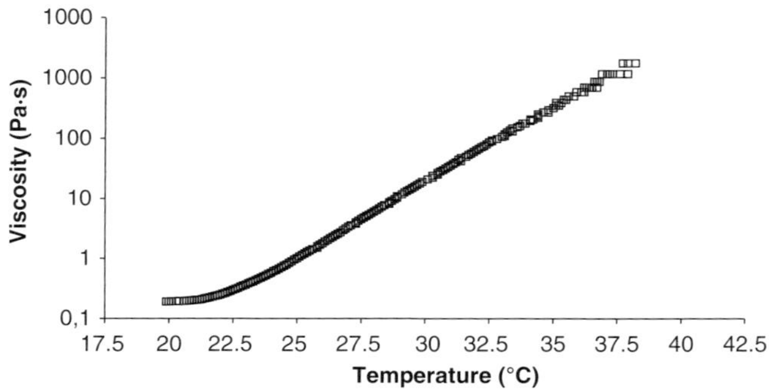

Figure 8.9 Variation of viscosity versus temperature for EMP Hydrogel (C = 4% in water) for a shear stress τ = 10 Pa. (From Grossiord and Seiller, 2001. STP Pharma.)

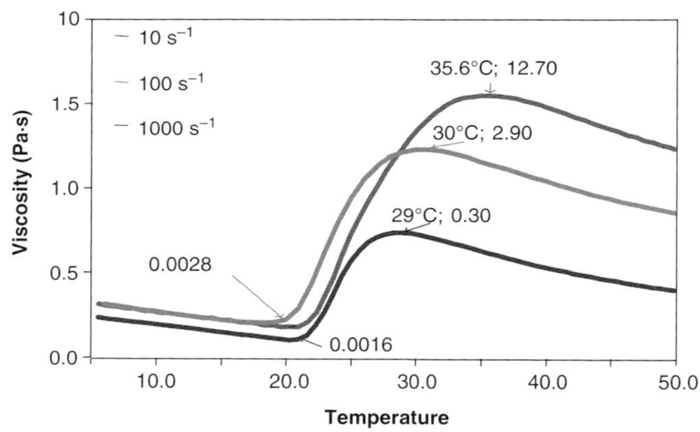

Figure 8.10 Variations of viscosity versus temperature for EG56 Hydrogel for different shear rates ($10\,s^{-1}$, $100\,s^{-1}$, $1000\,s^{-1}$). (From Djedour et al., 2006. CME 2006.)

- *EG 56 Hydrogel.* This is a hyperbranched poly(ethylene oxide-b-propylene oxide-b-ethylene oxide). It belongs to the ExpertGel™ family, which is synthesized by PolymerExpert (Pagnoux et al., 2003). The EG 56 Hydrogel is obtained by polyaddition. The poloxamer 407 molecules are linked together via urethane, allophanate, urea, and biuret links. As in the behavior of EMP Hydrogel, the viscosity increase is progressive (Figure 8.10). Interestingly, the viscosity of this sample does not decrease much under shear: at 30°C the viscosity of the 6.15% EG 56 solution varied from 13 Pa·s at $\dot{\gamma} = 10\,s^{-1}$ to 3 Pa·s at $\dot{\gamma} = 100\,s^{-1}$.

The two kinds of multiple emulsions considered contained in the first formulation the EMP Hydrogel in the external phase (called TRME7) and in the second formulation thickening agent the EG 56 Hydrogel (denoted EGME).

- *EMP Hydrogel/multiple emulsions systems.* Various formulations were prepared by changing mainly the nature and concentration of lipophilic and hydrophilic surfactants, the concentration of the EMP Hydrogel, as well as the composition of the primary emulsion and its volume fraction in the multiple emulsion. Finally, the TRME7 system was chosen to summarize the fragmentation and release properties under shear, because it seems to realize the best compromise between stability and release properties (Olivieri et al., 2001, 2003).

The release curves obtained for TRME7 submitted to different shear at 20°C and 35°C are presented in Figure 8.11 as a function of the shear rate. The total fraction released was much larger at 35°C than at 20°C for the same shear rate value. A shearing of $1000\,s^{-1}$, which is commonly reached during topical applications, led to a 60% release at 35°C, whereas a 20% release was obtained only at 20°C. This confirmed the interesting application expectations for this new W/O/W thermally reversible multiple emulsion. Indeed this emulsion was little sensitive to shear at low temperatures (during the fabrication process)

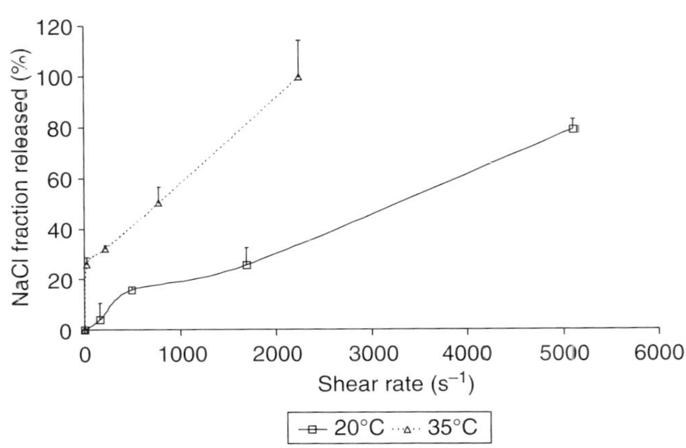

Figure 8.11 Variations of the released NaCl from TRME7 multiple emulsion versus shear rate at $T = 20°C$ and $T = 35°C$. (From Agnely et al., 2006. JDDST.)

and became more and more sensitive to shear as the temperature increased (e.g., when applied on skin).

- EGME multiple emulsion (Agnely et al., 2006). The aim of this work was to formulate a sprayable shear-sensitive W/O/W multiple emulsion using the EG 56 thermoassociating polymer. Thanks to the properties of this polymer, the multiple emulsion (EGME) is fluid at ambient temperature, in order to avoid droplet breakup during the fabrication process and thickens at skin temperature upon its topical application. This viscosity increase should allow the release of the active matter entrapped in the internal aqueous phase under the shear of application. After the formulation step, rheological, granulometric, and conductometric analysis were performed to assess the thermoresponsive behavior and the fragmentation release characteristics of this emulsion. Furthermore we tested the ability to spray EGME emulsion by simulating in rheological tests the shear conditions endured during its packaging and use. The emulsion was sheared at 20°C for 15 minutes at very high shear stresses, and then the amount of released $MgSO_4$ of multiple globules was evaluated by conductometric analysis. The results are illustrated by Figure 8.12.

Because the release under very high shear ($8600\,s^{-1}$) is not so important (maximum 35%), we can conclude that the EGME emulsion is stable under high shear and can be sprayed. In a second step, to mimic a topical application, the EGME emulsion was sheared for 15 minutes at 100 and $1000\,s^{-1}$ at 30°C. At $1000\,s^{-1}$, a high $MgSO_4$ released fraction (86%) was observed. These results show that EGME can be sprayed, thanks to its resistance to high shear stresses at room temperature. Thus high fragmentation and release can be obtained upon topical application.

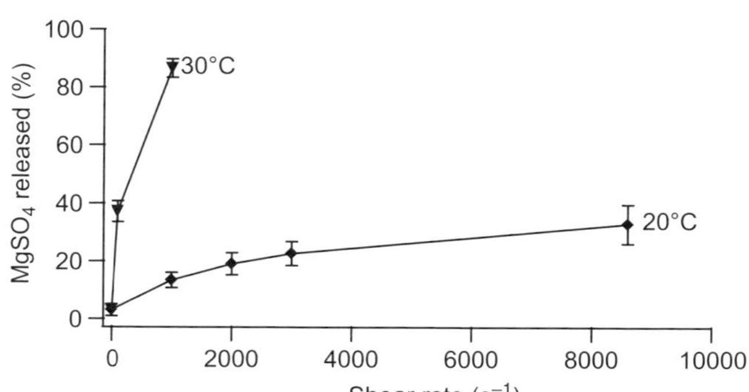

Figure 8.12 Variations of the released $MgSO_4$ from EGME multiple emulsion versus shear rate at $T = 20°C$ and $T = 30°C$. (From Agnely et al., 2006. JDDST.)

8.2.3 Conclusion

This review has focused on the release by breakup of the oily membrane, resulting either from a hypo-osmotic dilution or by shear application among the various release mechanisms of molecules initially entrapped in the internal aqueous phase of W/O/W multiple emulsions. Experimental studies have shown very interesting features that have practical application and allowed us to form some assumptions about the molecular mechanisms. In particular, it has been found that these mechanisms can be controlled by different emulsion formulations and/or application parameters as follows:

- For the swelling/breakdown kinetics: The concentration gradients of different hydrosoluble molecules and the concentration of the lipophilic surfactant.
- For the breakup induced by shear: The multiple globules' diameters, the interfacial tension, the rheological properties of the multiple emulsion, and the shear rate being applied.

It is possible to take advantage of this great variety of parameters and control the release from the internal phase by one of these mechanisms, in developing applications of multiple emulsions for pharmaceuticals, cosmetics, and the food industries.

8.3 REACTIVE DETOXIFYING EMULSIONS: ENVIRONMENTAL AND THERAPEUTIC APPLICATIONS

The reactive detoxifying emulsions (RDE) (Stambouli, 2006) are water-oil-water multiple emulsions. This detoxification emulsion technique is an emerging technology, and is directly based on a more known solvent extraction (SE) process that has been extensively applied for the selective recovery and refining of a wide variety of solutes (metal ions, pollutants, toxics, etc.) in hydrometallurgy and nuclear industry.

The environmental applications of the RDE process are especially attractive. The high surface area achieved with the RDE technology provides fast and complete removal of contaminants when aqueous phases are treated with a low toxic content, such as in industrial waste streams (Touati et al., 2006). Some more ingenious applications of the RDE have been recently investigated and extended to life sciences. Their suitability has been demonstrated in vitro for a large variety of molecules in drug overdose trials of both oral and topical intoxications (Assouly et al., 2003.; Hamoudeh et al., 2006; Frasca et al., 2006).

8.3.1 Reactive Detoxifying Emulsions Principle

The RDE action is illustrated in Figure 8.13. The incorporation of a suitable organic extractant carrier in the oily phase of the multiple emulsion enabled

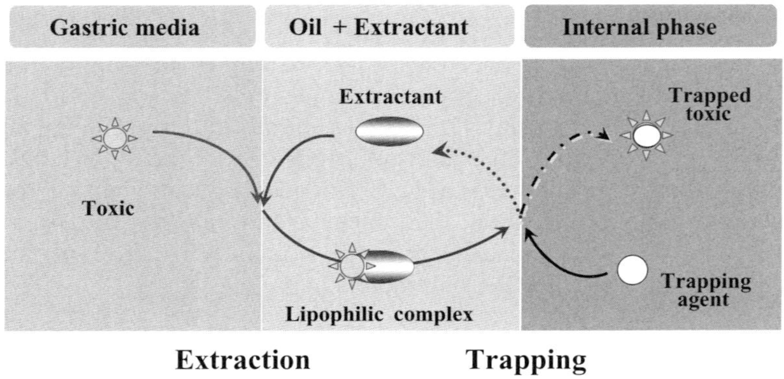

Figure 8.13 Schematic illustration of the extraction-trapping cycle.

TABLE 8.1 Guidelines for extractants classification versus toxics

Mechanism	Extractants	Solutes	Examples
Cationic exchange	Long-chain organic acids	Metallic cations, basic drugs	Cd^{2+}, Hg^{2+}, Cs^{+}, Nivaquine, paraquat, etc.
Anionic exchange	Amines or quaternary ammonium salts	Anionic metallic complexes, acidic drugs	Cyanide complexes, $Zn(CN)_4^{2-}$, $Ni(CN)_4^{2-}$, acetylsalicylic acid
Solvation	Long-chain alcohols and esters	Neutral solutes	Cyanhydric acid HCN, paracetamol

the formation of a lipophilic complex with the solute. This detoxifying complex diffuses through the oily phase to the internal aqueous interface where a trapping (stripping) agent captures the solute and releases the extractant for a new extraction–trapping cycle. Thanks to the strong similarity between the chemical mechanisms in solvent extraction (SE) and reactive detoxifying emulsions (RDE) techniques, easy transpositions can be drawn between SE and RDE configurations to and in the choice of appropriate "extractant–trapping agent" for a specific application.

Table 8.1 provides some general guidelines on the extractants, which are classified according to their extraction mechanism. The trapping agent is a mineral acid or base (typically HCl or NaOH), either alone or mixed with a specific complexing agent (NaOH + EDTA in the case of mercury).

8.3.2 Advantages/Drawbacks of RDE Detoxification

The main advantages of the RDE detoxification are as follows:

1. Only small amounts of extractant are required. So the use of expensive and selective extractant is possible for very specific and delicate detoxification.
2. High interfacial area are developed with the emulsions. The RDE detoxification is therefore efficient even for dilute aqueous solutions (<1 g/L), unlike the SE technique.
3. A concentration of the toxic (pollutant) in the internal phase is achieved because of the high volume ratio between external and internal aqueous phases. RDE detoxification could therefore suitably be applied as a preconcentration step in a more classical process.
4. Extraction and stripping steps are simultaneously combined. This presents an economic advantage.

The advantages (1) and (2) are quite suited for environmental application processes. However, unlike RDE technology, a scrubbing step can be added to the SE process when high extraction selectivity is desired. Only a highly selective extractant can be used to meet such a requirement for RDE.

Beside the RDE extraction performances, their stability represents a crucial point that needs to be elucidated. While long-term emulsion stability should be ensured for the therapeutic applications, only short-term stability has to be achieved for environmental applications. For the later applications an emulsion breakage (splitting or coalescence) step is required in the process in order to separate the toxic-loaded emulsion into a toxic-concentrated aqueous solution, which is then valorized and the free oily phase is recycled in the emulsification step. These opposite requirements need more extensive parametric study to determine any delicate compromises.

8.3.3 Environmental Applications

Thanks to the high interfacial area developed, RDE technology provides fast and complete removal of the contaminants in the treatment of aqueous phases with a low toxic content, as in industrial wastewaters. Li made the first technical application of this process in the 1960s (Li, 1978). In the beginning, RDE was used for production of metals, and in the 1980s, this technique was tested for applications in wastewater treatment (Draxler et al., 1988). A big advantage of this kind of treatment is good separation of metal ions in very dilute solutions. Some examples of environmental applications will be illustrated. The feasibility of the pollutant removal, such as removal of heavy metals (Touati et al., 2006; Wolfsberger, 2003), radioactive cesium (Pareau et al., 1999), and cyanide (Durand, 1996), has been demonstrated and tested at micro-pilot scale in our laboratory.

Removal of Heavy Metals

Removal of Cadmium from Phosphoric Acid Phosphoric acid has many uses in food and detergent industries and in agricultural applications as fertilizer. Almost all phosphate fertilizers contain small amounts of cadmium, a highly toxic contaminant affecting humans and a wide variety of living organisms. Its removal is absolutely essential. Various remedial processes were reported in the literature such as ion exchange, solvent extraction, and liquid membrane processes either supported or by RDE technology. The viability of RDE technology depends strongly on both the choice of an efficient extracting system and the formulation of stable emulsions in such "aggressive" acidic conditions (50–85% H_3PO_4). A variety of organophosphorus acids have been used as extractants for the recovery of cadmium with more or less success. To the best of our knowledge, only one paper deals with the use of di(2-ethylhexyl) dithiophosphoric acid (D_2EHDTPA) for cadmium removal.

1. *Solvent extraction experiments.* Solvent extraction (SE) batch experiments were conducted in order to identify the best "extractant-trapping agent" couple. The extraction mechanism of Cd^{2+} with D_2EHDTPA (noted AH) is a cationic exchange type:

   $$Cd^{2+} + 2AH \Leftrightarrow CdA_2 + 2H^+.$$

   In the aim to identify one trapping agent to incorporate in the internal aqueous phase of the emulsion and to determine the optimum conditions for trapping, several mineral acids were tested. The best performances were obtained with HCl (4 M). The equation of the stripping reaction is

   $$CdA_2 + 2Cl^- + 2H^+ \Leftrightarrow CdCl_2 + 2AH.$$

   However, such acidities are prejudicial with the stability of the emulsions. The optimal trapping phase was a mixture of NaCl (3 M) and HCl (1 M) whose stripping (trapping) performances were found quite comparable to those of HCl (4 M).
2. *RDE detoxification experiments.* As shown in Figure 8.14, the extraction of cadmium by RDE is fast and complete. Moreover the emulsion remains stable during all the duration of experiment (30 min). This duration is largely sufficient in the aim of an industrial implementation where average residence time in various technologies of extraction (mixers-settlers, pulsed columns, etc.) do not exceed 10 minutes. To verify the material balance and the trapping action, the loaded emulsion was splitted by electrocoalescence. The results were in excellent agreement with the predictions.

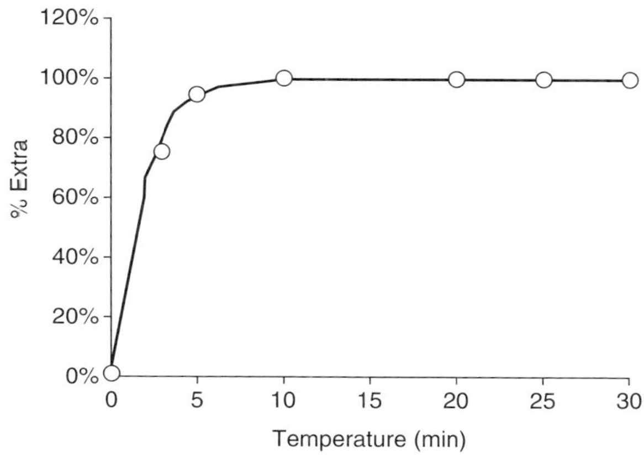

Figure 8.14 Kinetics of the cadmium extraction by the RDE detoxification method.

Extraction of Divalent Mercury from Hydrochloric Solutions Mercury is a naturally occurring element that appears in several inorganic and organic forms in the nature. Even in small concentration levels, inorganic mercury is naturally transformed to methyl-mercury, which quickly enters the aquatic food chain and accumulates in biological tissues. If ingested into the human body, mercury especially affects the nervous system. Recently mercury was ranked third on the CERCLA (Comprehensive Environment Response, Compensation, and Liability Act) priority list of hazardous substances, after arsenic and lead. One of our researches deals with the application of RDE technology for mercury extraction from hydrochloride acid media, using both trioctyl amine (TOA) and a quaternary ammonium salt (Aliquat 336) as the extraction reagents. Batch solvent extraction experiments were first conducted in order to choose the appropriate trapping (stripping) agent, namely a mixture of ethylene diamine tetraacetic acid (EDTA) and NaOH. Slope method analysis led to the identification of the extracted metallic complexes. The results clearly demonstrate the presence of both organic complexes $HgCl_3{\cdot}TOAH$ and $HgCl_4{\cdot}(TOAH)_2$ in the case of TOA and one complex 1:1 with Aliquat 336.

Kinetics experiments with RDE were then conducted, using either ECA 4360 or Abil EM90 as lipophilic surfactant. The effects of the concentration of the different reagents (TOA, NaOH, EDTA, and surfactant) on the extraction performance of mercury were investigated to optimize the emulsion formulation. An observed decrease of the internal phase pH, related to hydroxide consumption or HCl co-extraction, had an undesirable effect on the mercury stripping. Thus extraction time is a crucial parameter to be optimized for good stripping performances (Figure 8.15).

After optimization, continuous RDE experiments were carried out on a laboratory-scale pulsed column (1 m in height, 2 cm in diameter) equipped with alternate disks and rings. Extraction efficiency as high as 95% was obtained

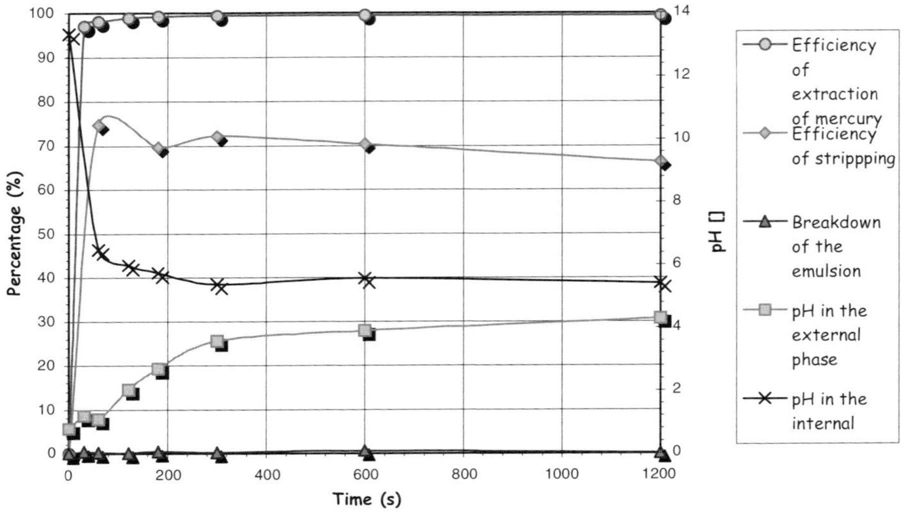

Figure 8.15 Kinetics profiles and extraction performance of mercury by the RDE method.

using a synthetic mercury solution (1 g/L). A concentrated internal phase of about 10 g/L of mercury was obtained after electrocoalescence splitting of the loaded RDE. Higher mercury depletion with respect to environment regulations (less than 50 ppb) would then require a pulsed column of 3.5 m height, worked under similar hydrodynamic conditions.

RDE Refining of Cyanide Industrial Wastewaters Cyanides are used in many industrial processes such as surface treatment and noble metal plating. After several cycles the baths must be renewed in order to maintain the same operating conditions. The conventional physicochemical process applied to eliminate cyanide from liquid effluents is mainly oxidation by hypochlorite or ozone. Depleted wastewaters contain 0.1 mg/L of free cyanide and may be rejected into the environment. The new environmental regulations are more restrictive and concern both free and total cyanide. RDE technology was applied in order to reduce both free and complexed cyanide levels.

As two toxics are concerned, two "extractant-trapping agent" couples have been identified and their performances optimized (Figure 8.16). A continuous RDE process has been successfully tested on a laboratory-scale mixer-settler battery, allowing the reduction of cyanide concentration by one order of magnitude.

Other Applications Nuclear fuel reprocessing operations generate different types of effluents. Chemical segregation of most or all the radioactive components from the inactive bulk constituents of the waste can reduce the

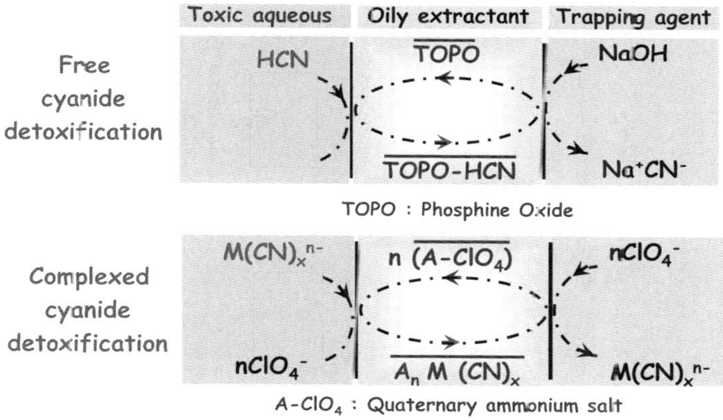

Figure 8.16 Schematic illustration of RDE extraction method of free and complexed cyanide "extractant-trapping agent" couples.

volume of waste requiring vitrification and subsequent disposal in a deep geologic repository. Because of their diversity and complexity, highly selective separation processes are required, and some new improved schemes have been recently developed for specific types of effluents. A RDE extraction process scheme was developed for the elimination of cesium (Cs) from an effluent containing traces of Cs along with high concentrations of sodium and hydroxide. Cesium is selectively extracted and concentrated. RDE technology was chosen because of its main advantages: possibility of treating much diluted solutions and possibility of concentration. The desired selectivity is obtained using a new extractant, synthesized in our laboratory: a calixarene formed by the acid-catalyzed condensation of resorcinol and octanal, named resorcarene or PR8 (Figure 8.17).

PARABAR 9220 (polyolefin alkyl anhydride) is used as surfactant. The internal aqueous phase of the emulsion is HNO_3 (1 mol/L) and the organic one contains PR8, PARABAR 9220 and dodecane. Surprisingly, by employing commonly used volume ratios of the emulsion (around 1), it is not possible to extract cesium. Emulsions containing a very low volume of internal phase (10% or less) are nevertheless very efficient. This is a very surprising and original behavior. The batch extraction studies show an extraction yield of 95% for cesium and less than 1% of sodium is co-extracted. The concentration factor of cesium is then 20. Continuous RDE extraction studies in mixer-settlers and a pulsed column are performed to test the viability of the continuous process. Because of the high RDE breakage observed in a mixer-settler, a pulsed column is then preferred. The experimental results are quite satisfactory (>90% Cs extracted, Cs concentration factor 20 and 1.2% sodium co-extracted). Theses results could then be improved by optimizing chemical and operating conditions.

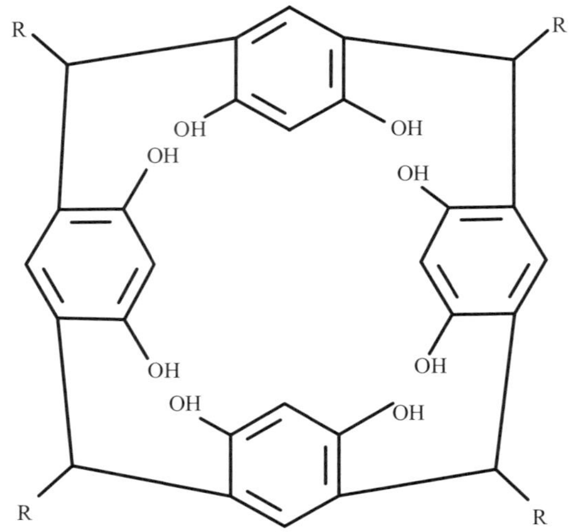

Figure 8.17 Chemical structure of resorcarene.

8.3.4 Therapeutic Applications (Assouly et al., 2003; Hamoudeh et al., 2006; Stambouli, 2006; Frasca et al., 2006; Grosber et al., 2006)

During the past decade, drug acute intoxications were recognized to be a serious problem threatening the public health sector. The clinicians still lack a fully comprehensive detoxification treatment for the daily accidents of drug overdose intoxications. Only a few means of detoxification are available to date but insufficiently applied: activated charcoal, ipecac-induced vomiting, gastric lavage, and so forth. However, they are continuously dismissed as being not much effective when used at a late stage, as causing many complications, and as having sometimes considerable contraindications that limit their applications. In the light of these problems, finding rapid, effective, and safe detoxifying treatments still remains a challenge.

Reactive detoxifying emulsion technology may be developed as an alternative and effective solution. The feasibility of RDE has been investigated in vitro on a large variety of toxics (paracetamol, aspirin, Nivaquine, paraquat, etc.) in treating oral or topic intoxications. An investigation of a suitable RDE could be targeted, after an oral ingestion of a drug overdose, to limit the drugs systemic passage and eliminate it by the faecal route in trapping the drug in the internal phase of the emulsion.

It is noteworthy that detoxification by multiple emulsions has been carried out in literature, though only so far by Morimoto et al. (Morimoto et al., 1979, 1982) to eliminate partially salicylic acid and quinine sulphate. However, these authors used "passive" emulsion systems without an extractant agent. In con-

TABLE 8.2 Typical RDE formulations

Toxic Nature	Studied Toxics	Extractant	Trapping Agent	Potential RDE Detoxification
Acidic	Aspirin	Long-chain amine (TOA)	Mineral base (NaOH)	Gastric
Basic	Nivaquine Zopiclone Paraquat	Organophosphoric acid (D_2EHPA, $D_2EHDTPA$)	Mineral acid or base (HCl, NaOH)	Gastric Gastric Gastric or cutaneous
	Nicotine			Cutaneous
Neutral	Paracetamol	Long-chain alcohol	NaOH	Gastric

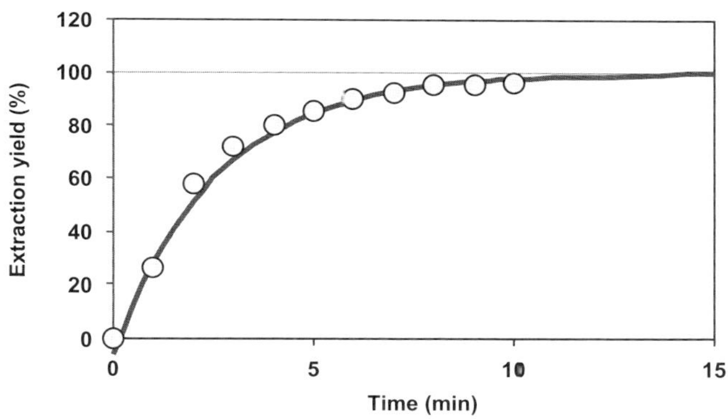

Figure 8.18 Extraction profile of aspirin by the percutaneous RDE detoxification.

trast, the aim of our work is to optimize the formulation of reactive emulsions including an "extractant-trapping agent" couple.

Typical RDE Formulations versus Toxics A large group of toxics was investigated for potential applications in gastric or percutaneous detoxification. An appropriate "extractant-trapping agent" couple was tested in vitro for each of the toxics (Table 8.2). For most toxics the RDE compositions was optimized to more than 80% of detoxification after 5 minutes. For example, Figure 8.18 illustrates the extraction performances for aspirin. For other dangerous toxics (Nivaquine, paraquat), the detoxification performances were more difficult to optimize because of various formulation parameters.

Nivaquine Detoxification Introduced in 1946, Nivaquine® (Chloroquine sulfate) was largely used in the chemoprophylaxis of malaria. Because of the narrow margin between its therapeutic and toxic dose, fatal chloroquine poisonings occur frequently in suicide and murder cases. Chloroquine (CQ) is extracted by di(2-ethylhexyl) phosphoric acid (D_2EHPA) according to a cationic exchange reaction:

$$CQH_n^{n+} + (n+p)\overline{AH} \rightarrow \overline{CQH_nA_n(AH)_p} + nH^+,$$

where $\overline{AH}$ represents the extractant monomer and $\overline{CQH_nA_n(AH)_p}$ the lipophilic complex. Classical study in solvent extraction allowed the determination of the stoichiometry of the lipophilic complex $\overline{CQH_2A_2(AH)_5}$. Thus chloroquine is extracted in a di-protonated form, confirming its basic character. So a mineral acid such as HCl appears to be a suitable trapping (stripping) agent.

The effects of the extractant amount as well as the emulsion type were mainly investigated:

- The presence of the extractant was confirmed to be absolutely necessary. Above 0.33%, its concentration has less significant effect (Figure 8.19).
- As it was expected, the white multiple emulsion ME1 (without an extractant nor a trapping agent) was not able to extract more than 10% of nivaquine in the first minutes. This weak extraction could be connected to the only phenomenon of passive diffusion. Figure 8.20 illustrates the influence of the emulsion type on the chloroquine extraction performance. With the multiple emulsions ME3 (with an extractant and a trapping agent), ME2 (only a trapping agent), and ME4 (only an extractant), it appears that the extraction efficiency is of the order ME4 > ME3 > ME2. The surprising performance of ME4 versus ME3 (and to a certain extent, ME2) can be related to proton leakage to the external aqueous phase.

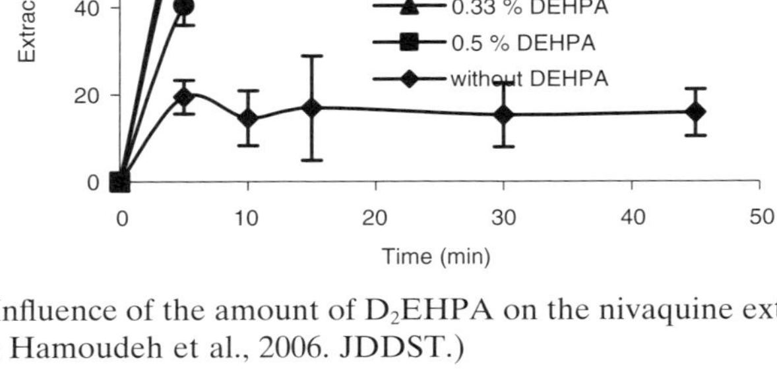

Figure 8.19 Influence of the amount of D_2EHPA on the nivaquine extraction performance. (From Hamoudeh et al., 2006. JDDST.)

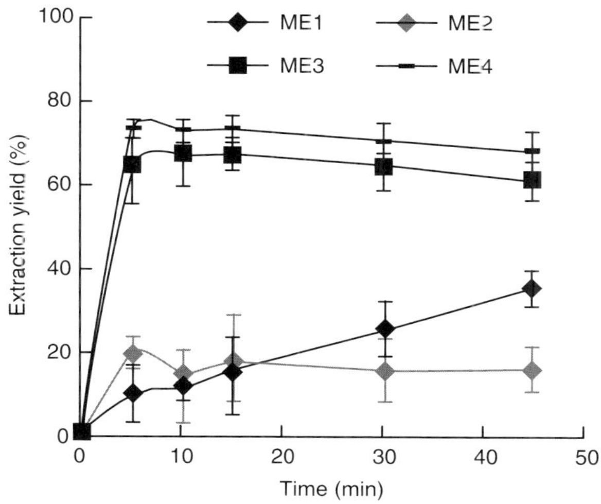

Figure 8.20 Influence of the emulsion type on the chloroquine extraction performance. (From Hamoudeh et al., 2006. JDDST.)

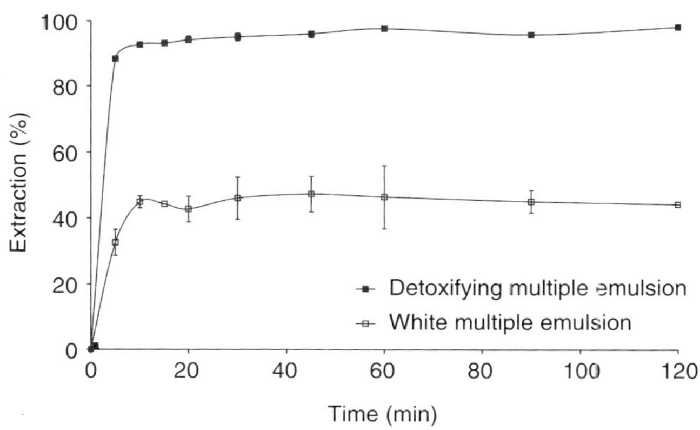

Figure 8.21 Multiple emulsion extraction performance.

Paraquat Detoxification Paraquat (1, 1′-dimethyl-4,4′-bipyridinium chloride) is a nonvolatile compound that is very soluble in water. Paraquat is a very toxic herbicide inducing pulmonary cell death in presence of oxygen. Paraquat is extracted by di(2-ethylhexyl) dithiophosphoric acid (D_2EHDTPA) according to a cationic exchange mechanism. NaOH is an appropriate trapping-agent. Figure 8.21 illustrates detoxification kinetics with either a reactive or a white (without "extractant-trapping agent" couple) As expected, RDE extraction is fast and complete. Surprisingly, about 45% of paraquat was extracted by the white emulsion, probably due to the passive diffusion through the oily membrane.

8.3.5 Conclusion

The RDE detoxification concept was validated for both environmental and therapeutic applications. While the environmental detoxification is a relatively known topic that has been tested at a micro-pilot scale, the therapeutic applications are more novel. The wide range of pollutants/toxics investigated contain very different types of molecules, so the potential for applying this innovative technology is high.

REFERENCES

Agnely F, Djedour A, Bochot A, Grossiord JL. 2006. Properties of various thermoassociating polymers: pharmaceutical and cosmetic application. *J Drug Deliv Sci Tech* 16(1): 3–10.

Assouly E, Grossiord JL, Pareau D, Seiller M, Stambouli M. 2003. Patent "Emulsions simples et multiples destinées à la détoxication de l'organisme ou de surfaces," *Demande internationale* PCT/FR03/01674, 4 juin.

Bromberg L. 1998a. Self-assembly in aqueous solutions of polyether-modified poly(acrylic acid). *Langmuir* 14: 227–234.

Bromberg L. 1998b. Polyether-modified poly(acrylic acid): Synthesis and properties. *Ind Eng Chem Res* 37: 4267–4274.

Bromberg L. 1998c. Properties of aqueous solutions and gels of poly(ethylene oxide)-b-poly(propylene oxide)-b-poly(ethylene oxide)-g-poly(acrylic acid). *J Phys Chem B* 102: 10736–10744.

Davis SS. 1981. Liquid membranes and multiple emulsions. *Chem Ind* 3: 683–687.

Dickinson E, Evison J, Gramshaw JW, Schwope D. 1991. Preparation of fine protein stabilized water-in-oil-in-water emulsions. *Food Hydrocolloids* 5: 481–485.

Djedour A, Agnely F, Mekhloufi G, Ponton A, Grossiord JL. 2006. W/O/W thermogelified multiple emulsions: Original forms for drug release after shear. *4th World Congress on Emulsions*, Lyon.

Draxler J, Fürst W, Marr J. 1988. Separation of metal species by emulsion liquid membranes. *J Membr Sci* 38: 281–293.

Durand G, Pareau D, Stambouli M, Coste M. 1996. Extraction of residual low concentration of cyanide from an industrial effluent by a surfactant liquid membrane process. *Proceedings of the International Solvent Extraction Conferences, Australia*, pp 1553–1558.

Frasca S. Chauvierre C, Seiller M, Pareau D, Du Halgouët B, Stambouli M, Grossiord JL. 2006. Multiple emulsions: In vivo paraquat detoxification. *4th World Congress on Emulsions*, Lyon.

Garti N. 1997. Double emulsions-scope, limitations and new achievements. *Colloids Surf A* 123–124: 233–246.

Geiger S, Tokgoz S, Fructus A, Jager-Lezer N, Seiller M, Lacombe N, Grossiord JL. 1998. Kinetics of swelling-breakdown of a W/O/W multiple emulsions: Possible mechanisms for the lipophilic surfactant effect. *J Controlled Release* 52: 99–104.

Geiger S, Jager-Lezer N, Tokgoz S, Seiller M, Grossiord JL. 1999. Characterization of the mechanical properties of a W/O/W multiple emulsion oily membrane by a micropipette aspiration technique. *Colloids Surf A* 157(1–3): 325–332.

Grosber M, Pareau D, Seiller M, Grossiord JL, Stambouli M. 2006. Rationalisation des méthodes de fabrication d'émulsions destinées à la désintoxication médicamenteuse. *4th World Congress on Emulsions*, Lyon.

Grossiord JL, Seiller M. 1998. Rheology of W/O/W multiple emulsions: Formulation, characterization and breakup mechanisms. In: *Multiple Emulsions, Structure, Properties and Applications*, Grossiord JL, Seiller M. eds. Paris: Editions de Santé, pp 169–192.

Grossiord JL, Seiller M, eds. 1998. *Multiple Emulsions, Structure, Properties and Applications*. Paris: Editions de Santé.

Grossiord JL, Seiller M, Puisieux F. 1993. Significance of rheological analysis in studies on W/O/W multiple emulsions. *Rheol Acta* 32: 168–180.

Grossiord JL, Seiller M. 2001. W/O/W multiple emulsions: a review of the release mechanisms by break-up of the oily membrane. *STP Pharma Scis* 11(5): 331–339.

Hamoudeh M, Seiller M, Chauvierre C, Auchere D, Lacour B, Pareau D, Stambouli M, Grossiord JL. 2006. Formulation of stable detoxifying W/O/W reactive multiple emulsions: In vitro evaluation. *J Drug Deliv Sci Technol* 16(3): 223–228.

Jager-Lezer N, Terisse I, Bruneau F, Tokgoz S, Ferreira L, Clausse D, Seiller M, Grossiord JL. 1997. Influence of lipophilic surfactant on the release kinetics of water-soluble molecules entrapped in a W/O/W multiple emulsion. *J Controlled Release* 45: 1–16.

Li NN. 1978. Metal extraction by combined solvent and liquid membrane extraction. US patent 4,086,163.

Matsumoto S, Inoue T, Kohda M, Ikura K. 1980. Water permeability of oil layers in W/O/W emulsions under osmotic pressure gradients. *J Colloid Interface Sci* 77: 555–563.

Morimoto Y, Subigayashi K, Yamaguchi Y, Kato Y. 1979. Detoxication capacity of a multiple (W/O/W) emulsion for the treatment of drug overdose: Drug extraction into the emulsion in the gastro-intestinal tract of rabbits. *Chem Pharm Bull* 27: 3188–3192.

Morimoto Y, Yamaguchi Y, Subigayashi K. 1982. Detoxication capacity of a multiple (w/o/w) emulsion for the treatment of drug overdose. Detoxication of quinine sulfate with the emulsion in the gastro-intestinal tract of rabbits. *Chem Pharm Bull* 30: 2980–2985.

Muguet V, Seiller M, Barratt G, Clausse D, Marty JP, Grossiord JL. 1999. W/O/W multiple emulsions submitted to a linear shear flow: Correlation between fragmentation and release. *J Colloid Interface Sci* 218: 335–337.

Muguet V, Seiller M, Barratt G, Hebel N, Marty JP, Grossiord. JL. 2000. Influence of the fragmentation of multiple globules on the rheological properties of W/O/W multiple emulsions. *J Rheol* 44(2): 379–395.

Muguet V, Seiller M, Barratt G, Ozer O, Marty JP, Grossiord JL. 2001. Formulation of shear rate sensitive multiple emulsions. *J Controlled Release* 70: 37–49.

Olivieri L, Seiller M, Bromberg L, Ron E, Couvreur P, Grossiord JL. 2001. Study of the break-up under shear of a new thermally reversible water-in-oil-in-water (W/O/W) multiple emulsion. *Pharmaceutl Res* 18(5): 689–693.

Olivieri L, Seiller M, Bromberg LE, Besnard M, Duong TNL, Grossiord JL. 2003. Optimization of the release under shear of a new thermally reversible W/O/W multiple emulsion. *J Controlled Release* 88: 401–412

Pagnoux A, Dolatkhani M, Chaffaux P. 2003. Heat-sensitive polymers and heat reversible gel. Patent WO 03,106,536.

Pareau D, Kaplan P, Stambouli M. 1999. Purification of a radioactive liquid effluent: Elimination of caesium by surfactant liquid membrane extraction. *Proceedings of the International Solvent Extraction Conferences,* Barcelona, pp 965–969.

Potier L, Raynal S, Seiller M, Grossiord JL, Clausse D. 1992. Study of state transitions within multiple W/O/W emulsion using calorimetry (DSC). *Thermoch Acta* 204: 145–155.

Raynal S, Grossiord JL, Seiller M, Clausse D. 1993. A topical W/O/W multiple emulsion containing several active substances: Formulation, characterization and study of release. *J Controlled Release* 26: 129–140.

Raynal S, Pezron I, Potier L, Clausse D, Grossiord JL, Seiller M. 1994. Study by differential scanning calorimetry, rheometry and electroconductimetry of mass transfers at subambient and ambient temperatures in multiple water/oil/water emulsions entrapping MgSO4. *Colloids Surf* 91: 191–205.

Srinivasan MP, Stroeve P. 1986. Subdrop ejection from double emulsion drops in shear flow. *J Membr Sci* 26: 231–236.

Stambouli M. 2006. Reactive detoxifying emulsions: Environmental and therapeutic applications. *4th World Congress on Emulsions*, Lyon.

Tadros TF. 1992. Review: Future developments in cosmetic formulations. *Int J Cosmet Sci* 14: 93–111.

Taylor GI. 1932. The viscosity of a fluid containing small drops of another fluid. *Proc R Soc London A* 138: 41–48.

Taylor GI. 1934. The formation of emulsions in definable fields of flow. *Proc R Soc London A* 146: 501–523.

Tomita M, Abe Y, Kondo T. 1982. Viscosity change after dilution with solutions of water/oil/water emulsions and solute permeability through the oil layer. *J Pharm Sci* 71: 332–334.

Touati M, Benna-Zayani M, Kbir-Ariguib N, Trabelsi-Ayadi M, Grossiord JL, Stambouli M. 2006. Extraction of cadmium from phosphoric acid media by di(2-ethylhexyl) dithiophosphoric acid. *4th World Congress on Emulsions*, Lyon.

Wolfsberger A. 2003. Elimination of mercury from wastewater by emulsion extraction. Diplomarbeit. IVIU Monatanuniversität Leoben, Austria.

CHAPTER 9

Surface-Modified Fine Multiple Emulsions for Anticancer Drug Delivery

AJAY J. KHOPADE and N. K. JAIN

Contents

9.1 INTRODUCTION

9.1.1 Multiple Emulsions

Multiple emulsions are complex systems consisting of a dispersed phase that itself contains globules dispersed within that are miscible with the continuous phase. This leads to water-in-oil-in-water (W/O/W) or oil-in-water-in-oil

Multiple Emulsions: Technology and Applications, Edited by Abraham Aserin

(O/W/O) type multiple emulsion, each with a miscible phase separated by an immiscible phase. In general, the two aqueous phases are identical, and therefore a $W_1/O/W_1$ emulsion is a two-component second-order system and an $O_1/W/O_2$ emulsion is a three-component second-order system. In this manner ternary, quaternary, and even higher order emulsions are possible. In principle, n-order emulsions can be prepared by remicellization of an $(n-1)$ phase into another continuous phase. Multiple emulsions are also called emulsions of emulsions, double or triple emulsions. Seifriz (1925) first published a photograph of a typical W/O/W type multiple emulsion almost 80 years ago. The multiplicity of the multiple emulsion is verified by light microscopy, which is typically visible as shown in Figure 9.1.

On the basis of size observed under inverted phase contrast microscope and images captured with a high-speed camera, Florence and Whitehill classified multiple emulsions into three types: *type A*, composed of relatively small multiple drops of mean diameter 8.5 μm containing a few relatively large internal droplets of mean diameter 3.3 μm; *type B*, composed of larger multiple drops of mean diameter 19 μm containing smaller but more numerous multiple droplets of mean diameter 2.2 μm; and *type C*, with vast number of very small internal droplets entrapped (Florence and Whitehill, 1982).

Over the last 25 years multiple emulsions have been extensively studied and investigated in the pharmaceutical and cosmeticological fields. Although there were limitations in their manufacturing on large scale and applications, numerous researches were carried on multiple emulsion systems avoiding limitations and harnessing advantages. As a result a few commercial preparations of multiple emulsions by Bioderma, Lancaster, Estee Lauder, and Rubinstein laboratories are available in the market (Yazan et al., 1993; Fox, 1986). The microsphere and liposome based depot formulations available in the market, Lupron depot® and Depocyte®, respectively, are not typical multiple emulsion systems, but yet a multiple emulsification process is used for their

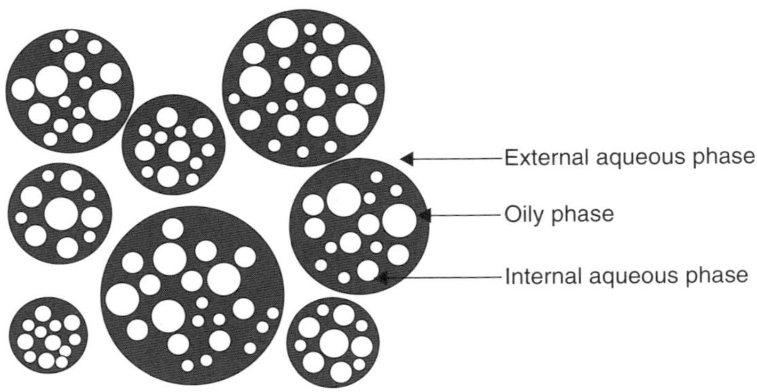

Figure 9.1 Schematic of a typical multiple emulsion system indicating internal, external, and oil phases.

manufacture. These successes raise hope for pharmaceutical scientists to explore this field further.

9.1.2 Multiple Emulsions Containing Anticancer Agents

The word *cancer* inspires fear despite improved modern treatment methods and health consciousness. Research pursuits in this field aim at achieving a better understanding of cancer etiology and growth, its treatment and possible cure by designing active agents with different targets. Nonspecificity of the active agents has led to the huge disappointments in clinics, and the search continues for techniques that considerably alleviate the problems of conventional time-tested chemotherapeutic agents. Among the many exciting and technologically challenging approaches being proposed including nanoparticles, liposomes, solid lipid nanoparticles, and multiple emulsions. Multiple emulsions are expected to allow smart, delivery of anticancer agents by sparing normal tissues and selectively attacking tumor tissues. Multiple emulsions are unique systems for drug delivery as they allow simultaneous encapsulation of two drugs in two distinct domains and release them slowly over a prolonged period. The studies on multiple emulsions encapsulating anticancer drugs are summarized below.

Fukushima et al. (1983) prepared multiple emulsions containing 5-fluorouracil and cytarabine by a two-step emulsification technique. They showed prolonged release of both drugs from the system. In a further study they used oily lymphographic agent as an oil phase to prepare a W/O/W emulsion containing anticancer agents (Fukushima et al., 1987). The release rate of 5-fluorouracil from the multiple emulsion was so fast that the release was completed within 1 h while the release of cytarabine and glucose was sustained and peplomycin was hardly released from the multiple emulsion.

The release rate of methotrexate encapsulated in the internal phase of W/O/W emulsions, stabilized by an interfacial interaction between albumin and sorbitan monooleate, was measured as function of two formulation variables—the oil phase and the secondary emulsifier composition. The release rate was significantly affected by the nature of the oil phase and surfactants with high HLB values (Omotosho et al., 1989). The influence of the oil phase of the W/O/W emulsions on the oral absorption of 5-fluorouracil in the rat was determined by measuring liver and lymphatic accumulations of the drug. The multiple emulsion system showed potential as a lymphotropic carrier to the mesenteric lymph nodes following oral administration (Omotosho et al., 1990).

Lipidiolized W/O and W/O/W multiple emulsions containing doxorubicin hydrochloride with different emulsifiers were prepared to evaluate in vitro sustained release behavior and the pharmacokinetic and tissue distribution function in Sprague-Dawley rats. The results showed sustained release of drug in vitro, which was correlated in vivo by prolongation of plasma drug levels and decreased clearance. The tissue distribution and retention of drug was dependent on the concentration of emulsifier and emulsion type. The lipidiol and hydrogenated castor oil played an important role in the prolongation and selective retention of W/O and W/O/W emulsion in vivo (Lin et al., 1992).

Methotrexate loaded multiple emulsion was prepared by using bovine serum albumin as stabilizer in internal phase to avoid internal droplet coalescence. By microwave technique the albumin in internal aqueous phase was solidified to form a microspheres-in-oil-in-water emulsion (S/O/W). The formulation and process variables were optimized and evaluated for physicochemical characteristics, such as microscopic structure, electrical charge, particle size distribution, rheological behavior, yield, entrapment efficiency, drug release and stability. The emulsions were found to be stable and showed prolonged release in vitro (Tao et al., 1992).

Ma et al. (1993) prepared multiple emulsions containing etoposide and evaluated its physicochemical properties. The release was found to be biphasic, a fast phase releasing 50% of the drug in 10 hours followed by slow phase in which up to 88% of the drug was released in next 38 hours.

Cytarabine loaded W/O/W multiple emulsions were prepared using nonionic surfactants (Tween® and Span®) by the emulsification-sonication technique and characterized by studying the osmotic behavior. The system exhibited prolonged release pattern (Kim et al., 1995).

From most of the studies described above, it can be noted that the multiple emulsions are proposed mainly for sustained release of anticancer drugs but not for drug targeting. One obvious reason is their large size, which is dependent on the size of internal aqueous phase droplets. If it was possible to reduce the size of the internal aqueous droplets to nanometers without coalescence, it may be possible to make multiple emulsions that are very small in size. A method for obtaining small-sized multiple emulsion with a solidified oil phase is reported by Morel et al. (1994) to avoid internal droplet coalescence and migration into external phase.

9.1.3 Targeting with Multiple Emulsions

The active and passive targeting of a multiple emulsion to the tumor site through surface modification is a promising and unexplored extension. The targeting is generally achieved with other colloidal carriers by coating the tumor-specific ligand or ligands capable of avoiding reticuloendothelial (RES) recognition. We have been trying multiple emulsion preparation and coating approaches in our laboratory with a fairly good success rate, albeit numerous problems still need to be addressed. The work was initiated with dextran-coated multiple emulsions bearing an antiinflammatory drug that showed enhanced accumulation of to the inflammatory sites (Khopade et al., 1995). We have also studied passive, size-based targeting of multiple emulsion to the lungs coupled with enhanced macrophage uptake by mannan coating (Khopade et al., 1996). To our knowledge, this was the first example in the literature on multiple emulsion targeting to the lungs through an intravenous route. Morel's (Morel et al., 1994) solidified multiple emulsion, the multiple emulsions we

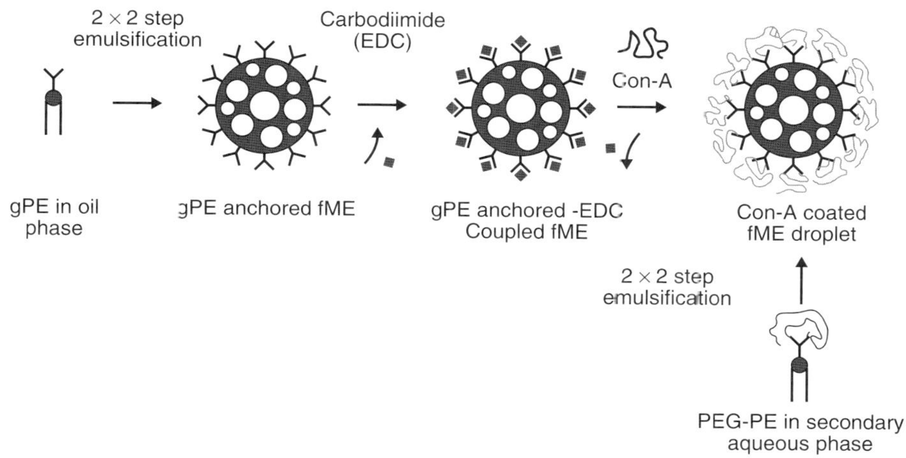

Figure 9.2 Schematic of the methods used for coating fMEs with Con-A and PEG.

used in the studies noted above, and some of the multicompartmental structures where the internal droplet sizes of the delivery system were in nanometric range (Khopade et al. 2007, unpublished results, Figure 9.2) while their droplet sizes varied have prompted us to classify multiple emulsions into three types: coarse multiple emulsions > 3 μm diameter, fine multiple emulsions (fME) of about 1–3 μm diameter, and micro-multiple emulsions < 1 μm diameter, all having internal droplets in nanometer range. The microscopy was used to prove the multiple nature of such emulsions and a confocal microscope was used to locate fluorescent probes in the internal phase (Khopade and Jain, 2002).

The following discussion will focus on the preparation and characterization of surface-modified fME for anticancer drug delivery. The fME bearing a model anticancer drug, 6-mercaptopurine (6-MP) were subjected to two types of surface-modification approaches for improved cancer therapy: (1) Concanavalin A (Con-A) coating for ligand-based targeting (active targeting) and (2) monosialogangliosides (GM_1), sphingomyelins (SM), and poly(ethylene glycol) (PEG) coatings to improve the circulation half-life of the drug (passive targeting).

9.2 PREPARATION OF fME

A two-step emulsification technique are used to prepare W/O/W fME. A careful selection of oil, surfactant, aqueous phase, and emulsification process variables is required to prepare fMEs. Typically the oil phase consists of corn oil containing palmitic acid and phosphatidylcholine as the emulsifier, which is dissolved with heating to 60°C. The internal aqueous phase was 1% w/v 6-

MP solution in phosphate buffer at pH 10. The primary W/O emulsion was prepared by using a mechanical stirrer (with specially designed blades that could be lowered into a small-volume glass vessel) followed by sonication at 55 W with a probe type ultrasonicator for 15 minutes. Basic pH 10 allows improved drug solubility in the internal phase. The free fatty acid was added for gelling the oil phase and to impart the potential (charge) This emulsion was added to the external phase consisting of phosphate buffer, pH 7.4, containing taurodeoxycholate and butyric acid at 60°C with stirring for 5 minutes followed by sonication.

Cosurfactant facilitates droplet size reduction in the fME range. The multiple emulsion was diluted with equal volume of ice-cold external phase (<4°C), stirred for 15 minutes, and immediately transferred to a refrigerator. The pH of the external phase was increased to about 8.0 to 8.9 and then restored to 7.4 with addition of an acid. The multiple emulsion was formed with a relatively high percentage of the secondary surfactant; further a cosurfactant had to be added to reduce the size of multiple emulsion. The formulation and process variables like amount of surfactants, phase volume, and time period of sonication for both emulsification steps were optimized by way of the appropriate evaluation parameters, mainly solidification, droplet size, yield, viscosity, and/or stability.

9.2.1 Coating of fME with Concanavalin-A

Phosphatidylethanolamine (PE) was reacted with glutaric acid anhydride in the presence of triethylamine to obtain an anchor product, *n*-glutaryl phosphatidyl ethanolamine, g-PE, after purification on a silica gel G column (1 × 30 cm) with an isotonic saline used as a mobile phase (Weissig et al., 1989). The anchors g-PE and PE were incorporated into the oil phase during the emulsification step and followed by Con-A coupling by the carbodiimide coupling method. The Con-A bound multiple the emulsion was separated by exclusion chromatography on a Sephadex G-50 (Pharmacia, Germany) column. The fME coating with the concanavalin is illustrated in Figure 9.3.

The proof of coating and the coating quality were evaluated in terms of a dextran-induced aggregation. The rate of aggregation depends on the coverage of Con-A over the surface and the free saccharide binding sites available for the dextran. The aggregation is reflected as increased turbidity, which is measured spectrophotometrically at 620 nm as a function of time with a suitably diluted sample. This method was used earlier on polysaccharide-coated emulsion systems to evaluate the Con-A coating qualitatively (Iwamoto et al., 1991). Uniform distribution of anchor (g-PE) over the surface of droplets results in uniform and efficient carbodiimde-mediated coupling. Other methods such as random cross-linking of Con-A showed loss of sugar binding sites.

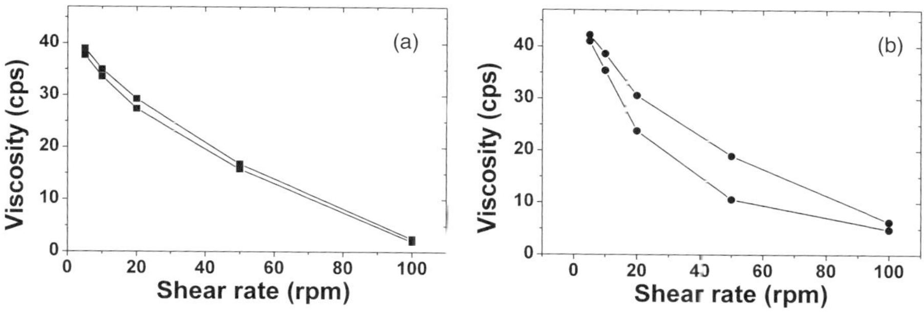

Figure 9.3 Rheological properties of fME formulations: (*a*) fME, *SM*-fME, and GM_1-fME; (*b*) PEG-fME and Con-A-fME.

9.2.2 Coating of fME with Stealth Lipids

The lipid-derivatized PEG was prepared by reacting PEG-2000 with carbonyldiimidazole in benzene and the residue (PEG-imidazole carbamate) with phosphatidylethanolamine in the presence of trimethylamine (Allen et al., 1991). The purified PEG-carbonyl phosphatidyl ethanolamine (PEG-PE) fractions on silica gel column chromatography were used for surface modification by incorporating it in the secondary aqueous phase (Figure 9.3). The fMEs coated with monosialoganglioside (GM_1) and sphingomyelin (*SM*) were prepared by incorporating them in the oil phase during the primary emulsification step.

9.3 CHARACTERIZATION OF FINE MULTIPLE EMULSIONS

9.3.1 Physicochemical Characterization

Droplet Size Droplet size was determined on a Var tan Coulter counter (Model 4000, Zymo Instruments, India) after diluting the multiple emulsion 100 times with an external phase. The droplet size ranged from < 1.0 to 1.93 μm. Table 9.1 shows the droplet size and the polydispersity of the multiple emulsion formulations. The particle sizes of 6-MP loaded, coated, and uncoated multiple emulsions was nearly the same. The mean diameter of PEG-coated multiple emulsions was found to decrease with time, and it reached its minimum size after 6 to 8 hours. The size was reduced by nearly 15%, and the polydispersity was decreased from 0.18 to 0.09. Kim et al. (1995) found the separation of the aggregated multiple emulsion droplets to occur over several days of equilibration. In our case, it was few hours because of steric stabilization of individual particles surface coated (Vanderhoff and El Asser, 1988).

Encapsulation Efficiency Encapsulation efficiency was determined by centrifuging multiple emulsion at 2500 rpm for 10 minutes. The bottom clear

TABLE 9.1 Physicochemical characteristics of multiple emulsion formulations

Formulation	Mean Size ± SD (μm)	PI	ζP (mV)	EE (%)	pH
fME	1.6 ± 0.13	0.08	−24.1 ± 1.7	92.5 ± 1.2	8.2
Coated					
Con-A-fME	1.3 ± 0.12	0.09	−22.8 ± 2.4	90.3 ± 1.3	8.7
PEG-fME	1.3 ± 0.12	0.09	−12.8 ± 3.4	89.9 ± 2.1	8.8
GM_1-fME	1.4 ± 0.11	0.07	−19.5 ± 1.5	90.2 ± 1.7	8.8
SM-fME	1.4 ± 0.19	0.13	−21.7 ± 2.1	91.5 ± 0.9	8.6

Note: ζ: Zeta potential; SD: standard deviation; EE: entrapment efficiency %; PI: polydispersity index = SD/Mean.

aqueous layer was filtered using 0.2 μm pore size membrane filter and analyzed spectrophotometrically at 318 nm. Encapsulation efficiency was calculated by

$$\text{Encapsulation efficiency}\,(\%) = \frac{\text{Total 6-MP} - \text{Free 6-MP}}{\text{Total 6-MP}} \times 100.$$

The entrapment efficiency was 88.5% to 93.7% (Table 9.1). This may be attributed to noncoalescence of internal droplets despite the second sonication step. Gelling of the oil phase upon congealing imparts stability to internal droplets and results in less expulsion/breaking of droplets and drug dumping in the external phase. The entrapment efficiency correlates with micro-multiple emulsion reported by Morel et al. (1994).

ζ-Potential The ζ-potential was determined on a Zetasizer (Malvern Instruments, UK). Multiple emulsion was diluted 1000 times with external phase and introduced into the cell of the instrument. The electrophoretic mobilities of the droplets were read and the mobilities (υ) were converted to the electrophoretic potential (ζ) using the Smoluchowski relation $\zeta = \upsilon\eta/\varepsilon$, where η and ε are the viscosity and permittivity of the solution, respectively. The charge was imparted on multiple emulsion due to the presence of free fatty acids and surfactants in oil phase, which also improve colloidal stability. A decrease in the ζ-potential with the addition of surface modifiers was in the order PEG-PE > GM_1 > *SM* > Con-A. The ζ-potential was not changed for Con-A-fME because of the presence of ionic groups on Con-A. Despite GM_1 and *SM* having a charge, their charge is masked by hydrophilic determinants in their structure reflected as lower ζ-potential. The ζ-potential decreased significantly in case of PEG-fME due to masking of charge by non-ionic PEG molecules. A decrease in ζ-potential suggests the possibility of aggregation however, negligible aggregation was observed due to steric entropic-enthalpic repulsion mechanism (Vanderhoff and El Asser, 1988). This effect was relatively less for Con-A, GM_1 and *SM* hence slight aggregation was observed after 7 d.

Rheology Rheological studies were performed using a cone and plate viscometer (BrookeField Engineering Labs, Massachusetts, USA). The shear rate was applied from 5 to 100 rpm. The flow curves (shear rate vs. viscosity) for multiple emulsion samples are shown in Figure 9.3*a* and *b*. Multiple emulsion samples containing GM_1 and *SM* showed overlapping of up and down curves immediately after preparation suggesting nearly the Newtonian flow (Figure 9.3*a*). Con-A and PEG coated multiple emulsions showed hysteresis and sometimes crossing of up and down curves (Figure 9.3*b*), suggesting hydrodynamic interaction due to Con-A and PEG chains during the shearing cycle (Deem, 1988). This behavior also provides an indirect proof of polymer coating.

***In vitro* Drug Release Profile** The drug-loaded multiple emulsion was placed in a dialysis bag (Spectrapore-2 Sigma, USA) and was kept in a rotating basket of USP type I dissolution apparatus. The basket was lowered in 900 ml of isotonic phosphate buffered saline (PBS, pH 7.4) at 37 ± 1°C and dialyzed (Hino et al., 1995). At appropriate time intervals, 5.0 ml aliquots were withdrawn and replenished with an equal volume of dialyzing fluid. The samples were analyzed spectrophotometrically. As illustrated in Figure 9.4, the multiple emulsions showed prolonged release of 6-MP. The drug release from the multiple emulsions in 24 hours was between 18% and 25%. If it is assumed that drug is solubilized in the reverse mixed micellar phase, the Magdassi and Garti release profile can be expected (Magdassi and Garti, 1995). But the zero-order release profile was obtained. Since the solubility of 6-MP in lipid is low, the release from a multiple emulsion is determined kinetically by permeation through a gelled lipid membrane into the surrounding medium. The protruding polymer chains and loops further provide a hydrophilic barrier due to the presence of relatively stagnant aqueous layer associated with coated multiple emulsions. The release profile of GM_1-fME and *SM*-fME almost overlapped with that of uncoated multiple emulsion.

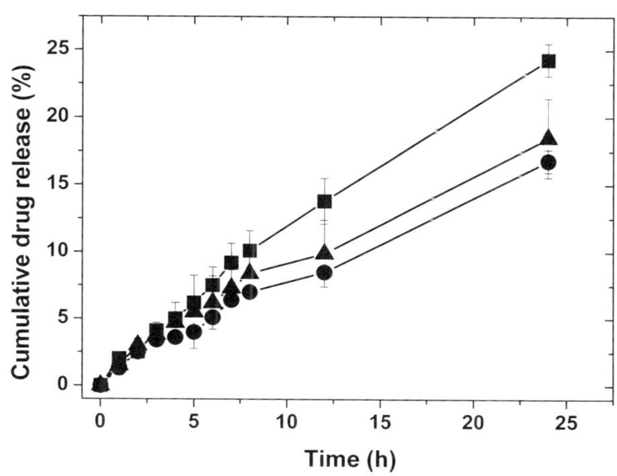

Figure 9.4 Release profile of 6-MP from fME formulations. ■: fME; ▲: Con-A-fME; ●: PEG-fME.

Stability The stability parameters studied included aggregation, creaming, change in droplet diameter, apparent viscosity, and leakage of 6-MP during 30 d storage by the methods described previously. Creaming was determined visually by measuring cream height in a 10 ml measuring cylinder kept at 4°C. Leakage of drug was monitored by UV spectrophotometric method. The stability characteristics are shown in Figure 9.5*a* through *e*. Most of the changes take place in 7 d, followed by a slow or no change. Slight aggregation was observed after 30 d, which is also reflected as the change in droplet size of <7% in case of Con-A-fME. Cream height was less than 5% v/v due to small droplet size. The leakage was less than 4% in all cases due to very low partition coefficient of drug. The apparent viscosities were slightly increased on storage and attained a plateau. The stability characteristics of GM_1-fME and SM-fME also overlapped with those of the uncoated multiple emulsion and hence are excluded from the figure for clarity. No phase separation was observed during 30 d period. The multiple emulsion was quite stable during the one month storage period.

9.3.2 Biological Characterization

Pharmacokinetics Sprague-Dawley albino rats of either sex weighing about 250 g were divided into groups of four rats each. Free drug, coated, and uncoated multiple emulsion formulations, equivalent to 2.5 mg 6-MP per kg

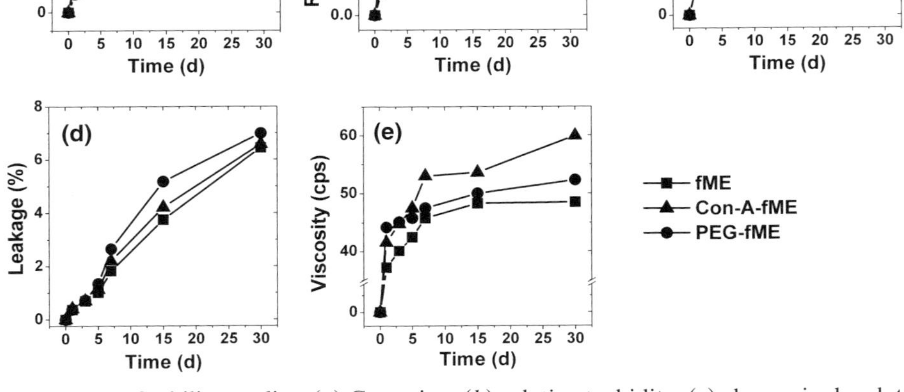

Figure 9.5 Stability studies: (*a*) Creaming, (*b*) relative turbidity, (*c*) change in droplet size, (*d*) leakage, and (*e*) viscosity. ■: fME; ▲: Con-A-fME; ●: PEG-fME.

body weight approximating to 1.25 to 2.0 ml injection volume, were administered separately to each group through caudal vein. At 0.5, 1, 2, 3, 4, 6, and 8 h post injection, 200 µl of blood was collected from the retro-orbital plexus of the rats. The samples were analyzed for drug content by a HPLC method (Sahnoun et al., 1990). The increase in half-life of the drug from multiple emulsion formulations compared to free drug were studied as a function of PEG-PE, GM_1, and *SM* concentrations. In the case of PEG-fME, the maximum $T_{1/2}$ was observed with 60 µg/ml of PEG-PE in external phase. While for multiple emulsions containing *SM* and GM_1, the maximum $T_{1/2}$ was obtained with 40% (w/w) of total lipid surfactants. Because the half-life was significantly higher for PEG-fME compared to *SM*-fME and GM_1-fME, only the emulsion was selected for further in vivo studies.

The in vivo clearance of coated and uncoated multiple emulsions was analyzed from blood level data of 6-MP after i.v. injection from various formulations. The plasma profile is shown in Figure 9.6. The pharmacokinetic parameters are shown in Table 9.2. The coated multiple emulsion showed

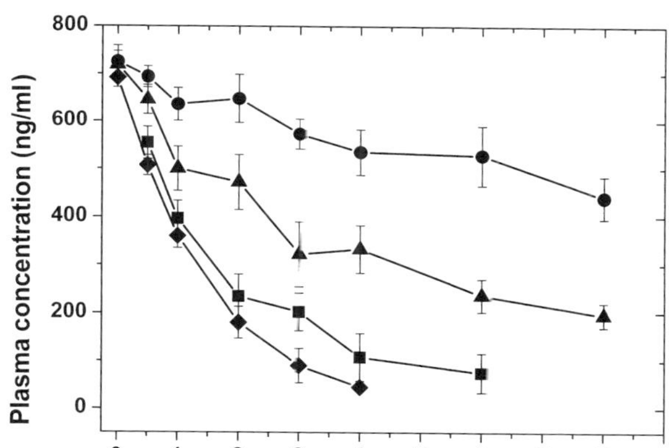

Figure 9.6 Plasma profile of 6-MP from multiple emulsion formulations. ◆: Free drug; ■: fME; ▲: Con-A-fME; ●: PEG coated.

TABLE 9.2 Pharmacokinetic parameters

Formulation	$T_{1/2}$ (min)	AUC (ng/ml × h)	Clearance (ml/h)	V_d (ml)	K_{el} (1/h)
Free	60.08 ± 2.9	740 ± 36	844 ± 34	1219 ± 108	0.69 ± 0.033
fME	102.35 ± 18.2	1114 ± 104	560 ± 85	1380 ± 273	0.41 ± 0.068
Con-A-fME	497.47 ± 36.8	1761 ± 96	354 ± 78	4250 ± 508	0.08 ± 0.006
PEG-fME	754.62 ± 22.1	4358 ± 116	143 ± 34	2602 ± 367	0.05 ± 0.004

TABLE 9.3 Tissue concentration of 6-MP in rats administered with fME formulations

	Liver		Lung		Spleen	
Formulation	1 h	24 h	1 h	24 h	1 h	24 h
Free	16.2 ± 1.20	3.5 ± 0.18	5.6 ± 0.23	0.7 ± 0.33	4.3 ± 0.11	0.5 ± 0.22
FME	15.6 ± 0.91	9.1 ± 1.20	7.4 ± 0.25	4.0 ± 0.40	8.2 ± 0.24	4.5 ± 0.31
Con-A-fME	21.7 ± 2.13	18.2 ± 1.10	17.7 ± 0.98	13.5 ± 0.23	12.9 ± 0.24	10.6 ± 0.16
PEG-fME	7.4 ± 0.16	8.5 ± 1.51	6.8 ± 0.50	4.9 ± 0.28	9.2 ± 0.10	7.0 ± 0.62

prolonged half-life compared to free 6-MP and uncoated fMEs. However, uncoated fMEs also showed about 1.80 times greater half-life than the plain drug. PEG-fME and Con-A-fME, respectively, showed 12.47 and 4.33 times greater half-life than plain drug and 6.92 and 2.40 times greater compared to fME. The AUC values, respectively, for PEG-fME and Con-A-fME were 4.57, 2.86, and 1.43 times greater than free 6-MP and 3.18 and 1.99 times greater than fMEs. This shows an increase in the bioavailability of the drug from the fME formulations. The clearance was decreased for coated fMEs free drug and uncoated fMEs. The volume of distribution was increased, which indicates that the drug is distributed in the peripheral compartment. The uptake was 2.72 and 1.51 times for PEG-fME and Con-A-fME compared to free drug and 2.16 and 1.20 times compared to fME, respectively. Overall, the parameters show that multiple emulsion based delivery of 6-MP improves its bioavailability by decreasing clearance and prolonging its half-life in rat plasma.

Organ Distribution Albino rats were treated similarly as above. Animals were sacrificed and the organs were removed at 1, 2, 4, 8, 12, and 24 hour intervals after administration of formulation. Blood was collected by cardiac puncture, and the lymph was collected from the thoracic lymph duct before removal of organs. They were homogenized and centrifuged, and the supernatant was subjected to analysis after suitable treatment. Drug localization was determined as tissue-to-blood AUC ratio. The percentage of drug depletion in 24 hours was calculated as

$$\text{Drug depletion} = \frac{\text{Drug conc. in tissue at 1h} - \text{Drug conc. in tissue at 24h}}{\text{Drug conc. in tissue at 1h}} \times 100.$$

The tissue concentration of 6-MP from various formulations are listed in Table 9.3. In general, high concentration values were observed for Con-A-fME

Kidney		Gut		Lymph		Skin	
1 h	24 h	1 h	24 h	1 h	24 h	1 h	24 h
5.6 ± 0.21	0.64 ± 0.19	3.1 ± 0.15	0.4 ± 0.14	7.10 ± 1.29	0.2 ± 1.16	0.56 ± 0.08	0.12 ± 0.09
8.8 ± 0.33	5.1 ± 0.32	4.5 ± 0.21	2.9 ± 0.25	22.5 ± 3.12	14.6 ± 1.40	0.99 ± 0.09	0.65 ± 0.09
13.5 ± 0.15	10.7 ± 0.11	5.0 ± 0.21	5.4 ± 0.12	26.7 ± 2.34	22.3 ± 2.50	1.50 ± 0.06	1.61 ± 0.08
9.9 ± 0.12	7.4 ± 0.18	3.5 ± 0.17	3.6 ± 1.95	20.8 ± 2.11	17.9 ± 0.08	1.2 ± 1.12	1.2 ± 0.07

in all the tissues at 1 hour due to uptake by RES system but low values for fMEs, which are preferentially cleared by lymphatic route. The values were still lower for PEG-fME. There were higher levels of 6-MP in tissues at 24 hours for Con-A coated multiple emulsion compared to the free drug. Moreover substantially greater amounts of drug were retained in tissues for coated fMEs at 24 hours compared to uncoated fME formulations.

The tissue-to-blood AUC ratios provides correlation between the availability of the drug in plasma to the tissue. In general. higher ratios were observed for Con-A-fME and fMEs than PEG-fMEs. The lymph remains the major extracting organ, followed by liver and spleen. Accumulation in other organs may be due to a sustained release effect from multiple emulsions. Higher values in the gut may be due to the intrinsic capability of a drug to reach the intestine coupled with the sustained release effect (Florey, 1978).

Drug depletion provides information about the residence time of the drug in a particular organ. Drug depletion from the free drug solution was 78% to 98% in all organs. There was 41% to 46% decrease in drug concentration in liver, lung, spleen, and kidney at 24 hours, but it was decreased to a lesser extent in gut, lymph, and skin (34–35%) with uncoated fMEs. Coated fMEs showed 16% to 24% decrease in 24 hours in all the organs except skin and gut where the drug level was greater than at 1 hour, which may be attributed to the slow accumulation of the drug in these tissues. The drug depletion from tissues in 24 hours was significantly slower for Con-A-fME than the uncoated fMEs. The less depletion for Con-A-fME compared to PEG-fME may be attributed to an increased retention in the tissues for prolonged periods (Shier, 1979) while PGE-fME had comparatively low tissue uptake.

Although the pharmacokinetics and biodistribution of 6-MP in fME administrated rats have demonstrated promising anticancer delivery system, their activity on tumor cells required deeper investigation. Hence we also compared the in vitro cultured cell toxicity and in vivo antitumor efficacy of fMEs and surface-modified fMEs with the free drug.

9.4 EVALUATION OF ANTICANCER ACTIVITY

9.4.1 In vitro Cytotoxicity

Cells Murine leukaemia L-1210 cell lines were obtained in plastic Roux culture flasks as a monolayer culture in Dulbecco's modified Eagle's medium (DMEM) supplemented with 100 μg/ml of streptomycin, 10% foetal calf serum, and 1000 U/ml of penicillin from the National Centre for Cell Science (NCCS), Pune, India.

Uptake Studies Cell lines were harvested by trypsinization and centrifugation and washed twice with a phosphate-buffered saline (PBS, pH 7.4) solution. Cells (1×10^8) were resuspended in 10 ml DMEM containing foetal calf serum in a 36-well culture plate and incubated for 2 to 3 hours at 37 ± 1°C. Free 6-MP, fME, Con-A-fME, and PEG-fME bearing 6-MP in a dose equivalent to 1000 ng was added to each well. At 60, 120, and 180 minutes intervals, 0.5 ml of the suspension was removed and placed in 2 ml ice-cold PBS. The samples were centrifuged at 200 g and washed with 5 ml cold PBS to remove non-cell associated 6-MP. Cell viability was >90% at the beginning and end of each assay as determined by trypan blue exclusion. The final pellet was analyzed by HPLC for 6-MP content (Sahnoun et al., 1990). The results of the cell uptake study demonstrate that there was no detectable amount of 6-MP in tumor cells supplemented with the free drug at 15 minutes; it reached maximum at 60 minutes and maintained same level for 180 minutes (Figure 9.7). Encapsulation in fME slightly increased the 6-MP association with the cells, possibly due to the surface properties imparted by the phospholipids and bile salts, used as emulsifiers. The activity of lipid-based systems like niosomes and liposomes has been attributed to surface activity enabling better permeation of

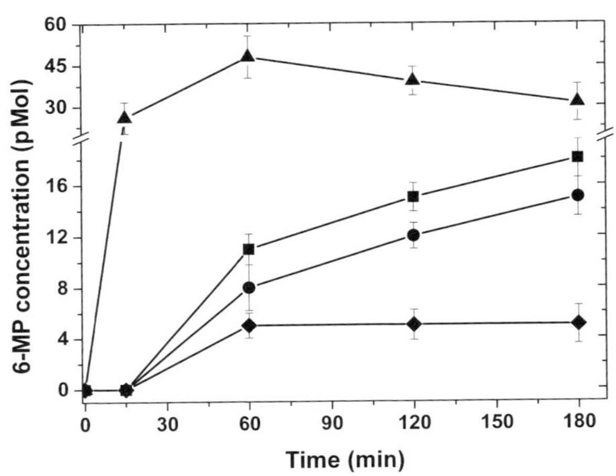

Figure 9.7 6-MP uptake by in vitro cultured leukemia L-1210 cells. ■: fME; ▲: Con-A-fME; ●: PEG-fME; ♦: Con-A.

the drug in the cells (Chandraprakash et al., 1993). There was slow uptake from fME, reaching a maximum by 180 minutes. The uptake by Con-A-fME reached a maximum in 60 minutes and decreased at 180 minutes. At 15 minutes 26 pMol uptake was noted, showing a rapid association of the drug in Con-A-fME; this may be due to endocytosis. Uptake by PEG-fME was less at all intervals studied, showing a slow association of the drug in PEG-fME to cancer cells, which was probably due to steric hindrance produced by hydrophilic PEG chains.

Cytostatic Assay Cells at an initial density level 2×10^5 cells/ml were placed in each well of multiwell plates. After 24 hours incubation the medium was removed by centrifugation at 200 g. One ml of fresh medium with the respective formulations at the indicated doses was added to each well. These plates were incubated for 4 days under the conditions mentioned above. On day 4, the final cell density in each well of the plate was determined. The cellular sensitivity to the drug was defined as the mean IC_{50}, which is the concentration of 6-MP required to inhibit cell growth to 50% of the untreated cell population. The cytostatic activity was assessed and found to be nearly same for the free drug and fME (insignificant at $p = 0.01$ but significant at $p = 0.05$ level). There was slight increase in the activity of the multiple emulsions, which was probably due to the increased uptake of the drug. IC_{50} was decreased for Con-A-fME compared to the free drug and fME (Table 9.4, Figure 9.7). This shows that cytostatic activity is directly correlated to the cell uptake of the drug from the formulation. The IC_{50} was $55.8 \pm 3.5\,\mu g$ for Con-A, which caused inflammatory reaction in mice when injected by the i.p. route. IC_{50} was decreased 1.64 times and 1.47 times for PEG-fME compared to the free drug and fME formulation, respectively, while it was decreased by 1.12 times for fME than for the free drug. Cytostatic activity showed direct correlation with the cell uptake of the drug from the formulations.

Cytotoxicity Cell lines were incubated in the 36 well culture plates at a seeding strength of 2×10^5 cells/ml. The respective formulations were added in a 100 ng concentration. The plates were incubated at 5% CO_2 and 95% humidified air at 37 ± 1°C. The cell viability was assessed by trypan blue after 12 hours and 24 hours. Cytotoxic activity as a function of cell viability revealed an insignificant difference at $p = 0.01$ levels (t-test) at 12 hours and 24 hours. The cytotoxic activity showed a similar profile among the formulations; that is, the order of toxicity was Con-A-fME > fME > PEG-fME > free drug > Con-A. The number of viable cells decreased with time (Table 9.4).

9.4.2 In vitro Anticancer Activity

Antitumor Activity Balb/c mice (20–25 g) were used throughout the study. They were housed in microlon boxes and kept on standard diet (Chakan Oil Mills, Pune, India) and water *ad libitum*. Murine leukaemia cells lines (L-1210) were maintained by weekly i.p. injection of 10^6 cells/mouse.

TABLE 9.4 In vitro cytostatic and cytotoxic activity of 6-MP fME formulations against leukemia L-1210 cell lines

		% Viability	
Formulation	IC_{50} (μM)	12 h	24 h
Free	2.8 ± 0.2	86.5 ± 3.0	77.3 ± 2.5
Con-A	55.8 ± 3.5*	94.3 ± 1.5	92.3 ± 3.5
fME	2.5 ± 0.1	78.6 ± 2.5	63.6 ± 4.4
Con-A-fME	0.7 ± 0.09	62.0 ± 3.6	50.5 ± 3.1
PEG-fME	1.7 ± 0.39	83.7 ± 5.1	70.3 ± 2.2

Note: *(μg) caused inflammatory reaction if injected i.p. in mice; IC50 = inhibitory concentration to reduce cell population by 50%; $n = 4$.

The animals were inoculated intraperitoneally (i.p.) and intravenously (i.v.) with 2 to 3 × 10^5 tumor cells/mouse on day 0, except for the control group. The treatment was given on days 1, 3, 5, and 7 with free 6-MP, Con-A, fME, Con-A-fME, and PEG-fME, 24 hours post-inoculation at an i.v. and i.p. dose of 2.5 mg/kg. The control group was treated with PBS. Median survival time (MST) for each group containing six mice was noted with reference to control. The animals surviving more than 40 days were considered to be cured. Survival times of treated groups (T) were compared with those of control groups (C) by the following equation:

$$\text{Increase in life (ILS)} = \text{T/C\%} = \frac{\text{MST of treated group}}{\text{MST of control group}} \times 100.$$

The antitumor activity fME formulations of 6-MP in terms of MST, ILS, and the number of survivors on day 40 is shown in Table 9.5. The MST in mice was found to depend on two factors: prolonged release of the drug from the formulation and direct interaction of tumor cells with the formulation. Thus the i.v.–i.v. and i.p.–i.p. tumor cells–formulation injection route showed greater MST and number of survivors on day 40 than the i.v.–i.p. and i.p.–i.v. injected tumor cells–formulation, respectively. There was no significant difference in MST either between the i.v.–i.v. and i.p.–i.p. or between the i.v.–i.p. and i.p.-i.v. combinations of tumor cells–formulation injection routes. The MST was found in the order Con-A-fME > PEG-fME > fME > free drug > Con-A > control (Table 9.5) in all cases, which reflected a direct correlation with the in vitro studies. The ILS was 208% to 235% with Con-A-fME. The ILS varied between 211.84% and 231.79% for PEG-fME. This shows the improved therapeutic efficacy of coated formulations. PEG-fME, however, had lower efficacy than Con-A-fME, which was due to the site specificity of the latter ligand. Also there was not much difference with respect to the route of the drug and tumor administration, as seen in case of Con-A-fME. Thus the superior activity of the PEG-coated emulsion may be due to slow and consistent release of the

TABLE 9.5 Antitumor activity of 6-MP multiple emulsion formulations in Balb/c mice bearing murine leukemia L-1210

Injection Route	Route of Tumor Administration	Formulation	MST (SD) (days)	ILS Survivors	T/C (%)
i.p.	i.p.	Control	17.4 (2.19)	100.00	0/6
		Free	23.8 (1.70)	136.78	0/6
		Con-A	17.8 (0.83)	102.29	0/6
		fME	28.6 (1.51)	164.36	2/6
		Con-A-fME	41.0 (3.18)	235.63	5/6
		PEG-fME	36.7 (2.32)	222.41	4/6
i.v.	i.p.	Control	15.2 (1.40)	100.00	0/6
		Free	20.5 (1.61)	134.86	0/6
		Con-A	16.0 (1.23)	105.26	0/6
		fME	25.4 (0 73)	167.10	2/6
		Con-A-fME	31.7 (2.95)	208.55	3/6
		PEG-fME	33.2 (2.23)	211.84	3/6
i.p.	i.v.	Control	15.3 (1.41)	100.00	0/6
		Free	24.1 (1.22)	157.51	0/6
		Con-A	15.6 (0.78)	101.96	0/6
		fME	29.5 (0.65)	192.81	2/6
		Con-A-fME	32.6 (1.60)	213.07	2/6
		PEG-fME	34.5 (2.62)	218.95	2/6
i.v.	i.v.	Control	17.3 (1.62)	100.00	0/6
		Free	21.5 (1.00)	124.27	0/6
		Con-A	18.0 (0.49)	104.04	0/6
		fME	31.2 (1.43)	180.34	2/6
		Con-A-fME	39.9 (1.81)	230.63	4/6
		PEG-fME	35.1 (1.50)	231.79	5/6

Note: SD: Standard deviation.

drug from the formulation; resulting in continuous perfusion of the drug to cancer cells.

Tumor Cell Growth Studies on in vivo tumor cell growth inhibition with fME formulations were carried out under identical experimental conditions, as described earlier. The animals inoculated i.p. with tumor cells were sacrificed 24 hours after the last treatment, and the tumor cells were collected by i.p. wash, thrice with 3 ml PBS. Cell viability was noted by the trypan blue exclusion test. The total number of viable cells/animal of the treated group was compared with that of the control group. Upon treatment, the number of cancerous cells decreased compared to the control group (Table 9.6). There was no significant difference between the free 6-MP and fME formulations, whereas the free Con-A treatment decreased the number of cells slightly (p = 0.05, student t-test). The number of tumor cells decreased significantly with the Con-A-fME formulation, indicating its specific uptake by the L-

TABLE 9.6 Effect of 6-MP fME formulations on cell counts in intraperitoneal wash fluid in Balb/c mice

Formulation	Injection Route	Number of Tumor Cells ($\times 10^8$)	Normal Peritoneal Cells ($\times 10^6$)
Control	i.p.	9.7 ± 0.87	4.8 ± 1.1
Free drug		7.3 ± 0.55	5.3 ± 0.9
Con-A		6.5 ± 0.91	11.7 ± 1.0
fME		4.9 ± 1.30	7.4 ± 0.7
Con-A-fME		0.2 ± 0.34	13.6 ± 1.4
PEG-fME		3.2 ± 0.80	5.6 ± 1.7
Control	i.v.	9.9 ± 1.10	—
Free drug		8.6 ± 0.68	—
Con-A		9.1 ± 0.42	—
fME		6.2 ± 1.21	—
Con-A-fME		6.9 ± 0.53	—
PEG-fME		5.9 ± 0.79	—

1210 cells. The number of tumor cells also decreased with the PEG-fME formulation, mainly due to slow release of 6-MP from the formulation. The rapid lymphatic uptake of fME and PEG-fME compared to Con-A-fME decreased the relative drug concentration at the tumor (i.p.) site. This was established by the fact that the number of cells on a single-dose treatment was nearly the same as that of control, whereas it decreased for the free drug and Con-A-fME. In the case of the multiple dose treatment, the drug concentration was maintained for a longer period compared to that of the free drug; hence better efficacy was noted. The effect of PEG-fME was independent of the route of administration, since the antitumor efficacy was due only to the prolonged release of the drug. Intravenous treatment showed comparatively less decrease in number of tumor cells for Con-A-fME. In contrast, with i.p. treatment the direct contact of formulation with the cells led to a significant decrease in the tumor cell population.

9.4.3 Toxicity in Mice

Effect of Formulations on Normal Peritoneal Cells The mice, each in groups of four, were treated i.p. with the respective formulations for two days. Peritoneal exudate cells were counted 24 hours after the last treatment and compared with the control group. The normal peritoneal cells were unaltered with the fME and PEG-fME multiple emulsion formulation treated free drug, whereas the number increased with the Con-A and Con-A-fME treated formulations due to the inflammatory response produced by Con-A (Table 9.6).

In vivo antitumor activity of the coated multiple emulsions was promising than the results of in vitro activity.

Several types of lipid-based systems for the effective delivery of anticancer drugs have been studied. Better delivery of methotrexate to the tumor site because of surface activity and better permeation of drug in presence of lipids has led to better tumoricidal activity (Chandraprakash et al., 1993). The therapeutically optimal exposure of the tumor to the drugs may be achieved by appropriately designing liposomes that can retain the drug concentration quantitatively for direct transport to accessible target cells and release the drugs at a predetermined rate (Kirby and Gregoriadis, 1982). The antitumor efficacy obtained for our multiple emulsions formulations may be attributed to similar reasons.

Haematological Studies The new surface-modified carriers are excellent candidates for antitumor application, but presence of relatively high levels of anticancer drug for prolonged period can lead to blood dyscrasias. Therefore, to evaluate the toxic effect of the fME formulation, haematological parameters were studied. Similar treatment was given as stated previously. On day 8 blood was withdrawn from each mice, and white blood cells (WBC), red blood cells (RBC), platelets, differential count, haemoglobin (Hb), and packed cell volume (PCV) were determined by standard methods. The average of all determinations was computed. The haematological parameters were found to differ from the control group. There was no significant difference in blood parameters with respect to the route of administration of formulation. The number of RBC was not changed by free 6-MP, fME, and PEG-fME treatments. However, in presence of Con-A-fME there was a slight decrease in RBC count, which may be due to the agglutination property of lectins on red cells. The total WBC count increased in the tumor-bearing animal but was restored after treatment. The platelet counts showed no significant differences among the various formulations. It was slightly decreased in tumor-bearing mice.

The differential WBC count of tumor-bearing mice showed a decrease in the number of circulating lymphocytes to approximately 50%, an increase in neutrophils to 206.25%, and an increase in monocytes to 147.8%. Treatment restored the counts. The mean percent recovery (MPR) of the cell count was calculated as

$$\text{MPR} = \frac{\text{Mean count in treated mice} - \text{Mean count in tumor-bearing mice}}{\text{Mean count in control mice} - \text{Mean count in tumor-bearing mice}}.$$

For free 6-MP, Con-A, fME, PEG-fME, and Con-A-fME treated mice, the mean recovery, respectively, of the lymphocyte count was 60.0, 6.6, 86.6, 93.3 and 110%; the neutrophils recovery was 76.4, 23.5, 94.2, 85.2, and 85.2%; and the monocyte recovery was 45.4, 81.8, 54.5, 109.09, and 63.6%. The coated multiple emulsion formulations were found to be safer and non-toxic.

9.5 CONCLUSION AND OUTLOOK

In summary, the formulations were complied to meet the requirements of small particle size and a stable drug-carrier association. The effects of Con-A-fME and PEG-fME on the pharmacokinetic and tissue distribution of 6-MP reveals that the drug disposition substantially depends on the properties of the ligands coated on fMEs. A long circulation effect was prominent with PEF-fME, while tissue retention was longer with Con-A-fME. Coating with PEG-PE increased the half-life of the drug despite the fMEs size of nearly 1000 nm. The system has an intrinsic advantage of being lymphotropic, and hence it may control tumor metastasis. The fMEs were also found to be effective against leukemia L-1210 cells both in vitro and in vivo in Balb/c mice. The antitumor effect of Con-A-fME was specific, while that of PEG-fME were non-specific. The toxicological profiles of both formulations were satisfactory. Despite the promise shown in present experiments, more decrease in the size of multiple emulsions is necessary. We have conducted a few experiments using dendrimers and other hyperbranched polyelectrolytes with oppositely charged surfactants to support the internal aqueous phase. Using a polyamidoamine (PAMAM) dendrimer/ dicetylphophate complex system, we were able to obtain micro-multiple emulsions with the mean particle size of 450 nm (Khopade et al., unpublished results). We could also observe an internal sponge-like phase in this system. It is arguable whether such small-sized solidified particles with internal arrangement can be classified under classic mesophasic systems or modified multiple emulsions. Nevertheless, from the applied pharmaceutical viewpoint, it is important that a multiple emulsion in its classic or modified form shows promise as a potential drug delivery system, in general, and for anticancer drugs, in particular.

REFERENCES

Allen TM, Hansen C, Martin F, Redeman C, Young YA. 1991. Liposome containing synthetic lipid derivatives of polyethylene glycol show prolong circulation half-lives in vivo. *Biochim Biophys Acta* 1066: 29–36.

Chandraprakash KS, Udupa N, Uma Devi P, Pillai GK. 1993. Effect of niosome encapsulation of methotrexate, macrophage activation on tissue distribution of methotrexate and tumor size. *Drug Deliv* 1: 133–137.

Deem DE. 1988. Rheology of Dispersed Systems. In: *Pharmaceutical Dosage Forms: Disperse Systems*, Vol. 1, Lieberman HA, Reiger MM, Banker GS eds. New York: Marcel Dekker, pp 422–424.

Fox C. 1986. Introduction to multiple emulsions. *Cosmet Toilet* 101: 101–106.

Fukushima S, Juni K, Nakano M. 1983. Preparation of and drug release from W/O/W type double emulsions containing anticancer agents. *Chem Pharm Bull* 31: 4048–4056.

Fukushima S, Nishida M, Nakano M. 1987. Preparation of and drug release from W/O/W type double emulsions containing anticancer agents using an oily lymphographic agent as an oil phase. *Chem Pharm Bull* 35: 3375–3381.

Florence AT, Whitehill D. 1982. The formulation and stability of multiple emulsions. *Int J Pharm* 11: 277–308.

Florey K. 1978. *Analytical Profiles of Drug Substances*, Vol. 7. New York: Academic Press, pp 344–355.

Hino T, Takeuchi HT, Niwa T, Kitagawa M, Kawashima Y. 1995. The analysis of drug release from diluted water/oil/water emulsion by a model of the rupture of oil membrane. *J Pharm Pharmacol* 47: 1–7.

Iwamoto K, Kato T, Kawahara M, Koyama N, Wanatabe S, Miyaki Y, Sunamoto J. 1991. Polysaccharide coated oil droplets in oil-in-water emulsions as targetable carriers for lipophilic drugs. *J Pharm Sci* 80: 219–224.

Kim CK, Kim SC, Shin SJ, Kim KM, Oh KH, Lee YB, Oh IJ. 1995. Preparation and characterization of cytarabine-loaded W/O/W multiple emulsion. *Int J Pharm* 124: 61–67.

Kirby C, Gregoriadis G. 1982. The effect of lipid composition of small unilamellar liposomes containing melphalan and vincristine on drug clearance after injection into mice. *Biochem Pharmacol* 32: 609–615.

Khopade AJ, Jain NK. 2002. Multiple emulsions. In: *Novel and Controlled Drug Delivery*. Jain NK, ed. New Delhi: CBS Publishers, pp 381–407.

Khopade AJ, Mahadik KR, Jain NK. 1996. Targeting of multiple emulsions to the lungs. *Pharmazie* 51: 558–562.

Khopade AJ, Paradkar AR, Mahadik KR. 1995. *Stealth Multiple Emulsion System for passive Targeting of an Antiinflammatory Drug*. International Seminar on Recent Trends in Pharmaceutical Sciences. Ootacamund, India: JSS College of Pharmacy.

Khopade AJ, Khopade SA, Tripathi PK, Jain NK. 2007. Spongesomes: Vesicles from poly(amidoamine) dendrimer doped phospholipids with a sponge-like internal structure. (Unpublished manuscript).

Lin SY, Wu WH, Lui WY. 1992. In vitro release, pharmacokinetic and tissue distribution studies of doxorubicin hydrochloride (Adriamycin HCl) encapsulated in lipiodolized W/O emulsion and W/O/W multiple emulsion. *Pharmazie* 47: 439–443.

Magdassi S, Garti N. 1995. Release of markers from the inner water phase of w/o/w emulsions stabilized by silicon based polymeric surfactants. *J Controlled Release* 33: 1–12.

Ma JL, Xiaong QM, Tao T, Chen RZ. 1993. Physicochemical properties and its release in vitro of multiple emulsion containing etoposide. *Chin J Pharm* 24: 357–360.

Morel S, Gasco MR, Cavalli R. 1994. Incorporation in lipospheres of [D-Trp-6] LHRH. *Int J Pharm* 105: R1–R3.

Omotosho JA, Whateley TL, Florence AT. 1989. Methotrexate transport from internal phase of W/O/W emulsion. *J Microencapsul* 6: 183–192.

Omotosho JA, Florence AT, Whateley TL. 1989. Release of 5-fluorouracil from intramuscular W/O/W multiple emulsions. *Biopharm Drug Dispos* 10: 257–268.

Omotosho JA, Florence AT, Whateley TL. 1990. Absorption and lymphatic uptake of 5-fluorouracil in the rat following oral administration of multiple emulsion. *Int J Pharm* 61: 51–56.

Seifriz W. 1925. Studies in emulsions. *J Phy Chem* 29: 738–749.

Sahnoun L, Serre-Debeauvais F, Lang J, Faucon G, Gaven M. 1990. Determination of 6-mercaptopurine and its metabolites in plasma or serum by high performance liquid chromatography. *Biomed Chromatogr* 4: 144–147.

Shier WT. 1979. Lectins as drug carriers. In: *Drug Carriers in Biology and Medicine*, Gregoriadis G. eds. London: Academic Press, pp 43–70.

Tao T, Ma JL, Xiong QM, Chen RZ. 1992. Study on preparation of methotrexate-loaded multiple emulsion. *Chin J Pharm* 23(9): 393–396.

Vanderhoff JW, El Asser M. 1988. *Theory of colloids in pharmaceutical dosage forms.* In: *Pharmaceutical Dosage Forms: Disperse Systems* Vol. 1, Lieberman HA, Reiger MM, Banker GS. eds. New York: Marcel Dekker, pp 119–120.

Weissig V, Lasch J, Gregoriadis G. 1989. Covalent coupling of sugars to liposomes. *Biochim Biophys Acta* 1003: 54–57.

Yazan Y, Seiller M, Puisieux F. 1993. Multiple emulsions. Boll Chim. *Farmaceutico* 132: 187–196.

CHAPTER 10

Application of Emulsion Technology to Transarterial Injection Chemotherapy for Hepatocellular Carcinoma Using Double-Emulsion Enclosing Vesicles of Anticancer Drug Solution

SHUSHI HIGASHI

Contents

Multiple Emulsions: Technology and Applications, Edited by Abraham Aserin

10.1 INTRODUCTION

10.1.1 Clinical Aspect of Hepatocellular Carcinoma (HCC)

Hepatocellular carcinoma (HCC) is as a malignant tumor known to be common in Japan. The tumor arises from hepatocytes following a long-term hepatitis B or C virus infection (Bisceglie, 1997). In past its prevalence was restricted to countries in Asia and Africa, but it is now moving gradually to other countries including the United Kingdom and the United States because of large-scale immigration from countries contaminated with the virus.

In the early stage of the disease, HCC is detected as a single nodule clearly defined against normal liver tissue, and within a year, it tends to increase in size and number (see Figure 10.1). Distant metastases including the lung or lymph node is rare. Therefore the prognosis of a patient is determined mainly by the state of the lesions in the liver.

Surgery is the most effective treatment for the tumor. But approximately 80% of the patients are excluded from indication of surgical resection because

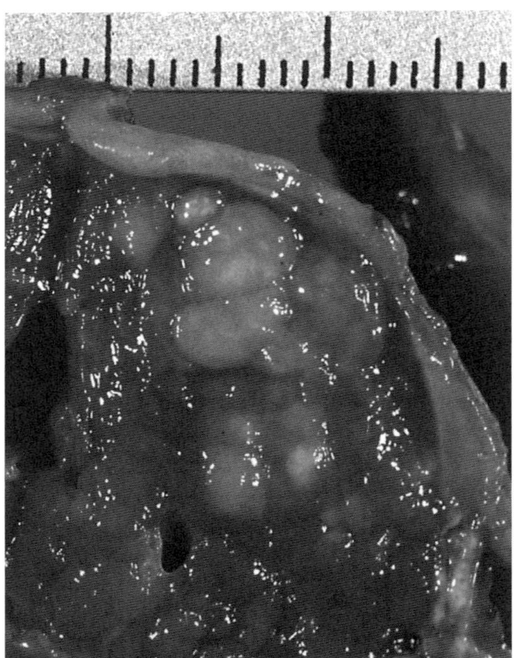

Figure 10.1 Cross section of a hepatocellular carcinoma arising in the posterior part of the right lobe of the liver. Before hepatectomy, the patient was diagnosed as having early stage HCC, a solitary nodule 20 × 12 mm in diameter. However, as shown, it was revealed that the lesion consists of many tumor nodules.

of multiple cancer nodules in the liver, or because of impaired liver function due to liver cirrhosis or chronic hepatitis.

10.1.2 Transarterial Chemotherapy for HCC

For the inoperable case of HCC, arterial injection or transcatheter arterial embolization (TAE) was invented (Pannuti, 1986; Doci, 1988). In this treatment, the hepatic artery entering HCC is occluded with an embolic substance, such as gelatin sponge particles, on the basis that HCC is pre-dominantly fed by the hepatic artery (Burgener, 1984).

The initial report on TAE was published as a treatment for renal cell carcinoma. In 1978 Yamada et al. reported a modification of TAE (Matsumura, 1978). They injected gelatin sponge particles soaked with an anticancer drug and were able to attain a better anticancer effect; their term for this method was transcatheter arterial chemoembolization (TACE). Later, it was revealed that the therapeutic efficacy of TACE is not sufficient in the case of HCC. This is because the collateral pathway of the hepatic artery for the tumor frequently appears within a year, so there is high incidence of recurrence. Furthermore the therapy brings about severe tissue damage in the liver (Belli, 1997).

Recently iodinated poppy-seed oil (IPSO, Lipiodol Ultra-Fluid: Laboratoire Guerber, France), the ethyl ester of iodinated fatty acids obtained by hydrolysis of poppy-seed oil, has been used for injection chemotherapy (Miller, 1987). Initial application of IPSO was reported by Idezuki et al. (1966). They used IPSO as a diagnostic material in injecting the oil into the portal vein. According to this report, a plain abdominal X-ray film was taken after the injection and revealed a negative shadow of HCC surrounded by thick staining of nontumorous liver tissue. Afterward, it was revealed that IPSO selectively accumulates in HCC tissue when it is injected into the hepatic artery (Iwai, 1984; Kobayashi, 1992; Nakakuma, 1983; Tsai, 1986).

Kanematsu et al. (1984) reported a new injection material for TACE; a mixture of IPSO and an aqueous solution of hydrophilic anticancer drug, doxorubicin. This mixture was prepared by sonication or agitation. Its antitumor effect was inadequate because the mixture easily separates into oil and water.

To avoid separation, they used a 22% urografin solution as the aqueous phase and equal to IPSO in specific gravity. When the volume ratio of oil and water was three to one, the mixture showed the best stability, though at most for 24 hours.

A magnified image of the mixture shows these characteristics well (see Figure 10.2). The aqueous solution of anticancer drug is not enclosed in the IPSO microdroplets. Therefore the aqueous solution, when injected into the hepatic artery, gets expelled the from HCC tissue, leaving the IPSO behind. Furthermore the micodroplets in the mixture are multifarious in diameter. The result is heterogeneous accumulation of oil in the liver. Larger IPSO microdroplets tend to remain in nontumorous liver tissue for more than two weeks.

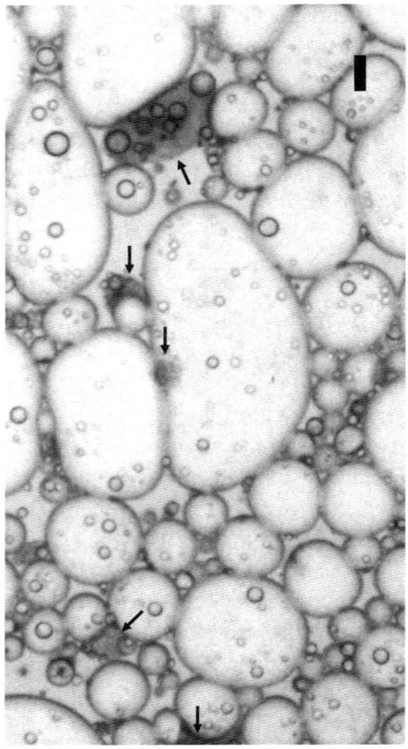

Figure 10.2 Magnified image of conventional emulsion for a use of transcatheter arterial chemoembolization (TACE), prepared by agitating a mixture of IPSO and an aqueous solution of epirubicin. The IPSO microdroplets are multifarious in size, ranging from 0.1 to 200 μm, and the aqueous solution of epirubicin (arrows) is not enclosed in the microdroplets. Indicator: 30 μm.

10.2 PREPARATION OF W/O/W EMULSION

10.2.1 Invention of a New Injection Material, W/O/W, for TACE

To reduce shortcomings of the conventional IPSO mixture, we invented a double emulsion as a form of water-in-oil-in-water (W/O/W) emulsion (Higashi et al., 1993, 1995, 1996). In the emulsion, numerous spherical IPSO microdroplets, enclosing minute vesicles of an aqueous solution of epirubicin, are suspended in a normal saline liquid. The IPSO microdroplets are almost equal in size, and the epirubicin solution appears in the IPSO microdroplets as multiple vesicles. When injected into the hepatic artery, both IPSO microdroplets and epirubcin accumulate exclusively in the HCC tissue and deliver a good antitumor effect.

10.2.2 Preparation of Porous-Glass Membrane

The porous-glass membrane invented by our colleagues has become the core technology for preparing double emulsions (Nakashima, 1991). A magnified view of the membrane shows numerous pores that penetrate through the entire thickness of the glass membrane (see Figure 10.3).

The method used to produce porous-glass membranes is as follows: Raw materials—SiO_2, Al_2O_3, B_2O_3, Na_2O, K_2O, CaO, and MgO—are heated to be made into glass at a temperature of 1200°C for 60 minutes. The glass is then subjected to heat treatment at a temperature of 700°C–750°C. This treatment causes the homogeneous primary glass to decompose into two different types of glass phases, the Al_2O_3/SiO_2-rich phase and the CaO/B_2O_3-rich phase. The size of each separated microphase is approximately 1 nm in the beginning of the thermal treatment at approximately 700°C and gradually increased to 10,000 nm as the heating temperature increases or the heating intensifies. Next a hydrochloric acid treatment causes the CaO/B_2O_3 phase to leach out and form a porous network of equal size in the glass. The glass membrane with a pore size of 20 μmm is prepared as a tube, which is sized to fit the emulsifying module (see Figure 10.4).

10.2.3 Preparation of W/O/W Emulsion with Porous-Glass Membrane: Membrane Emulsification

Sixty milligrams of epirubicin were dissolved in 5 ml of a 5.8% (w/v) glucose solution and were mixed with 10 ml of IPSO and 500 mg of polyglycerol esters of polycondensed fatty acids of castor oil. The mixture was sonicated for 10 minutes to form a water-in-oil (W/O) emulsion that was injected into an

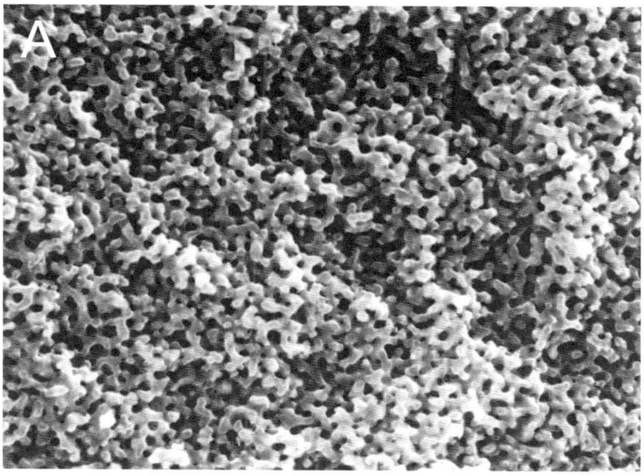

Figure 10.3 Magnified view of the surface of a porous-glass membrane. Numerous pores penetrating through the membrane are observed.

Figure 10.4 Porous-glass membrane prepared to fit an emulsifyng module. The pore size of the membrane is 19.3 μm for preparing IPSO microdroplets 70 μm in diameter.

emulsifying module in which W/O passed through the glass membrane with pores into 0.4% saline containing 1% (w/v) polyoxyethylene 60 stearate (HCO-60). In the water-in-oil-in-water (W/O/W) emulsion obtained, IPSO micodroplets, enclosing numerous minute vesicles of an aqueous solution of epirubicin, were suspended in saline (see Figure 10.5). We call this method membrane emulsification.

10.3 CHARACTERISTICS OF W/O/W EMULSION

10.3.1 Confirmation of Selective Accumulation of IPSO Microdroplets in HCC Immediately after TACE

Long-term accumulation of the IPSO microdroplets in nontumorous tissue is hazard to patients because the microdroplets firmly enclose the anticancer drug. We quantitatively analyzed IPSO accumulation in the tumors and non-tumorous tissue by assessing CT values immediately after and 7 days after TACE with the W/O/W emulsion (see Figure 10.6).

Design of Research Ten successive patients were included in the study. The W/O/W emulsion, consisting of IPSO microdroplets 70 μm in diameter, was injected through a catheter inserted into the hepatic artery or a segmental branch thereof. The mean amount of the W/O/W emulsion per one injection was 10 ml. The mean CT values for the tumor (Hounsfield unit, HU/pixel) were calculated from the CT values of all pixels in the whole area corresponding to the largest cross section of the main tumor. The mean CT values for the

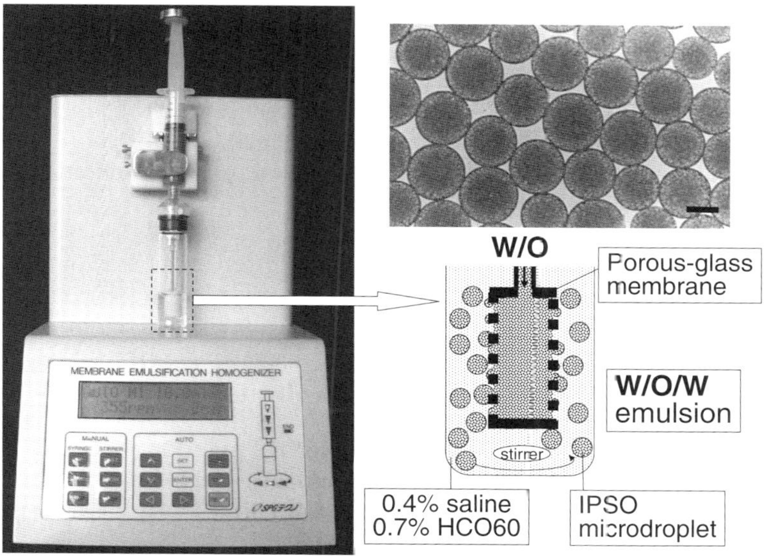

Figure 10.5 Preparation of water-in-oil-in-water (W/O/W) emulsion by the membrane-emulsifying technique. Water-in-oil (W/O) emulsion, prepared by sonication and consisting of numerous vesicles of an aqueous solution of epirubicin, is passed through the porous-glass membrane into saline containing polyoxyethylene 60 stearate. IPSO microdroplets in W/O/W emulsion are almost equal in size and firmly enclose vesicles of epirubicin solution. Indicator: 30 μm.

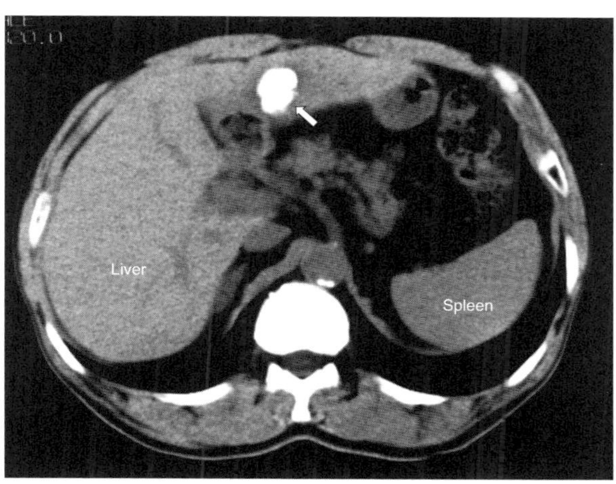

Figure 10.6 Computed tomogram (CT) taken 1 week after TACE with W/O/W emulsion. Thick accumulation of IPSO (arrow) is clearly noted only in HCC tissue.

juxta-tumor nontumorous tissue were calculated from the CT values of 250 randomly selected pixels within an area up to 10 mm from the edge of the tumor. The mean CT values for the nontumorous tissue distant from the tumor were calculated from the CT values of 250 randomly selected pixels approximately 10 mm from the edge of the tumor. The mean CT values for the control part of the liver were obtained from the CT values of 250 randomly selected pixels in an area where the emulsion had not been injected.

Results The mean CT values in the tumor were 332 HU/pixel on day 0 and 291 on day 7 (see Figure 10.7). There was no significant difference. The mean CT values in the juxta-tumor nontumorous tissue decreased significantly from 166 to 60. The mean CT values in the nontumorous tissue 10 mm distant from the tumor decreased significantly from 142 to 62. The mean CT values in the control, noninjection part of the liver, were 65 on day 0 and 61 on day 7.

According to the study, the 70 μmm sized IPSO microdroplets disappeared from the nontumorous tissue within a week after the injection. However, the IPSO remained in the tumor as a thick accumulation.

10.3.2 Serum Concentration of Epirubicin after TACE with W/O/W

Cancer chemotherapy often delivers adverse effects to patients, such as bone marrow depression, loss of hair, or digestive organs disorder. Among the drugs, anthracyclines such as doxorubicin and epirubicin often pose the threat of

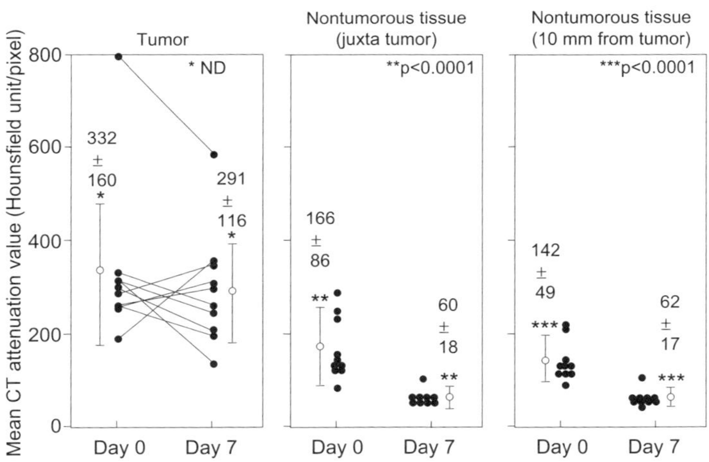

Figure 10.7 Mean CT attenuation values (Hounsfield unit, HU/pixel) immediately and 7 days after injection with W/O/W emulsion. The values were calculated by using numeric data of randomly selected 250 pixels in a main tumor, juxta-tumor nontumorous tissue within an area up to 10 mm from the edge of the tumor, and nontumorous tissue 10 mm distant from the tumor.

death to patients. In TACE with W/O/W emulsion, we further attempted to prevent drug-induced side effects by injecting a total of 60 to 90 mg of epirubicin, which was almost equal to the dose in intravenous use.

We measured the serum levels of epirubicin after TACE with W/O/W emulsion containing 60 mg of epirubicin. Venous blood samples were taken 1, 8, and 24 hours after the injection, and serum concentrations of the drug were measured using high-performance liquid chromatography.

The levels were 24 ng/ml immediately after injection, 5 ng/ml at 8 hours after injection, and 3 ng/ml at 24 hours after injection (unpublished data). These values are approximately one-third of those after intravenous injection with an aqueous solution of epirubicin (60 ng/ml for immediately after injection, 19 ng/ml for 8 hours, and 11 ng/ml for 24 hours after injection).

10.3.3 Influence of Diameter of IPSO Microdroplets on Their Outflow to the Lung

The IPSO microdroplets injected into the hepatic artery flow into the capillary vessels of the liver (see Figure 10.8), and some parts inevitably flow into the lung via the hepatic vein. The lung deposition of oil is fatal. It causes pulmonary disorders such as pulmonary edema, so, it is important to estimate the total outflow of the IPSO microdroplets into organs other than the liver. The outflow of IPSO from the liver may be closely related to the size of the microdroplets. Nevertheless, it has not been revealed how the size of IPSO microdroplets affects the amount of expelled IPSO from the liver into the lung.

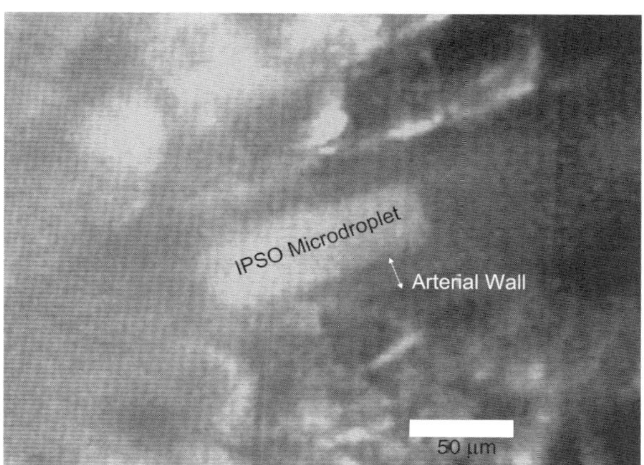

Figure 10.8 Fluorescent microscopic observation of an IPSO microdroplet flowing in the peripheral artery of the rabbit. The injected IPSO microdroplets change their form according to the caliber of the artery and rapidly flow through the vessels without being broken into fragments. Indicator: 50 μm.

Experiment Studies were performed using six female New Zealand White rabbits weighing 2.6 to 3.4 kg. The rabbits underwent laparotomy under general anesthesia using 33 mg/kg body weight of sodium pentobarbiturate. A portion of an oil-in water (O/W) emulsion, which had been prepared by passing 180 μl of ^{125}I-IPSO through the glass membrane and dripped into a physiological saline solution containing 1% of HCO-60 comprising IPSO microdroplets with diameters of either 10 or 60 μm, was injected into the proper hepatic artery of rabbits via a catheter (0.2 mm in internal diameter). The diameter of the IPSO prepared microdroplets, which appeared to be 10 or 60 μm, was actually 12.1 ± 0.1 or 61.0 ± 0.1 μm, respectively. The animals were sacrificed 6 hours after the injection by total exsanguinations. Organs including the liver, lungs, spleen, and kidneys were extirpated, weighed, and minced. Bile and urine were collected 6 hours after the injection. The radioactivities of these samples were measured by a gamma well scintillation counter (Auto Well Gamma System, ARC-1000M, Aloka Corp., Tokyo). The amount of IPSO in each organ was calculated by comparing the specific radioactivity of the samples with that of stocked ^{125}I-IPSO.

Results The amount of IPSO per gram wet weight of the lungs 6 hours after the injection was significantly larger in the animals treated with the 12 μm diameter microdroplets than those treated with the 61 μm diameter microdroplets (see Figure 10.9). The total IPSO amount excreted in bile and urine

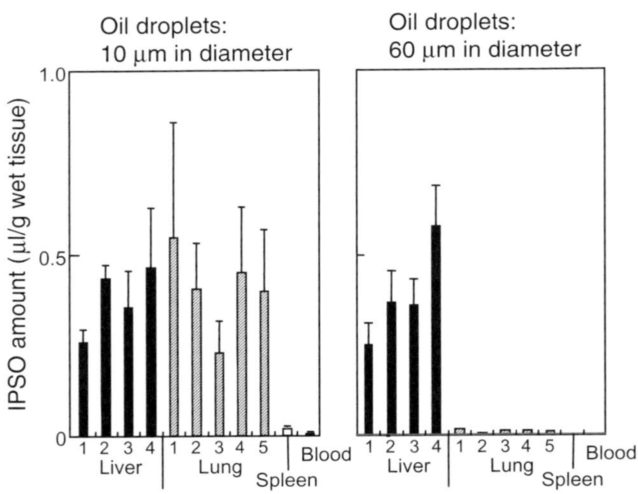

Figure 10.9 IPSO amount in organs 1 hour after hepatic arterial injection with a W/O/W emulsion consisting of ^{125}I-IPSO microdroplets that are 10 or 60 μm in diameter. The emulsion was injected into the proper hepatic artery of the rabbits after laparotomy under general anesthesia. The IPSO amount was calculated by measuring the radioactivity of the sample taken from the liver (each 4 lobe), lung (each 5 lobe), spleen, and blood.

during 6 hours were 2.7 ± 2.0 and 1.9 ± 2.2 μl/each animal, respectively, with the small microdroplets, and 0.94 ± 0.61 and 0.69 ± 0.78 μl/ each animal, respectively, with the large microdroplets. No significant differences were noted among the values.

Thus the diameter of the IPSO microdroplets injected into the hepatic artery affects the flux of IPSO through the liver to the lungs. To avoid a pulmonary complication in TACE with W/O/W emulsion, the IPSO microdroplets prepared for injection should all have the same diameter that is large enough not to pass through the liver.

10.3.4 Size of IPSO Microdroplets for Antitumor Effect in TACE with W/O/W Emulsion

At the beginning of our clinical trial, we used IPSO microdroplets designed to be 30 μm in diameter. During the clinical trials, there was some doubt about the size of microdroplets because the treatment sometimes failed to attain an adequate result. We wondered if the 30 μm diameter of IPSO microdroplets was appropriate. The result in animal experiments described above, which showed size of IPSO microdroplets should be large enough for preventing pulmonary embolism, encouraged us to attempt next clinical research.

Design of Research Thirty-two patients of HCC was randomly assigned to receive W/O/W emulsion of small IPSO microdroplets (30 μm in diameter) containing 60 mg of epirubicin (group-S, n = 16) or W/O/W emulsion of large IPSO microdroplets (70 μm) containing the same amount of epirubicin (group-L, n = 16). There were no significant differences between the groups in age, number of tumors, the maximum size of tumors, 15-min value of indocyanine green clearance test, or serum alfa-fetoprotein (AFP) levels before treatment. W/O/W emulsion containing 60 mg of epirubicin was injected via either the proper hepatic artery or its branch at a rate of 5 ml/min. No embolic substances were used. Serum AFP level was measured immediately before and 7 days after treatment.

Results Serum AFP levels decreased in 26 of 32 patients, and the remaining six patients whose serum AFP level increased or unchanged belonged exclusively to group-S (see Figure 10.10). There was a significant difference in the percent decline of serum AFP levels in seven days between the S and L groups: 18.9 ± 33.1% (mean ± SD) in group-S and 50.5 ± 19.8% in group-L, $p < 0.005$.

From the results it is clear that the size of IPSO microdroplets injected into the hepatic artery determines the decrease of serum AFP levels in patients with HCC. So the indication is that the size of IPSO microdroplets should be regulated to maintain an appropriate diameter (Higashi, 1999).

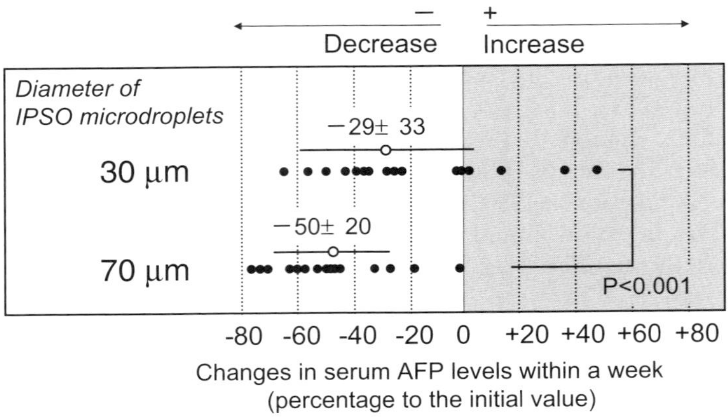

Figure 10.10 Relative changes in serum α-fetoprotein (AFP) levels seven days after TACE with W/O/W emulsion. Two kinds of emulsion consisting of IPSO microdroplets 30 or 70 μm in diameter were used in the trial. All plots indicate percentage to the initial value of AFP level. The therapeutic effect was remarkable in patients treated with large IPSO microdroplets.

10.4 MECHANISMS OF ACCUMULATION OF W/O/W EMULSION IN HCC TISSUE

The IPSO microdroplets injected into the hepatic artery inevitably flow into the peripheral branch of the artery, which runs through bile duct wall as a network called peribiliary plexus and finally connects to the portal vein or sinusoid (Kan, 1989) (see Figure 10.11). In the case of the liver bearing HCC, the tumor tissue is supplied with blood only by the hepatic artery, and not by the portal vein. Therefore immediately after the TACE with the W/O/W emulsion, the IPSO microdroplets accumulate in the hepatic artery along bile duct wall in nontumorous tissue and in tumor vessels in HCC tissue.

Afterward, the microdroplets in nontumorous tissue flow via the sinusoid into the hepatic vein or Disse's space, which lies outside of sinusoidal space. The microdroplets entering Disse's space inevitably flow into the bile canaliculi or lymphatic channels, finally out of the liver. In the HCC tissue the microdroplets remain there for a long time, probably because tumorous tissue is not equipped with a biliary or lymphatic drainage system.

The fate of the oil droplets depositing in the tumor is mystifying. We examined HCC tissue resected 30 days after a hepatic arterial injection of chemotherapy with a W/O/W emulsion containing a low dose (8 mg) of epirubicin. As shown in Figure 10.12, the IPSO microdroplets appearing red in color due to the oil-O-stain are seen only in the cytoplasm of the cancer cells, and not in the arterioles or the tumor sinusoids. Electron microscopy of the same specimen revealed that the IPSO microdroplets accumulated in the cytoplasm

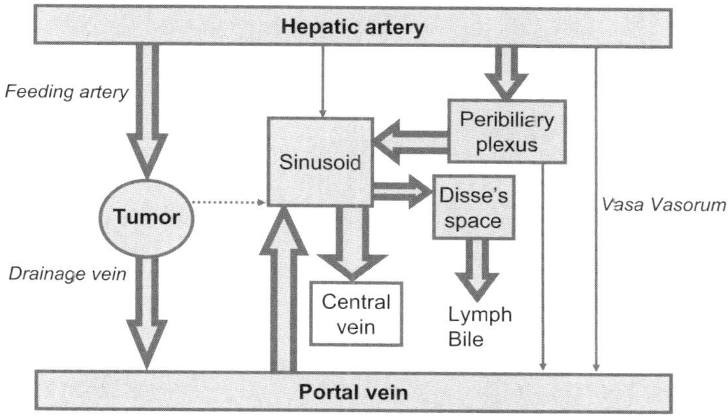

Figure 10.11 Diagram showing microarchitecture of vessels near HCC tissue. The hepatic artery branches off from feeding the tumor tissue to feeding other arteries as well as the sinusoid, the bile duct, and the portal vein. Blood flow through the tumor is drained more directly into the portal vein than into the sinusoids. A portion of serum in the sinusoids flows into Disse's space through which oil droplets is drained off as lymphatic flow.

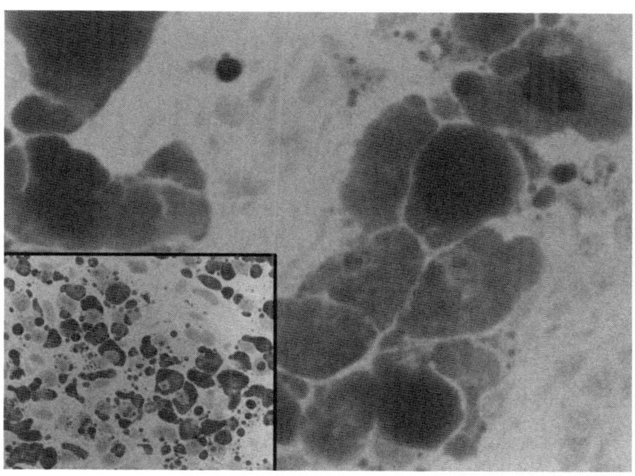

Figure 10.12 Microscopic view of human HCC tissue resected one month after a TACE with a W/O/W emulsion. The IPSO microdroplets, seen as red in color, remain only in the cytoplasm of the cancer cells, and not in the arterioles or tumor sinusoids. Nuclei are stained as thin purple area.

of the HCC cells (see Figure 10.13). The unique structure of the microvesicles of epirubicin is clearly visible because of their markably increased size. From the results we believe that almost all the IPSO microdroplets in the tumor vessels eventually enter into the cytoplasm through the cytoplasmic membrane of the cancer cells.

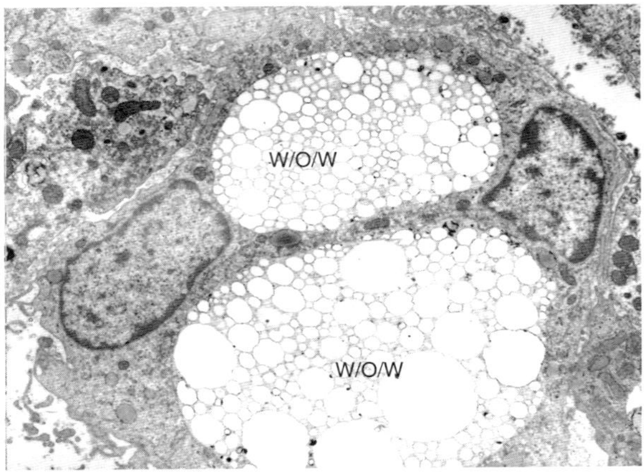

Figure 10.13 Electron micrograph of the same specimen as in Figure 10.12 showing the IPSO microdroplets that accumulate in the cytoplasm of HCC cells. The microvesicles of the aqueous solution of epirubicin have greatly increased in size.

10.5 CLINICAL TRIALS

10.5.1 Retrospective Study

The clinical trials were conducted successively and involved 63 patients (49 men and 14 women with a mean age of 65 years) who had been diagnosed as HCC and consented to the therapy; they were exclusively treated by the therapy. The patients comprised 29 cases of primary occurrence of HCC (group PRIM) and 34 cases of recurrence after partial hepatectomy for the disease (group RECR). Patients with 60% or more of tumor-occupying rate to the whole liver, persistent ascites, or hyperbilirubinemia exceeding 3 mg/dl were excluded from the study. The W/O/W emulsion was injected through a catheter located in either of the proper hepatic artery or its branches. The dose of epirubicin per one injection was 60 mg. For patients with inadequate IPSO deposition, subsequent injections were repeated at intervals of three months. A total of 172 injections were performed. No treatment included embolization with gelatin sponge particles, surgery, percutaneous ethanol injection therapy, or systemic anticancer chemotherapy.

In all the patients treated with W/O/W, a low fever, less than 38.5°C, was observed for three to seven days. One third of the patients complained of nausea or gastric discomfort, which disappeared within 6 hours after the injection. There were no fatal side effects, including myelosupression or loss of hair due to the anticancer agent, and pulmonary edema, gastric ulcer, or acute cholecystitis due to the oil emboli.

In groups PRIM and RECR, 25 patients died during their follow-up periods. Twenty-three of the patients died of hepatic failure or cachexia caused by HCC, and two others were free from tumors immediately before their deaths, which were brought on by other causes.

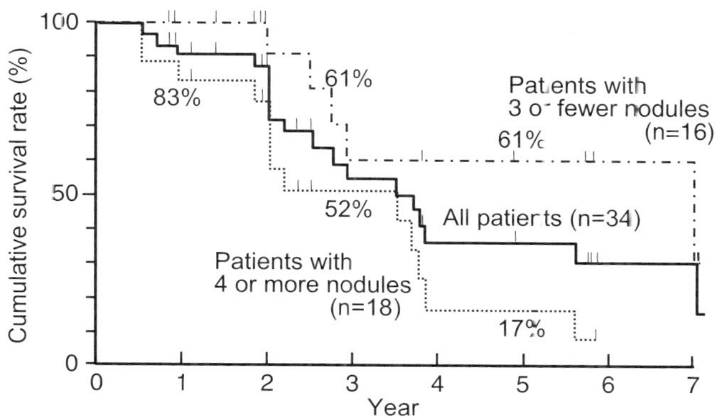

Figure 10.14 The cumulative survival rates of patients with recurrent HCC after a partial hepatectomy for the disease. The patients were exclusively treated by TACE with a W/O/W emulsion containing 60 mg of epirubicin. The 1-, 3-, and 5-year cumulative survival rates for all patients were 90%, 55%, and 37%, respectively.

The five-year cumulative survival rates (see Figure 10.14) were 34% in group PRIM and 37% in group RECR. As for the patients with three or fewer tumors, the rates were 44% in group PRIM ($n = 10$) and 61% in group RECR ($n = 16$). As for the patients with four or more nodules, the rates were 24% in group PRIM ($n = 19$) and 17% in group RECR ($n = 18$).

10.5.2 Case Presentation

An 80-year-old male, who had undergone partial resection of his liver for HCC in segment 4, was found to have multiple recurrent masses in the remaining liver one year after the resection (see Figure 10.15a). For the lesions a W/O/W emulsion containing 60 mg of epirubicin was injected via the left hepatic artery (the first injection; 12 months after the resection) and the right hepatic artery (the second injection; 15 months after the resection). The computed tomogram taken 30 months after the resection (18 months after the initial injection therapy) revealed no viable mass in the liver except for several scars (see Figure 10.15b). Forty-two months after the hepatic resection, the patient is still surviving without any discomfort or sign of recurrence.

10.6 CONCLUSION

In the W/O/W emulsion the vesicles of an aqueous solution do not separate from the IPSO and do not accumulate in the nontumorous tissue after TACE. This crucial finding can help attain longer survival for patients bearing multiple nodules who might otherwise be forced to undergo a long-term hospitalization. We hope to develop the W/O/W injection as a new therapeutic regimen for HCC.

(a)

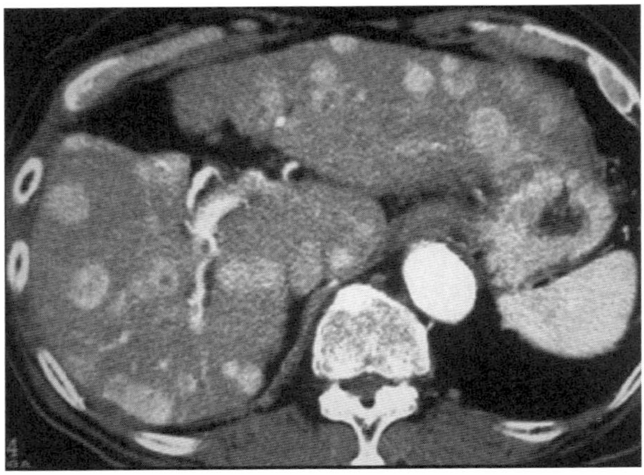

(b)

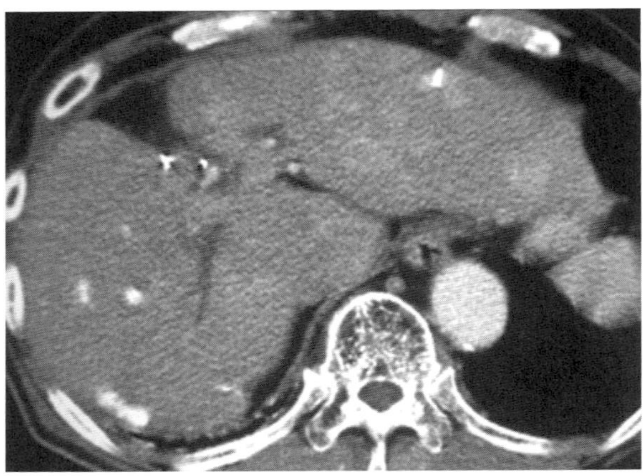

Figure 10.15 Clinical course of an 80-year-old male who underwent partial resection of the liver for HCC in segment 4. (*a*) An enhanced computed tomogram taken 12 months after the partial resection (on detection of recurrence). Numerous HCC nodules are observed in the remaining liver. The serum tumor marker (PIVKA-II) level was 18,569 MAU/ml. (*b*) An enhanced computed tomogram (early phase) taken after the second injection therapy. Almost all nodules disappeared and serum PIVKA-II level decreased to 19 MAU/ml.

Because our clinical trials are promising but preliminary, we intend to continue our research into the efficacy of chemotherapy injection with W/O/W. We hope to confirm our results with more strictly constructed trials.

REFERENCES

Belli L, Magistretti G, Puricelli GP, Damiani G, Colombo E, Cornalba GP. 1997. Arteritis following intra-arterial chemotherapy for liver tumors. *Eur Radiol* 7: 323–326.

Bisceglie AMD. 1997. Hepatitis C and hepatocellular carcinoma. *Hepatology* 26: 34S–38S.

Burgener FA. 1984. Peritoneal hepatic artery embolization in rabbits with VX2 carcinoma of the liver. *Cancer* 46: 56–63.

Doci R, Bignami P, Bozzetti F, Bonfanti G, Audisio R, Colombo M, Gennari L. 1988. Intrahepatic chemotherapy for unresectable hepatocellular carcinoma. *Cancer* 61: 1983–1987.

Higashi S, Shimizu M, Setoguchi T. 1993. Preparation of new Lipiodol emulsion containing water soluble anticancer agent by membrane emulsification technique. *Drug Deliv Sys* 8: 59–61 (in Japanese with English abstract).

Higashi S, Shimizu M, Nakashima T, Iwata K, Uchiyama F, Tateno S, Setoguchi T. 1995. Arterial injection chemotherapy for hepatocellular carcinoma using monodispersed poppy-seed oil microdroplets containing fine aqueous vesicles of epirubicin. *Cancer* 75: 1245–1254.

Higashi S, Maeda Y, Kai M, Kitamura T, Tsubouchi H, Tamura S, Setoguchi T. 1996. A case of hepatocellular carcinoma effectively treated with epirubicin aqueous vesicles in monodispersed iodized poppy-seed oil microdroplets. *Hepato-Gastroenterology* 43: 1427–1430.

Higashi S, Tabata N, Kondo KH, Maeda Y, Shimizu M, Nakashima T, Setoguchi T. 1999. Size of lipid microdroplets effects results of hepatic arterial chemotherapy with an anticancer agent in water-in-oil-in-water emulsion to heoatocellular carcinoma. *Pharmacol Exp Thr* 289: 816–819.

Idezuki Y, Sugiura M, Hatano S, Kimoto S. 1966. Hepatography for detection of small tumor masses in liver: Experiences with oily contrast medium. *Surgery* 60: 566–572.

Iwai K, Maeda H, Konno T. 1984. Use of oily contrast medium for selective drug targeting to tumor: Enhanced therapeutic effect and X-ray image. *Cancer Res* 44: 2115–2121.

Kan Z, Ivancev K, Hagerstrand I, Chuang VP, Lunderquist A. 1989. In vivo microscopy of the liver after injection of Lipiodol into the hepatic artery and the portal vein in the rat. *Acta Radiol* 30: 419–425.

Kanematsu T, Inokuchi K, Sugimachi K, Furuta T, Sonoda T, Tamura S, Hasuo K. 1984. Selective effects of Lipiodolized antitumor agents. *J Surg Oncol* 25: 218–226.

Kobayashi S, Narimatsu Y, Ogawa K, Hashimoto S, Nakatsuka S, Miura H, Ohzono H, Ka WJ, Ido K, Hiramatsu K. 1992. Transcatheter hepatic arterial chemoembolization using epirubicin-Lipiodol: Experiment and pharmacological evaluation. *Cancer Chemother Pharmaco* 31: S45–S50.

Matsumura T, Kawamura M, Yamamoto K, Yamaguchi T, Kawakita J, Maeda T, Nishio S, Hayahara N, Nakanishi J, Kishimoto T, Maekawa M, Yamada R, Minaguchi K, Nakatsuka H, Funai K, Tsujita M. 1978. Transcatheter embolization of renal cell carcinoma. *Nippon Hinyokika Gakkai Zasshi* 69: 1094–1103. (in Japanese)

Miller DM, O'Leary TJ, Girton M. 1987. Distribution of iodized oil within the liver after hepatic arterial injection. *Radiology* 162: 849–852.

Nakakuma K, Tashiro S, Hiraoka T, Uemura K, Konno T, Miyauchi Y, Yokoyama I. 1983. Studies on anticancer treatment with an oily anticancer drug injected into the ligated feeding hepatic artery for liver cancer. *Cancer* 52: 2193–2200.

Nakashima T, Shimizu M, Kukizaki M. 1991. Membrane emulsification by microporous 2nd International Conference of Inorganic Membranes, Montpellier, France 511–516.

Pannuti F, Camaggi CM, Strocchi E, Comparsi R, Rossi AP, Angelli B, Franchini A. 1986. Intrahepatic aterial administration of 4′-epidoxorubicin (epirubicin) in advanced cancer patients: A pharmacokinetic study. *Eur J Cancer Clin Oncol* 22: 1309–1314.

Tsai C, Kusumoto Y, Harada R, Shima M, Nakata K, Kono K, Sato A, Ishii N, Koji T, Nagataki S. 1986. Effect of intrahepatic arterial infusion of 131I-labelled lipiodol on hepatocellular carcinoma of rat. *Ann Acad Med Singapore* 15: 521–524.

CHAPTER 11

Lipiodol W/O/W Emulsion for Transcatheter Arterial Embolization Therapy Prepared with Two-Step Pumping Emulsification Method

TOMOAKI HINO and TAKAYUKI OHWAKI

Contents

11.1 INTRODUCTION

In Japan more than 30 thousand people die of hepatocellular carcinoma annually (Figure 11.1). About 6 thousand people die of viral hepatitis and 16 thousand die of hepatism (Table 11.1). Most of the patients of hepatocellular carcinoma have been afflicted with viral hepatitis and/or hepatism previously. Liver functions of the patients of hepatocellular carcinoma are damaged by not only carcinoma but also by hepatitis or hepatism.

The current methods for treating hepatocellular carcinoma are hepatectomy (Minagawa and Makuuchi, 2006), percutaneous ethanol injection therapy (PEIT) (Kumada et al., 1992; Jang et al., 2004; Koiwa et al., 2007), transcatheter

Multiple Emulsions: Technology and Applications, Edited by Abraham Aserin

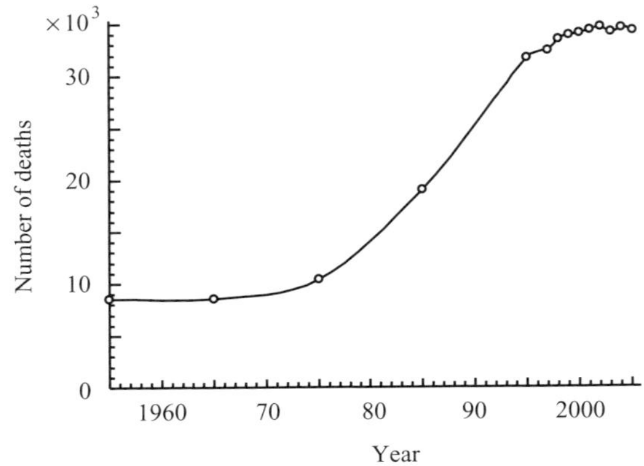

Figure 11.1 Annual number of deaths due to hepatocellular carcinoma in Japan. Derived from the annual data of Japan's Ministry of Health, Labor, and Welfare.

TABLE 11.1 Annual deaths due to hepatocellular carcinoma, viral hepatitis, and hepatism in Japan

		Viral hepatitis			Hepatism	
Year	Hepatocellular carcinoma[a]	Total	Hepatitis B	Hepatitis C	Total	Hepatocirrhosis[b]
1995	31707	5029	880	3542	17018	11301
1997	32359	4843	833	3449	16599	10876
1998	33434	4869	792	3574	16133	10333
1999	33816	5184	830	3831	16585	10315
2000	33979	5120	885	3756	16057	9837
2001	34311	5478	823	4194	15848	9538
2002	34637	5663	829	4411	15490	9220
2003	34089	5805	856	4530	15737	9220
2004	34510	5888	836	4648	15885	9150
2005	34268	6042	786	4855	16430	9387

Source: Derived from the annual data of Japan's Ministry of Health, Labor, and Welfare.
[a]Including cholangioma.
[b]Excluding alcoholic hepatocirrhosis.

arterial embolization (TAE) (Shepherd et al., 1987; Marrache et al., 2007), percutaneous microwave coagulation (Cui et al., 2005), grafting (Ohira et al., 2006), chemotherapy, and so on. The choice of treatment depends on the extent of damage of the liver function as well as the extent of the carcinoma. The extent of the carcinoma is specified by TNM scores (Table 11.2) and evaluated as in Table 11.3. The damage of liver function is evaluated as in Table 11.4. The treatment methods are selected as in Table 11.5.

TABLE 11.2 TNM scores for classification of hepatocellular carcinoma

Primary tumor (T)
- T1: Solitary tumor 2 cm or smaller in greatest dimension without vascular invasion
- T2: Solitary tumor 2 cm or smaller in greatest dimension with vascular invasion, or multiple tumors limited to one lobe, none more than 2 cm in largest dimension without vascular invasion, or a solitary tumor more than 2 cm in largest dimension without vascular invasion.
- T3: Solitary tumor more than 2 cm in largest dimension with vascular invasion, or multiple tumors limited to one lobe, none more than 2 cm in greatest dimension, with vascular invasion, or multiple tumors limited to one lobe, any more than 2 cm in greatest dimension, with or without vascular invasion.
- T4: Multiple tumors in more than one lobe, or tumor(s) at a major branch or portal or hepatic vein(s).

Regional lymph nodes (N)
- N0: No regional lymph node metastasis.
- N1: Regional lymph node metastasis.

Distant metastasis (M)
- M0: No distant metastasis.
- M1: Distant metastasis.

Source: Classified by the Liver Cancer Study Group of Japan.

TABLE 11.3 TNM staging system in hepatocellular carcinoma

Stage Grouping	T Score	N Score	M Score
Stage I	T1	N0	M0
Stage II	T2	N0	M0
Stage IIIA	T3	N0	M0
Stage IIIB	T1-3	N1	M0
Stage IVA	T4	N0-1	M0
Stage IVB	T1-4	N0-1	M1

Source: Classified by the Liver Cancer Study Group of Japan.

TABLE 11.4 Clinical staging system in liver function

Items	Stage I	Stage II	Stage III
Ascites	None	Easily controlled	Poorly controlled
Serum bilirubin (mg/dL)	<2.0	2.0–3.0	>3.0
Serum albumin (g/dL)	>3.5	3.0–3.5	<3.0
ICG R_{15}[a] (%)	<15	15–40	>40
Activity of prothrombin (%)	>80	50–80	<50

Source: Classified by the Liver Cancer Study Group of Japan.

[a]The retention rate of indocyanine green at 15 minutes.

TABLE 11.5 Treatment for hepatocellular carcinoma

	Clinical Stage		
TNM stage	Stage I	Stage II	Stage III
Stage I	Hepatectomy PEIT TAE	Hepatectomy PEIT TAE	PEIT TAE
Stage II	Hepatectomy PEIT TAE	Hepatectomy PEIT TAE	PEIT TAE
Stage III	Hepatectomy TAE	Hepatectomy TAE	TAE
Stage IV	Hepatectomy TAE	Hepatectomy TAE	TAE

Hepatectomy cannot be used on a patient whose liver function is severely damaged. PEIT cannot be used on a patient whose carcinoma is large. TAE therapy is the widely used treatment for hepatocellular carcinoma.

The mechanism of TAE therapy is as follows: Normal tissues are nourished mainly by portal veins, whereas malignant tissues are nourished by arteries. Consequently, if the hepatic artery, which nourishes malignant tissues is embolized, the tumorous tissues become severely damaged, but the damage to normal tissues is not particularly severe. The embolization is carried out with gelatin sponges, lipiodol, coils, or degradable starch microspheres by way of a catheter inserted from a femoral artery and monitored by X rays. Gelatin sponges and lipiodol are widely used as the embolizing materials.

Chemoembolization therapy is one of the improved methods of TAE, whereby anticancer agents delivered to the tumorous cells (Nishioka et al., 1993). The use of drug carriers can often enhance the efficiencies of antitumor agents (Ohnishi et al., 1984).

Lipiodol, for example, which is an ethyl ester of iodinated poppy-seed oil fatty acids, is widely used as an oily contrast medium and an embolizing material for TAE therapy (Raoul et al., 1992; Boucher et al., 2007). The deposition of lipiodol in the malignant tissue is detectable even one week after administration (Figure 11.2). Its potency as a contrast medium is owed to a large absorbance of X rays. This is due to the presence of iodine, and its affinity to malignant cells by an enhanced permeability and retention (EPR) effect (Maeda et al., 2000). Lipiodol penetrates into the malignant tissues because the tissues have hypervascularity. Lipiodol is not drained out from the malignant tissues because the tissues do not have lymphatic capillaries. Consequently lipiodol is accumulated in the malignant tissues (EPR effect).

Most anticancer drugs are water soluble, so they are used as emulsion with lipiodol. Conventional emulsion systems are oil-in-water (O/W) and water-in-

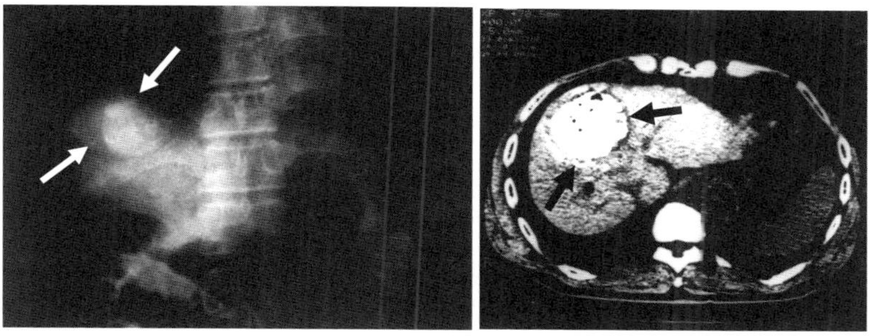

(a) (b)

Figure 11.2 Deposition of lipiodol one week after hepatic arterial infusion. (*a*) X-ray photograph; (*b*) helical CT-photograph. The patient was a 64-year-old male. The arrow shows the deposition of lipiodol in the hepatocellular carcinoma.

oil (W/O) systems. However, these two systems have some problems. Delivery of the agent to the tumor cells is not possible in the former system because the agent cannot be encapsulated in the emulsion droplets. The latter system has generally such a large viscosity that infusion of the emulsion to arteries/capillaries via catheters is difficult. A water-in-oil-in-water multiple emulsion (W/O/W emulsion) system is now being developed as a drug carrier that combines the features of encapsulation of the agent by oil membranes and the low viscosity in the external aqueous phase.

W/O/W emulsions are characterized by the vesicular structures in which internal and external aqueous phases are separated by oil membranes. Owing to the liquid membrane structures of oil phases, W/O/W emulsions have many potential applications in such diverse fields as pharmaceutics, cosmetics (Lee et al., 2004; Carlotti et al., 2005), food (Benichou et al., 2001; Su et al., 2006; Surh et al., 2007), and separation technologies (Salazar et al., 1992; Hirai and Kondo, 2007). The potential pharmaceutical applications of W/O/W include use for red blood cell substitutes (Zheng et al., 1993), treatment of drug overdosages (Morimoto et al., 1979), immobilization of enzymes (Chang and Lee, 1993), masking the taste of drugs (Vaziri and Warburton, 1994), and sustained release by carriers (Ghosh et al., 1997; Vasiljevic et al., 2006), lymphatic uptake (Omotosho et al., 1990), transdermal delivery (Piemi et al., 1998), and enhancement of enteral absorption (Shima et al., 2005). In addition to trials with dosage forms, W/O/W emulsions have been used as intermediates for the preparation of microspheres (Giovagnoli et al., 2007).

We developed a lipiodol W/O/W emulsion that encapsulates a water-soluble anticancer agent by using a two-step pumping emulsification procedure (Hino et al., 1995). We were able to investigate the acute toxicity and deposition of the agent and lipiodol in liver after i.v. and hepatic arterial administration of the emulsion to rats (Hino et al., 1997, 1999).

11.2 PREPARATION OF LIPIODOL W/O/W EMULSION WITH TWO-STEP PUMPING EMULSIFICATION

Before practical uses of the W/O/W emulsions especially for TAE can be established, two big problems have to be solved, namely the biocompatibilities of oil and the surfactant, and their sterilization and instability during storage. We designed a new way to prepare a W/O/W emulsion using a two-step pumping method (Figure 11.3) (Hino et al., 1995). The procedure is simple and easy. An injection syringe containing an oil and an injection syringe containing an aqueous solution of the drug are connected by means of a three-way cock. The W/O emulsion is created by pushing and pulling the contents alternately in the syringes. This emulsion is re-emulsified in the same way with the addition of an outer aqueous solution, resulting in the W/O/W emulsion. This two-step procedure can be easily carried out even during a surgical operation under sterilized conditions.

The formulations of the lipiodol W/O/W emulsion developed by us and the conventional emulsions for TAE (Nakamura et al., 1989) are shown in Table 11.6, together with their physical properties. All the materials used are injectable. The anticancer agent used was Farmorubicin® (FARM), which is a powder mixture of epirubicin hydrochloride (EPI) and lactose. The aqueous contrast medium used was Omnipaque 240®, which is an aqueous solution of 517.7 mg/ml iohexol, 0.1 mg/ml ethylenediaminetetraacetic acid disodium calcium, and 1.21 mg/ml tris(hydroxymethyl)aminomethane. The chemical structures of EPI

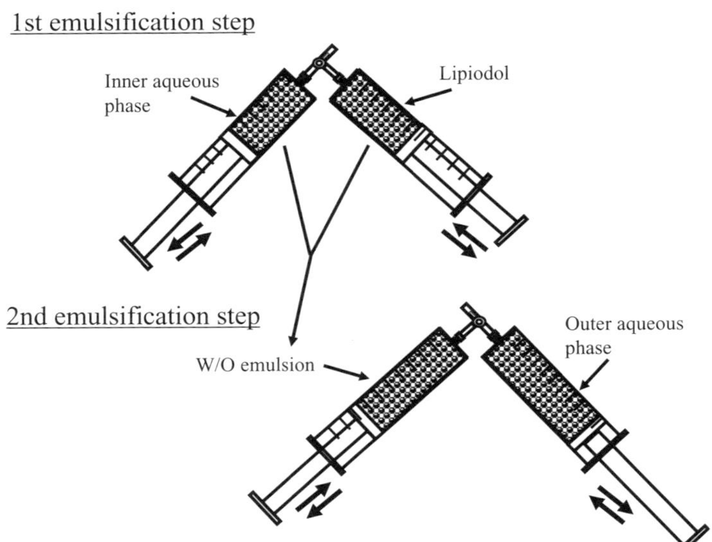

Figure 11.3 Preparation procedure of W/O/W emulsion with two-step pumping emulsification method.

TABLE 11.6 Formulations and physical properties of conventional O/W and W/O emulsions and of prepared W/O/W emulsion

Emulsion	Conventional O/W Emulsion		Conventional W/O Emulsion		W/O/W Emulsion	
Formulation						
Inner aqueous phase	—		Farmorubicin®	10 mg	Farmorubicin®	80 mg
			Urografin 76%®		Omnipaque240®	2 ml
				0.73 ml	HCO-60	240 mg
			Water	0.27 ml		
Oil phase	Lipiodol	4 ml	Lipiodol	3 ml	Lipiodol	3 ml
Outer aqueous phase	Farmorubicin®	80 mg	—		Omnipaque240®	5 ml
	Omnipaque240®	4 ml			Gelatin	150 mg
Physical property						
Viscosity (Pa·s)	0.0392		0.0512		0.0750	
Droplet size (μm)	8.5 ± 3.1		11.4 ± 6.8		69.2 ± 25.9	
Epirubicin hydrochloride trapped in the emulsion (%)	12.5		80.2		62.6	

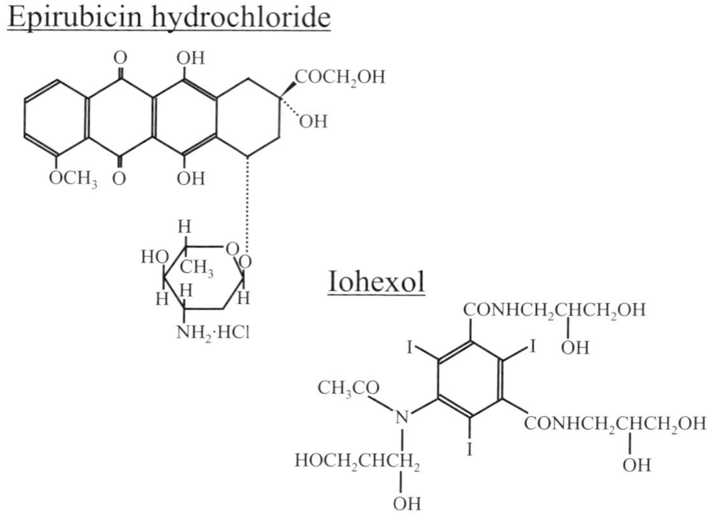

Figure 11.4 Chemical structures of epirubicin hydrochloride and iohexol.

(anticancer agent) and iohexol (x-ray contrast medium) are shown in Figure 11.4. The potential toxicities of EPI are an increase in plasma glutamic oxaloacetic transaminase (GOT) and also glutamic pyruvic transaminase (GPT) activities, ulcerations in the stomach, a size reduction of the thymus gland, leucopenia, anemia, and so on. The potential toxicity of lipiodol is edema, which is caused by the embolization.

An optical microphotograph of our prepared W/O/W emulsion is shown in Figure 11.5. The same W/O/W emulsion could be prepared by a two-step pumping emulsification procedure under sterilized conditions to avoid contamination.

The encapsulation efficiency of the emulsions is evaluated by a dialysis method that uses a seamless cellulose tube. The viscosities of the emulsion were measured by a cone-and-plate viscometer. The encapsulation efficiency of the W/O/W emulsion was high and its viscosity was so low (<0.1 Pa·s) that it was easy to handle.

The droplet sizes of an emulsion are determined from optical microphotographs. At first the volume mean diameter of the W/O/W emulsion droplets was somewhat large at 69.6 μm. Another W/O/W emulsion was prepared with the same formulation, and the number of droplets pumped at the second emulsification stage could be increased from 10 to 20, because the diameter of the droplets was reduced to 15.5 μm. This result indicates that the droplet size can be easily controlled by the pumping method and the obtainable size depends on the width of the embolization site during the operation.

The drug release profiles of the emulsions are shown in Figure 11.6. The W/O/W emulsion showed a sustained release pattern and was comparable to the release pattern from the W/O emulsion up to 7 hours.

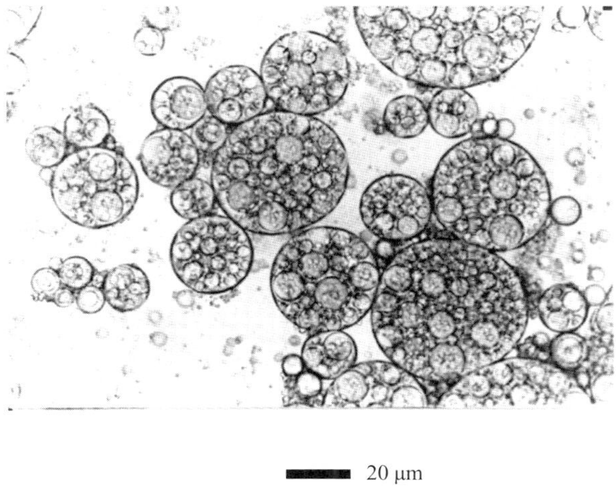

Figure 11.5 Optical microphotograph of W/O/W emulsion prepared by two-step pumping emulsification method.

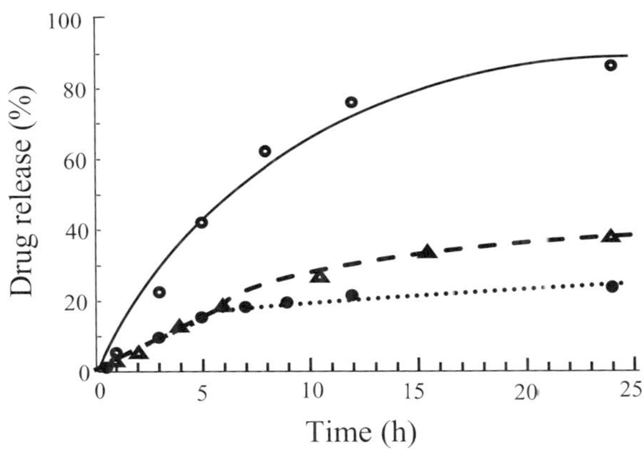

Figure 11.6 Drug release profiles of prepared W/O/W emulsion and conventional emulsions. ○: Conventional O/W emulsion; ▲: prepared W/O/W emulsion; ●: conventional W/O emulsion.

The correlation between hemolysis and drug release after incubation of an intermixed solution of a human erythrocytic suspension and the emulsion, or drug solution, are shown in Table 11.7. The drug solution and O/W emulsion caused significant hemolysis, whereas the hemolysis percentage of the W/O/W emulsion was very small (13.8%). The larger percentage of hemolysis was due to the larger release of the drug. The hemolysis was primarily caused by EPI existing in the outer aqueous phase. As the results show, the encapsulation of EPI with the W/O/W emulsion liquid membrane reduced the toxic effect of EPI on the erythrocytes.

TABLE 11.7 Percentage of hemolysis and drug release after incubation of an intermixed solution of erythrocytic suspension and the two different emulsions and drug solution

	Hemolysis (%)	Drug Release (%)
Drug solution	84.2	94.8
W/O/W emulsion	13.8	60.1
Conventional O/W emulsion	70.1	89.8

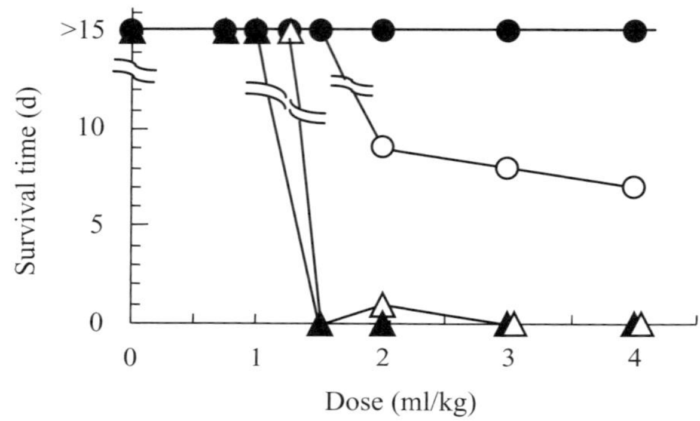

Figure 11.7 Survival time of rat after i.v. administration of emulsion. ○: W/O/W emulsion with Farmorubicin®; ●: W/O/W emulsion without Farmorubicin®; □: O/W emulsion with Farmorubicin®; ▲: O/W emulsion without Farmorubicin®.

11.3 ACUTE TOXICITY OF LIPIODOL W/O/W EMULSION AFTER INTRAVENOUS ADMINISTRATION TO RAT

When an excessive amount of emulsion is administered to the hepatic artery, the excess can be leaked to the vein. The W/O/W and O/W emulsions with and without FARM were administered via the tail veins of male Wistar rats and the acute toxicities of the emulsions were evaluated (Hino et al., 1997). Fifteen days after administration or after they died, the rats were dissected. Macroscopic pathological appearances and optical microscopic changes in hematoxylin-eosin (HE)-stained tissues of lung, stomach, and thymus gland were observed.

The survival times of the rats after i.v. administration of the emulsions are shown in Figure 11.7. The survival time of the rat administered the W/O/W emulsion is longer than that of the rat with the O/W emulsion. The rats administered the O/W emulsion (≥1.5 ml/kg) died within two days. Dyspnea was observed in these rats after administration of the O/W emulsion. The rats administered W/O/W emulsion encapsulating drug (≥2.0 ml/kg) died between seven and nine days. Diarrhea, bloody stool, and anemia were observed in these rats at four to five days after administration.

TABLE 11.8 Macroscopic pathological finding of lung after i.v. administration of emulsion

Dose (ml/kg)	0.7	1	1.5	2	3
W/O/W emulsion (with Formorubicin®)	—	—	—	—	a
W/O/W emulsion (without Formorubicin®)	—	—	—	—	—
Conventional O/W emulsion (with Formorubicin®)	—	—	b	b	b
Conventional O/W emulsion (without Formorubicin®)	—	—	b	b	b

Note: a: Bleeding in the lung was observed; b: pulmonary edema; —: no significant change was observed.

TABLE 11.9 Size reduction of thymus gland after i.v. administration of emulsion

Dose (ml/kg)	0.7	1	1.5	2	3
W/O/W emulsion (with Formorubicin®)	—	—	+	+	+
W/O/W emulsion (without Formorubicin®)	—	—	—	—	—
Conventional O/W emulsion (with Formorubicin®)	+	+	a	a	a
Conventional O/W emulsion (without Formorubicin®)	—	—	—	—	—

Note: +: Size reduction of thymus gland was observed; —: no significant change was observed; a: size reduction of thymus gland could not be observed because rat died in two days after administration.

TABLE 11.10 Macroscopic pathological finding of stomach after i.v. administration of emulsion

Dose (ml/kg)	0.7	1	1.5	2	3
W/O/W emulsion (with Formorubicin®)	—	—	a	b	b
W/O/W emulsion (without Formorubicin®)	—	—	—	—	—
Conventional O/W emulsion (with Formorubicin®)	c	d	e	e	e
Conventional O/W emulsion (without Formorubicin®)	—	—	—	—	—

Note: a: Gastric paries became thin; b: bleeding in the stomach was observed; c: brown spots were scattered in the inner mucosa; d: ulcer was observed; e: rats died in two days after administration; —: no significant change was observed.

The macroscopic pathological changes in the lung, thymus gland, and stomach are shown in Tables 11.8, 11.9 and 11.10, respectively. Pulmonary edema was observed in the lungs of rats given the large dose (≥1.5 ml/kg) of the O/W emulsion regardless of whether the drug was in the solution. Edema and bleeding were observed in the HE-stained tissues of the lungs of these rats. The deaths of these rats within two days after administration of the O/W emulsion was from the pulmonary edema due to the embolism caused by the fine oil droplets. Edema could not be detected in the lung tissues of the rats that had been administered the W/O/W emulsion.

The reduced of thymus gland and pathological changes in the stomach, namely a thinning of gastric paries and bleeding, were observed in the tissues of rats administered the large dose of the W/O/W emulsion (≥1.5 ml/kg) containing the drug. The deaths of these rats at seven to nine days after administration of the W/O/W emulsion containing the drug were due to the weakness of the anticancer drug. Similar pathological changes were observed in the tissues of rats administered a low dose of the O/W emulsion with FARM. Apparently, the encapsulation of the drug by the oil membranes of the W/O/W emulsion reduced the toxic effects of EPI.

11.4 ACUTE TOXICITY AND DEPOSITION IN LIVER OF RAT AFTER HEPATIC ARTERIAL ADMINISTRATION OF LIPIODOL W/O/W EMULSION AND O/W EMULSION

The acute toxicity and deposition in liver of the lipiodol W/O/W emulsion encapsulating anticancer agent after hepatic arterial administration to rats were investigated and compared to the O/W emulsion used conventionally for TAE therapy (Hino et al., 1999).

The hepatic arterial administration of W/O/W and O/W emulsions to rats were performed as follows: Nine-week-old male Sprague-Dawley rats were anaesthetized intraperitoneally with sodium pentobarbiturate solution. The gastroduodenal and splenic arteries of each rat were ligated with clips. The emulsion was infused to a left gastric artery with a catheter and then the ligatures were removed. The dose was 400 μl/kg. The hepatic artery was ligated with a thread at 5 minutes after administration of the emulsion.

X-ray CT photographs of the rats administered the emulsions are shown in Figure 11.8. The livers of rats administered the emulsions show wide regions of white (Figures 11.8*a* and *b*). No such white regions appear in the X ray of the control rat, which was not administered the emulsions; the bright areas in the X-ray image show only bones (Figure 11.8*c*). The wide white regions in the first two images of the rat livers are due to the deposition of lipiodol. The region of the O/W emulsion is wider than that of the W/O/W emulsion is

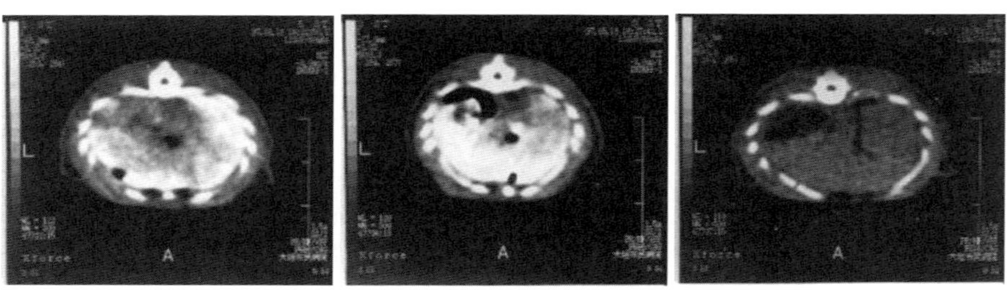

(a) (b) (c)

Figure 11.8 Helical X-ray CT photographs of rats after hepatic arterial administration of emulsions. (*a*) Rat administered W/O/W emulsion; (*b*) rat administered O/W emulsion; (*c*) control rat without administration of the emulsion.

because a larger dose of lipiodol was added to the same volume of emulsion. The distribution in the liver of the lipiodol present in the W/O/W emulsion is detectable even at 8 hours after administration. This result indicates that with suitable monitoring using an X-ray CT scanner during the surgical operation, the W/O/W emulsion could be successfully infused to a hepatocellular carcinoma.

The concentrations of EPI in the liver at 1, 8, and 24 hours are shown in Figure 11.9. The concentration of the agent in the liver of the rat administered the W/O/W emulsion was remarkably increased at 8 hours compared with that of the rat administered the O/W emulsion or drug solution. The concentration was increased and its residence was prolonged by encapsulation of the agent in the W/O/W emulsion. A more therapeutic effect of the W/O/W emulsion than that of the conventional O/W emulsion can be expected from the sustained release of the lipiodol to the malignant cells and its drug-targeting potency.

Figures 11.10 and 11.11 shows changes in the plasma GOT and GPT activities of rats after administration of the different emulsions. Both GOT and GPT activities temporarily increased within a single day in all rats. The increase due to the O/W emulsion was much larger than that due to the W/O/W emulsion. The possible cause of the larger increase is the toxic effect of EPI and embolization of the arteries by the emulsion droplets. If the former effect is dominant, the initial damage to normal hepatic cells by the aqueous drug solution should be larger than that by the W/O/W emulsion. However, the contrary effect was obtained: the increase in the values was owing mainly to the embolization by the lipiodol droplets.

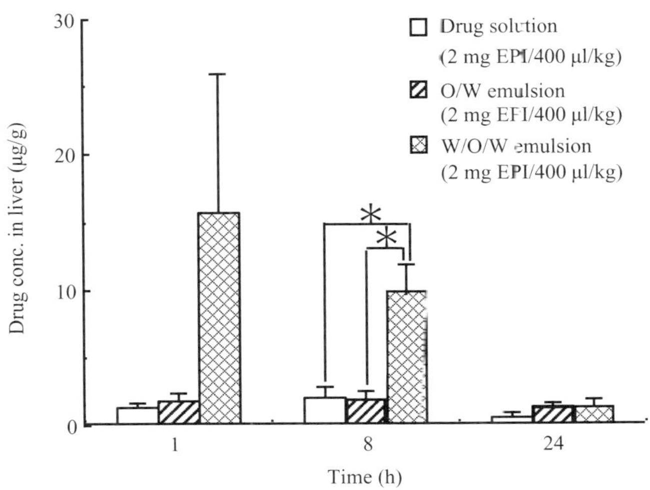

Figure 11.9 Concentration of epirubicin hydrochloride in the liver after hepatic arterial administration of emulsions ($n = 3$, $*p < 0.05$).

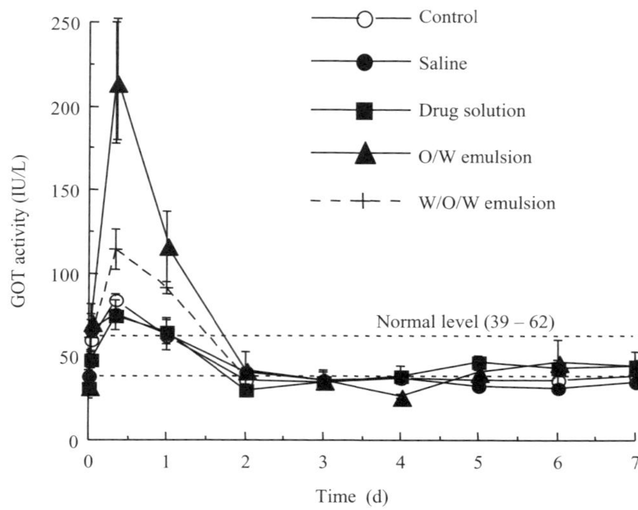

Figure 11.10 Changes in plasma GOT activities of rats after hepatic arterial administration of emulsions. ○: Control; ●: saline; ■: aqueous drug solution; ▲: O/W emulsion; □: W/O/W emulsion ($n=3$).

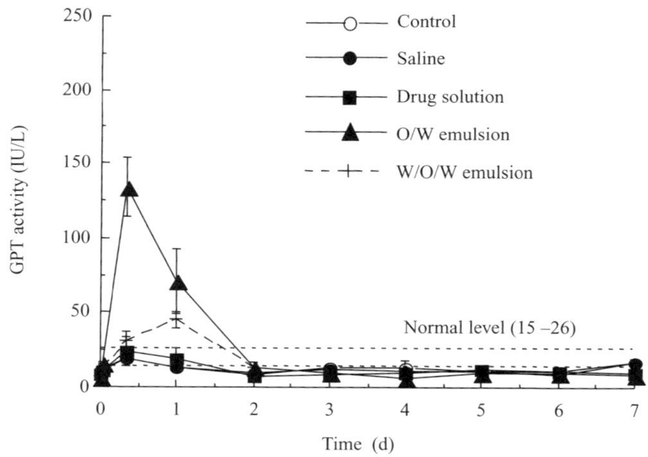

Figure 11.11 Changes in plasma GPT activities of rats after hepatic arterial administration of emulsions. ○: Control; ●: saline; ■: aqueous drug solution; ▲: O/W emulsion; □: W/O/W emulsion ($n=3$).

From the pathological findings after dissection, extensive necrosis was observed in the liver of the rat administered the O/W emulsion. By contrast, the area was small and slight in the organ of the W/O/W emulsion-administered rat. The arteries were embolized tightly with the many fine lipiodol droplets of the O/W emulsion, which induced the spread of necrosis. The embolization with the large multiple droplets of the W/O/W emulsion was loose, so the necrosis was not as remarkable and the damage to the normal hepatic cells by the lipiodol oil droplets of W/O/W emulsion was much smaller. These changes in the GOT and GPT are consistent with the values shown in Figures 11.10 and 11.11.

The effectiveness of the conventional O/W emulsion is considered only due to the embolizing effect of lipiodol, and the first shot of the anticancer agent independently of the droplets. No drug-carrying and drug-targeting effect, with a sustained release of the drug, can be expected of the O/W emulsion, whereas that effect is the aim of the W/O/W emulsion. Moreover the W/O/W can be expected to be more therapeutic with a less toxic effect. The weakness of the embolizing effect of a W/O/W emulsion can further be conquered by administration of gelatin sponge powder after infusion of the emulsion (Sasaki et al., 1987).

The W/O/W emulsion we prepared appear to be a desirable formulation for TAE therapy. Our investigation is currently proceeding to include tests on animals with implanted malignant cells.

11.5 CONCLUSION

The lipiodol W/O/W emulsion used for TAE therapy can be prepared by a two-step pumping emulsification procedure. This procedure is simple enough to be carried out even in a surgical operation under sterilized conditions. We found that a large amount of the aqueous drug solution becomes encapsulated in the resultant W/O/W emulsion and the drug-release profile from the emulsion shows a sustained release pattern. The hemolysis induced by the W/O/W emulsion was much more depressed compared to the EPI solution or the conventional O/W emulsion used for TAE therapy. The toxicity of the W/O/W emulsion after i.v. and hepatic arterial administration to rats was investigated. The survival time of the rat administered the W/O/W emulsion was longer that of the rat administered the O/W emulsion. The toxic effects of EPI and lipiodol on the normal hepatic cells were reduced. The concentration of EPI in the liver was increased and its residence was prolonged by encapsulating it in the W/O/W emulsion. Deposition of the emulsion in the liver was detected by means of an X-ray CT scanner. These trials suggest that the dose of the emulsion is controllable if monitored during a surgical operation. The W/O/W emulsion prepared by us turned out to be a desirable formulation for TAE therapy.

REFERENCES

Benichou A, Aserin A, Garti N. 2001, Polyols, high pressure, and refractive indices equalization for improved stability of W/O emulsions for food applications. *J Dispersion Sci Technol* 22: 269–280.

Boucher E, Garin E, Guylligomarc'h A, Olivie D, Boudjema K, Raoul JL. 2007. Intra-arterial injection of iodine-131-labeled Lipiodol for treatment of hepatocellular carcinoma. *Radiother Oncol* 82: 76–82.

Carlotti ME, Gallarate M, Sapino S, Ugazio E, Morel S. 2005. W/O/W multiple emulsions for dermatological and cosmetic use, obtained with ethylene oxide free emulsifiers. *J Dispersion Sci Technol* 26: 183–192.

Chang JH, Lee WK. 1993. Modeling of enzymatic reaction with emulsion liquid membrane. *Chem Eng Sci* 48: 2357–2366.

Cui Jun, Dong BW, Liang P, Yu XL, Yu DJ. 2005. Construction and clinical significance of a predictive system for prognosis of hepatocellular carcinoma. *World J Gastroenterol* 11: 3027–3033.

Ghosh LK, Ghosh NC, Thakur RS, Pal M, Gupta BK. 1997. Design and evaluation of controlled-release W/O/W multiple-emulsion oral liquid delivery system of chlorpheniramine maleate. *Drug Dev Ind Pharm* 23: 1131–1134.

Giovagnoli S, Blasi P, Schoubben A, Rossi C, Ricci M. 2007. Preparation of large porous biodegradable microspheres by using a simple double-emulsion method for capreomycin sulfate pulmonary delivery. *Int J Pharm* 333: 103–111.

Hino T, Takeuchi H, Niwa T, Nagaya K, Kawashima Y, Nakano S, Yamazaki F, Kumada T. 1995. Formulation of W/O/W emulsion encapsulating epirubicin hydrochloride for the transcatheter arterial embolization therapy for heparocellular carcinoma. *Drug Deliv Sys* 10: 115–119.

Hino T, Isozaki M, Yamamoto H, Takeuchi H, Kawashima Y, Nakano S, Yamazaki F, Kumada T, Sassa T. 1997. Evaluation of acute toxicity of epirubicin hydrochloride-encapsulating W/O/W emulsion administered to rat for the transcatheter arterial embolization therapy for heparocellular carcinoma. *Drug Deliv Sys* 12: 161–165.

Hino T, Shimabayashi S, Kawashima Y, Takeuchi H, Yamamoto H, Isozaki M, Nakano S, Yamazaki F, Kumada T, Sassa T. 1999. Comparison of lipiodol water-in-oil-in-water emulsion with oil-in-water emulsion: Acute toxicity and deposition in liver after hepatic arterial administration in rats. *Pharm Pharmacol Comm* 5: 355–359.

Hirai T, Kondo Y. 2007. Preparation of $Y_2SiO_5:Ln^{3+}$ (Ln = Eu, Tb, Sm) and $Gd_{9.33}(SiO_4)_6O_2$: Ln^{3+} (Ln = Eu, Tb) phosphor fine particles using an emulsion liquid membrane system. *J Phys Chem C* 111: 168–174.

Jang JW, Park YM, Bae SH, Choi JY, Yoon SK, Chang UI, Nam SW, Kim BS. 2004. Therapeutic efficacy of multimodal combination therapy using transcatheter arterial infusion of epirubicin and cisplatin, systemic infusion of 5-fluorouracil, and additional percutaneous ethanol injection for unresectable hepatocellular carcinoma. *Cancer Chemother Pharmacol* 54: 415–420.

Koiwa F, Kakuta T, Tanaka R, Yumita S. 2007. Efficacy of percutaneous ethanol injection therapy (PEIT) is related to the number of parathyroid glands in haemodialysis patients with secondary hyperparathyroidism. *Nephrol Dialysis Transplant* 22: 522–528.

Kumada T, Nakano S, Takeda S, Sugiyama K, Osada T, Kiriyama S, Yamada M, Okabe H. 1992. Treatment of small hepatocellular carcinoma. *Cancer Chemother Pharmacol* 31: S25–S29.

Lee JS, Kim JW, Han SH, Chang IS, Kang HH, Lee OS, Oh SG, Suh KD. 2004. The stabilization of L-ascorbic acid in aqueous solution and water-in-oil-in-water double emulsion by controlling pH and electrolyte concentration. *J Cosmet Sci* 55: 1–12.

Maeda H, Wu J, Sawa T, Matsumura Y, Hori K. 2000. Tumor vascular permeability and the EPR effect in macromolecular therapeutics: a review. *J Controlled Release* 65: 271–284.

Marrache F, Vullierme MP, Roy C, El Assoued Y, Couvelard A, O'Toole D, Mitry E, Hentic O, Hammel P, Levy P, Ravaud P, Rougier P, Ruszniewski P. 2007. Arterial phase enhancement and body mass index are predictors of response to chemoembolization for liver metastases of endocrine tumours. *Br J. Cancer* 96: 49–55.

Minagawa M, Makuuchi M. 2006. Treatment of hepatocellular carcinoma accompanied by portal vein tumor thrombus. *World J Gastroenterol* 12: 7561–7567.

Morimoto Y, Sugibayashi K, Yamaguchi Y, Kato Y. 1979. Detoxication capacity of a multiple (W/O/W) emulsion for the treatment of drug overdose: drug extraction into the emulsion in the gastro-intestinal tract to rabbits. *Chem Pharm Bull* 27: 3188–3192.

Nakamura H, Hashimoto T, Oi H, Sawada S. 1989. Transcatheter oily chemoembolization of hepatocellular carcinoma. *Radiol* 170: 783–786.

Nishioka Y, Kyotani S, Okamura M, Ohnishi S, Yamamoto Y, Tanada S, Nakamura T. 1993. A study of embolizing materials for chemo-embolization therapy of hepatocellular carcinoma: Effects of chitin concentration on cis-diamminedichloroplatinum (II) albumin microsphere properties and antitumor effect in VX_2 hepatocellular carcinoma model rabbits. *Biol Pharm Bull* 16: 1136–1139.

Ohira M, Ohdan H, Mitsuta H, Ishiyama K, Tanaka Y, Igarashi Y, Asahara T. 2006. Adoptive transfer of TRAIL-expressing natural killer cells prevents recurrence of hepatocellular carcinoma after partial hepatectomy. *Transplantation* 82: 1712–1719.

Ohnishi K, Tsuchiya S, Nakayama T, Hiyama Y, Iwama S, Goto N, Takashi M, Ohtsuki T, Kono K, Nakajima Y, Okuda K. 1984, Arterial chemoembolization of hepatocellular carcinoma with mitomycin C microcapsules. *Radiology* 152: 51–55.

Omotosho JA, Florence AT, Whateley TL. 1990. Absorption and lymphatic uptake of 5-fluorouracil in the rat following oral administration of W/O/W multiple emulsions. *Int J Pharm* 61: 51–56.

Piemi MPY, de Luca M, Grossiord JL, Seiller M, Marty JP. 1998. Transdermal delivery of glucose through hairless rat skin in vitro: Effect of multiple and simple emulsions. *Int J Pharm* 171: 207–215.

Raoul JL, Helesbach D, Bretagne JF, Ferrer DB, Duvauferrier R, Bourguet P, Messner M, Gosselin M. 1992. Chemoembolization of hepatocellular carcinomas: A study of the biodistribution and pharmacokinetics of doxorubicin. *Cancer* 70: 585–590.

Salazar E, Ortiz MI, Urtiaga AM, Irabien JA. 1992. Kinetics of the separation-concentrtation of chromium (VI) with emulsion liquid membranes. *Ind Eng Chem Res* 31: 1523–1529.

Sasaki Y, Imaoka S, Kasugai H, Fujita M, Kawamoto S, Ishiguro S, Kojima J, Ishikawa O, Ohigashi H, Furukawa H, Koyama H, Iwanaga T. 1987. A new approach to chemoembolization therapy for hepatoma using ethiodized oil, cisplatin, and gelatin sponge. *Cancer* 60: 1194–1203.

Shima M, Tanaka M, Kimura Y, Adachi S, Matsuno R. 2005. Enhancement in transport of a hydrophilic marker through intestinal epithelial cell (Caco-2) monolayer by W/O/W multiple emulsion containing C8TG. *Food Hydrocolloids* 19: 321–328.

Shepherd FA, Evans WK, Blackstein ME, Fine S, Heathcote J, Langer B, Taylor B, Habal F, Kutas G, Pritchard KI, Kuruvilla P. 1987. Hepatic arterial infusion of mitoxantrone in the treatment of primary hepatocellular carcinoma. *J Clin Oncol* 5: 635–640.

Su J, Flanagan J, Hemar Y, Singh H. 2006. Synergistic effects of polyglycerol ester of polyricinoleic acid and sodium caseinate on the stabilisation of water-oil-water emulsions. *Food Hydrocolloids* 20: 261–268.

Surh J, Vladisavljevic GT, Mun S, McClements DJ. 2007. Preparation and characterization of water/oil and water/oil/water emulsions containing biopolymer-gelled water droplets. *J Agr Food Chem* 55: 175–184.

Vasiljevic D, Parojcic J, Primorac M, Vuleta G. 2006. An investigation into the characteristics and drug release properties of multiple W/O/W emulsion systems containing low concentration of lipophilic polymeric emulsifier. *Int J Pharm* 309: 171–177.

Vaziri A, Warburton B. 1994. Slow release of chloroquine phosphate from multiple taste-masked W/O/W multiple emulsions. *J Microencaps* 11: 641–648.

Zheng S, Zheng Y, Beissinger RL, Wasan DT, McCormick DL. 1993. Hemoglobin multiple emulsion as an oxygen delivery system. *Biochim Biophys Acta* 1158: 65–74.

CHAPTER 12

Multiple Emulsions: Delivery System for Antigens

ASUMAN BOZKIR and ONGUN MEHMET SAKA

Contents

12.1 INTRODUCTION

12.1.1 Vaccination and Immunization

Vaccination is one of the most successful achievements of medical science. The number of cases of infectious diseases is a serious problem only in countries where vaccination is not systematically carried out (Saatci and Bozkir, 2003; Almeida and Alpar, 1996). Administration of vaccines routinely is considered to have been very effective in preventing a number of infectious diseases. The

Multiple Emulsions: Technology and Applications, Edited by Abraham Aserin

success of this is illustrated by the fact that mass vaccination resulted in worldwide eradication of smallpox in the 1970s. Moreover, because of the vaccination of children, a number of infections such as diphtheria, pertussis, tetanus, poliomielitis, measles, mumps, and rubella are under control in countries where such vaccinations are being systematically carried out (Klegerman, 1992).

The principle behind a vaccine is to mimic an infection in such a way that the natural specific defense mechanism of the host against the pathogen is activated, and the host remains free of the disease that normally results from a natural infection. The success of vaccination relies on the induction of a long-lasting immunological memory. Vaccination is also referred to as active-immunization because the host's immune system is activated for an acquired immunity against the particular pathogen. Apart from active immunization, the administration of specific antibodies can be utilized for short-lived immunological protection of the host, which is termed as passive immunization (Figure 12.1) (Ciardi et al., 1992; Klegerman, 1992; Jiskoot et al., 2002).

Vaccines are pharmaceutics that have traditionally consisted of live attenuated pathogens, whole inactivated organisms, inactivated toxins, or toxoids to prevent infectious diseases by immunization. Moreover some vaccines include recombinant protein subunits, synthetic peptides, protein polysaccharide conjugates, or plasmid DNA (O'Hagan et al., 2001). In the rapidly evolving field of new vaccine technologies there can be discerned improvements of existing vaccines and, secondly, the development of vaccines for diseases against which there are no vaccines currently available (Klegerman 1992).

Modern biotechnology has had an enormous impact on vaccine development. The elucidation of the molecular structures of pathogens and the tremendous progress made in immunology during the past few decades have led to the identification of protective antigens. Major goal of modern vaccine technology is to fulfill all requirements of the ideal vaccine, as summarized in Table 12.1 by expressing antigen epitopes and/or isolating those antigens that confer an effective immune response (Montagne and Rabinovich, 1992; Klegerman 1992; Jiskoot et al., 2002).

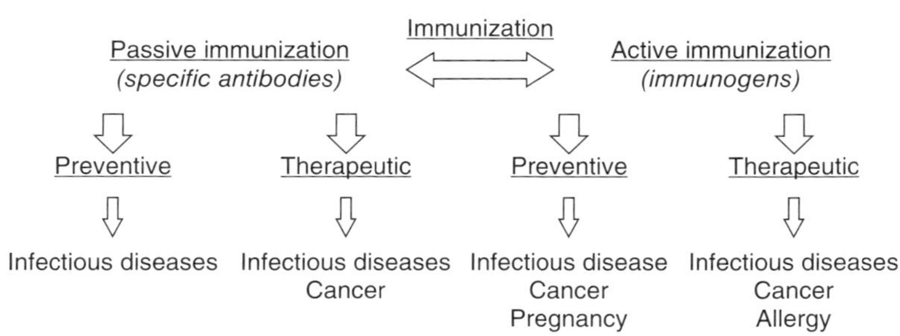

Figure 12.1 Schematic of active immunization (through vaccination) and passive immunization and examples of their fields (Adapted from Jiskoot et al., 2002).

TABLE 12.1 Characteristics of the (hypothetical) ideal vaccine

- Is 100% efficient in all individuals of any age
- Provides lifelong protection after single administration
- Does not evoke an adverse reaction
- Is stable under various conditions (temperature, light, transportation)
- Is easy to administer (preferably orally)
- Is available in unlimited quantities
- Is cheap
- Provides immunization for more infections
- Stimulates both mucosal and systemic immunization

TABLE 12.2 Mechanisms by which adjuvants improve the immune response

- Increase the immunogenicity of weak antigens
- Enhance the speed and duration of the immune response
- Modulate antibody avidity, specificity, isotype, or subclass distribution
- Stimulate cell mediated immunity
- Promote the induction of mucosal immunity
- Enhance immune responses in immunologically immature, or senescent individuals
- Decrease the dose of antigens in the vaccine and reduce costs
- Help overcome antigen competition in combination vaccines

Modern vaccines that are biosynthetic, recombinant, or made by other modern technology have a general problem, which is that these vaccines alone are often poorly immunogenic. So there is an urgent need for the development of potent and safe adjuvants that can be used with modern vaccines. Delivery systems are injectable devices that allow the multimeric presentation of antigens and may also contain adjuvants.

12.1.2 Adjuvants

Adjuvants are defined as any material that can increase the humaral and cellular response against an antigen. The success of immunization depends not only on the nature of the protective components but also on their presentation form, the presence of adjuvants, and the route of administration (see Sections 12.3 and 12.4).

Better insight into adjuvant action can aid in the rational design of vaccine formulations. The mechanisms proposed for adjuvant action include (1) slow release of the antigen, (2) attraction and stimulation of macrophages and lymphocytes, and (3) delivery of the antigen to regional lymph nodes (O'Hagan et al., 2001).

Adjuvants can be used to improve the immune response against to the vaccine antigens in several different ways, as summarized in Table 12.2 (Rabinovich et al., 1994; O'Hagan et al., 2001; Jiskot et al., 2002; Saatci and Bozkir, 2003). Table 12.3 lists some well-known adjuvants and delivery systems.

TABLE 12.3 Summary of adjuvants and antigen delivery systems

Adjuvant	*Characteristics*
Aluminum salts	Antigen adsoption is crucial (e.g., aluminum hydroxide
Lipid A and derivatives	fragment of lipopolysaccharide, a bacterial endotoxin)
Muramyl peptides	Active fragments of bacterial cell walls (Keitel et al., 1993)
Saponins	Plant triterpene glycosides (e.g., QS21)
Nonionic block copolymer	Synthetic amphiphiles (Todd et al., 1997; Newman et al., 1998)
Cytokines	Interleukins (1, 2, 3, 6, 12), interferon-γ, tumor necrosis factor (Rabinovich et al., 1994)
Cholera toxin, B subunit	Mucosal adjuvant (Hirabayashi et al., 1990; Tamura et al., 1997; Tomasi et al., 1997)
Delivery system	*Characteristics*
Emulsions	W/O, O/W, and W/O/W emulsions (Lidgate et al., 1992; Poland and Couch 1999; Bozkir and Hayta, 2004)
Liposomes	Phospholipid membrane vesicles (Allison and Gregoriadis, 1974; Michalek et al., 1992)
Iscom's	Micellar lipid-saponin complex
Nano-/micro particles	Biodegradable polymeric spheres (Bozkir et al., 2004)
Niosomes	Nonionic surfactant vesicles (Eratalay et al., 2004)

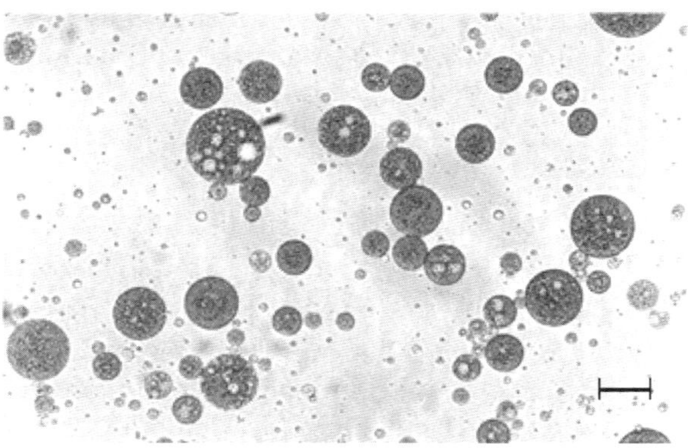

Figure 12.2 Picture of W/O/W multiple emulsions (bar = 10 μm).

12.2 EMULSION FORMULATIONS IN IMMUNIZATIONS

Water-in-oil emulsions, consisting of a aqueous antigen solution dispersed in a mineral oil phase with the aid of lipophilic emulsifier, have been prepared to act as efficient immunological adjuvants (Figure 12.2) over a long period of time. It is suggested that in addition to the depot effect, the oil attracts mononuclear cells about the antigen that take part in the antibody production and give a high antibody level (Garti and Aserin, 1996). On the other hand, because the mineral oil's physical properties cannot be metabolized, the mineral oil emulsion adjuvants will persist at the injection site. They cannot save their labil properties from fallowing in a few months, although they are kept at 4°C. Indeed there is limited use of mineral oil emulsions in humans. Therefore researchers have studied on alternative oil phases like hexadecane, squalene (Hilgers et al., 1999), sesame oil (Hilleman, 1966), and peanut oil (Yarkoni and Rapp, 1978).

Freund's incomplete adjuvant (FIA), a mineral oil emulsion that contains Arlacel A (2-dianhydro-D-mannitol-mono-9-octadecenoate), shows adjuvant activity. It was modified to give a greater antigenic response by incorporating dried, killed Mycobacterium tuberculosis in the mineral oil phase. These emulsions are termed as Freund's complete adjuvants (FCA) (Klegerman, 1992). Although far more effective than aluminum salts, Freund's adjuvants are used exclusively in animal trials. The granulomas and tuberculin hypersensitivity induced by FCA make it unacceptable for use in human and veterinary vaccines. FCA is also a potential hazard for laboratory personnel. Accidental self-inoculation can result in tuberculin sensitization followed by chronic local inflammation that responds poorly to antibiotic treatment. Accidental splashing of FCA in the eye, a risk for unguarded eyes during preparation of the emulsion or during injection of the animal, can result in severe ocular irritation and even blindness (Chapel and August, 1976). Allison and Byars (1992) replaced the components of FCA with alternatives and introduced Syntex adjuvant formulation (SAF), which was also shown to be very toxic for widespread use in humans. SAF consists of a peptide related to muramyl dipeptide (threonyl-*N*-acetylmuramyl-L-alanyl-disoglutamine) in a vehicle comprising squalane, Pluronic L121, and Tween 80. This vehicle has sometimes been used as an adjuvant without the inclusion of muramyl dipeptide Unlike Freund's complete adjuvant (FCA), which is viscous and difficult to work with, SAF is fluid and is easily mixed with antigen solutions (Hjorth et al., 1997). Multiple W/O/W emulsions contain both W/O and O/W simple emulsions. Because of the presence of a reservoir phase inside the droplets of another phase that can be used to prolonged release of active ingredients, multiple emulsions can avoid the booster (second or third vaccination) dose of the vaccine. Therefore the adjuvant emulsion (MF59), which contains 0.5% polysorbate 80 (Tween 80), 0.5% sorbitan trioleate, and 4.3% squalene, can be used to enhance the effectiveness of vaccination. MF59 (adjuvanted influenza vaccine, or FLUAD) was confirmed in elderly subjects in clinical trials, and these data allowed the approval of this product in Italy in 1997 (Minutello et al., 1999; O'Hagan et al., 2001; Podda, 2001). The addition of MF59 adjuvant emulsion to control the vaccine appeared to increase the frequency of reports of some postimmunization reactions. However, these reactions did not affect routine daily activities of the vaccines, nor the need for second and third vaccinations (Minutello et al., 1999). Hilgers et al. (1999) compares adjuvant effect of squalane/water emulsion with multiple emulsions composed of 10 mg synthetic sulfolipo derivatives of polysaccharide, 20 mg Tween 80 and 80 mg oil per ml. They found multiple emulsions show higher activity and stability than any single emulsion. As a result there is still a need for adjuvant formulations with lower toxicity and better biocompatibility.

12.3 MULTIPLE-EMULSION SYSTEMS USED IN IMMUNIZATION

12.3.1 Applications of Multiple Emulsions in Immunizations

The term "multiple emulsions" has been used to describe emulsion systems in which drops of the dispersed phases contain smaller droplets that have the same composition as the external phase (Figure 12.2) (Bozkir and Hayta, 2004). The water-in-oil-in-water (W/O/W) and oil-in-water-in-oil (O/W/O) types of multiple emulsions for entrapping drugs are pharmaceutically considered to be the most useful in controlled releasing and targeting drug delivery systems (Miyakawa et al., 1993). W/O/W emulsions have been studied as prolonged release delivery systems for such drugs as pentazocine, vancomycine, and methotrexate (Omotosho et al., 1989; Mishra and Pandit, 1990; Okochi and Nakano, 2000). Since 1960s it has been known that multiple emulsions are effective adjuvants. However, it was reported that because of water-in-oil emulsions' high viscosity, injections required high gauge needles and were difficult to apply. Also, even though emulsions were stored at 4°C, they were not very stable and would break down within a few months (Herbert, 1965). W. J. Herbert incorporated antigen into the inner water phase of a multiple emulsion to produce a high and prolonged antibody response. These antigenic multiple emulsions could be stored without evidence of breakdown for several months at 56°C and at room temperature (Herbert, 1965). Although W/O/W emulsions have many of the attributes of W/O emulsions, their lower viscosity, derived from water as the external phase, makes them easier to inject.

Many compounds with adjuvant activity are presently known (shown in Table 12.2) but only a few are applied routinely in human and veterinary vaccines. The application of the novel adjuvants are limited for several reasons, such as disappointing efficacy in the target animal species, insufficient safety, problems with large-scale preparations, and limited stability of the final formulations. Many studies have reported on O/W and W/O emulsions used as adjuvants and delivery systems for immunization (Hilgers et al., 1994a, b, 1999). The adjuvanticity of the W/O/W makes this type of multiple emulsion a suitable delivery system for immunization with prolonged release, and its interesting preparation is presented below.

12.3.2 Formation of Multiple Emulsions

W/O/W emulsions are today considered to be a convenient and reliable method for encapsulation of antigens (Bozkir and Hayta, 2004). The W/O/W emulsion is also a pro forma step in microparticular immunization formulations (Couvreur et al., 1997; Vila et al., 2002; Bozkir et al., 2004). First, homogenization is done to obtain a fine, homogeneous, and stable W/O emulsion using an Ika homogenizer (T25 basic Ika Labortechnic, Germany). Second, the continuous phase of the emulsion is poured into a dispersed phase to prepare the W/O/W emulsion. This is a reverse of the general method whereby a dispersed phase is poured into a continuous phase, and it allows for a more homogeneous particle size to be obtained this way than by the conventional method. The multiple emulsions are then prepared by emulsifying 2 ml of an aqueous dispersion of influenza antigen (in PBS with or without 0.2% hydroxypropyl-β-cyclodextrin) with 8 ml of squalane/soybean oil containing a mixture of Span 80 (2.1 g) and Cremophor RH-40 (HCO-40) (1.05 g) by a homogenizer at 10,000 rpm for 3 minutes per 10 ml of PBS (pH 7.4) containing 0.5% Pluronic F-68 (Poloxamer 188; block copolymers of ethylene oxide and propylene oxide; 80:20), which is added to the W/O emulsion and further homogenized at 8000 rpm for 5 minutes (Figure 12.3).

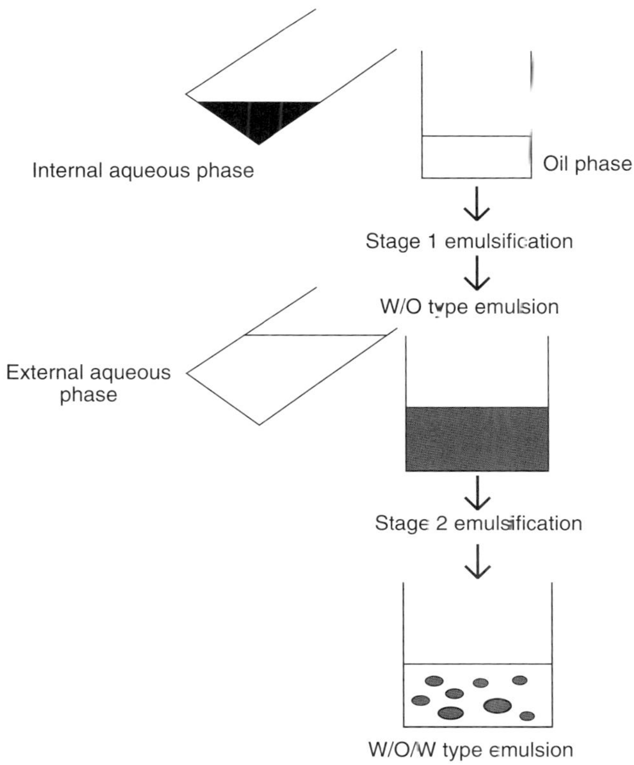

Figure 12.3 Schematic presentation of the antigen contained multiple emulsion preparation.

Hilgers et al. (1994, 1999) have prepared sulfolipo-polysaccharides (SLP)/squalene/water emulsions for immunization. In sum, their method was to dissolve the SLP in a minimal volume of methyltertiary butyl ether to which they added 2 g Tween 80 (per g SLP). The methyltertiary butyl ether was removed to yield viscous solutions of SLP in Tween 80. Then 80 ml of water (per g SLP) was added, slowly giving transparent or slightly opalescent solutions of SLP/Tween 80 in water. Subsequently 9 g of PBS and 8 g squalene (per g SLP) were added. The mixtures were emulsified by a microfluidizer. The emulsions were supplemented with thiomersal at a final concentration of 0.1 mg per ml and stored 4°C, until use within 30 months. The final composition of the adjuvant emulsion formulations was 10 mg SLP, 20 mg Tween 80, and 80 mg oil (squalene) per ml (Hilgers et al., 1999).

12.3.3 Multiple-Emulsion Stability

Surfactant migration and osmotic gradients are largely responsible for the poor physical stability of multiple emulsions. The breakdown of the oil layer and consequent loss of the internal droplets can result if the concentration of the primary emulsifier is decreased by passing from the oil layer to the external aqueous component. Such emulsifier migration is responsible for a decrease in the effective HLB of the second emulsifier, which means an increase in the optimal HLB proportional to the concentration of primer emulsifier. At a fixed concentration of primary emulsifier, the HLB shift is inversely proportional to the concentration of secondary emulsifier. It is possible to predict the optimal HLB if the two emulsifier concentrations and the required HLB of the oil are known. Briefly, the structure of the surface-active substances and their concentrations in the formulation were found to be effective on the emulsion type and its stability. For example, a mixture of Span 80 (HLB 4.3) with HCO-40 (HLB 15) at a certain ratio (1:2.3) was added to the oil phase and found to prevent phase separation of the emulsions as compared to the formulations prepared with Span 80 alone (Bozkir and Hayta, 2004). Pluronic F68 (Poloxamer 188) and Pluronic F127 (Poloxamer 407; block copolymers of ethylene oxide and propylene oxide; 101:56) were added at concentrations of 0.5–1–5% during external aqueous phase and 0.5% w/v Pluronic F68 and found to yield the most stable formulations (Okochi and Nakano, 2000). An inversion of W/O/W emulsions to O/W emulsion occurs if the HLB of the total emulsifiers approaches the required HLB of the oil. As a result of the increasing secondary emulsifier concentration, the total droplet size becomes too small to contain the internal water droplets. This situation also causes the inversion of the emulsion stability. Droplet size and distribution of parenteral emulsions have an impact on both toxicity and stability (Floyd, 1999). The droplet size decreases with the increasing stirring speed. The duration of stirring has also an effect on droplet size. With stirring at high speed over longer periods of time, the droplets are broken and phases separate (Floyd, 1999; Okochi and Nakano, 2000). Bozkir and Hayta (2004) prepared stable droplets for a primary water-oil emulsion with stirring at high speed over a short stirring duration and for multiple emulsions with a lower speed over a longer stirring duration. Temperature is another important parameter in the preparation of emulsion formulations (Floyd, 1999). Generally, a working temperature of 40°C to 70°C is established. However, immunization structures are all proteinic structures. Protein structures cannot tolerate this temperature range even for a short period (Yannarell et al., 2002). This problem can be solved in working at a lower temperature over a short period of time. It was shown that active and stable formulations can be prepared in 8 minutes at between 30°C and 40°C (Bozkir and Hayta, 2004).

12.4 MULTIPLE-EMULSION APPLICATIONS

Multiple emulsions have been widely studied as means of delivering drugs via oral, topical, and parenteral routes. The applications include protein delivery (Cournarie et al., 2004), delivery of antibiotics to the vagina (Tedajo et al., 2005), sustained delivery (Vaziri and Warburton, 1994), and vaccine delivery (Bozkir and Hayta, 2004). The immunological response to a vaccine also depends on the route of administration. Most current vaccines are administered intramuscularly, which induces immunization as a systemic immunity. However, the live polio vaccine and the live typhoid vaccine are administered orally. Local immunization (oral, intranasal, or intravagina) may be preferred, since mucosal surfaces are the common entrance to many pathogens. Moreover local immunization induces both mucosal and systemic immunity. Ease of administration and avoidance of systemic side effects are additional advantages of local immunization (Walker, 1994; Shalaby, 1995). Nevertheless, successful local immunization has only been achieved with a limited number of oral vaccines. Also there are very few studies on multiple emulsions used in the immunization process, especially on parenteral and oral administration.

12.4.1 Parenteral Administration

Most proteinic substances are poorly absorbed via the intestinal membrane. This is because of the extensive proteolytic degradation by intestinal enzymes and insufficient membrane permeability of such substances, which have high molecular weight and low lipophilicity. Multiple emulsions that contains influenza hemagglutinin in the internal aqueous phase, and show suitable particle size, are used for intramuscular (IM) administration. Bozkir and Hayta (2004) observed a fourfold increase in the immune response when multiple-emulsions formulations were compared to the conventional vaccine. Also it was shown that multiple-emulsions formulations have the adjuvant property of a drug vehicle, with a potential to entrap an antigen in the internal phase. This property increases the efficacy of multiple emulsions for vaccine delivery. A single administration of entrapped influenza hemagglutinin was shown to stimulate a more effective immune response in rats than the conventional vaccine (Bozkir et al., 2004).

12.4.2 Oral Administration

Multiple emulsions are systems of increasing interest for the oral delivery of hydrophilic drugs, since these drugs are unstable in the gastrointestinal tract. However, there are very few studies on oral tolerance that is enhanced by using multiple emissions as the delivery vehicle. Antigen incorporation into W/O/W emulsions has been used for the induction of immunity by parenteral injection (Herbert, 1965; Bozkir and Hayta, 2004). Besides the problem of oral route, even normal animals have limitations of T cells specific for any given antigen. This problem can be overcome by using transgenic animals whose T cells are reactive with a defined antigen (Elson et al., 1996). In this case the antigen was dissolved in microdispered oil particles and then delivered into the gut. The oil phase was a mixture of squalene and emulsifiers that are biodegradable and nontoxic. When animals were fed, the multiple emulsions passed through the upper gastrointestinal tract to the colon intact. Within 20 minutes, the emulsion particles were seen over Peyer's patches in the mice. Most of the emulsions were broken down in the colon with the feces' dehydration, but the promote particles were uptaken by the domes of Peyer's patches (Hearn et al., 1996). Also poloxamer, which has adjuvant activity, was used as an emulsifier to increase the immunization (Elson et al., 1996).

After three infusions of multiple-emulsions formulations, the animals produced both intestinal secratory IgA and circulating IgG antibody responses. The IgG titers persisted for one year and were boosted by a further oral immunization. Oral immunization using multiple emulsions also uniformly primed the animals for an increased response following parenteral immunization.

12.5 COMPARISION OF MULTIPLE-EMULSION DELIVERY SYSTEMS FOR ANTIGENS WITH OTHER SYSTEMS

Liposomes are phospholipid-based vesicles that have been evaluated both as adjuvants and delivery systems for slow release of antigens. The adjuvant effect can be observed in the membrane's charge and liposome type, in the position of the encapsulated or adsorbed antigens, in the phospholidid structure and their glass transition temperature, and in the ratio of phospholipids to antigen (Gregoriadis et al., 1989). Because biodegradable and biocompatible micro-/nanoparticles can be used to stimulate an inflammation reaction, they can be prepared as a delivery system for antigens (Maloy et al., 1994). The release of biological material from the biodegradable particles appears to be affected by the properties of the polymers and the morphology and the structure of the particle (Bozkir and Saka, 2004). Polymeric particles also have the potential to be adjuvants for DNA vaccines (Briones et al., 2001). The cationic particles have been shown to enhance both antibody and lymphocyte responses in a range of animal and cell models. They efficiently adsorb DNA and can deliver several plasmids simultaneously on the same formulation and at a range of different loading levels (Briones et al., 2001, Bozkir and Saka, 2004).

Comparisons of antigen carrier systems have been the subject of many studies. The reported antigenic activity depends on the properties of the carrier systems such as their adjuvant effect and/or particle size, and viscosity. For a comparison to be valid, the systems being compared must show similar properties; otherwise, the studies are misleading. Based on their properties, already it is evident that emulsion and liposome formulations offer more advantageous results. In our studies we found that multiple emulsions induce more antigenic activity than any cationic particle formulation with the classic subunit vaccine, and they enhance the immunization's efficacy (Bozkir et al., 2004).

12.6 CONCLUSION

Over the last 30 years considerable progress has been made in the preparation of well-characterized vaccine formulations loaded with a variety of antigens. Despite this considerable amount of work, it has still not reached agreement on a beneficial vaccine dosage form. The properties of multiple emulsions, which have the advantages of low viscosity, easy injection, and the stable formulations, appear to offer significant improvement as delivery vehicles for immunizations. Our multiple emulsion formulation of a single administration of entrapped influenza hemagglutinin proved to stimulate a more effective immune response in rats than the conventional vaccine.

REFERENCES

Allison AC, Byars NE. 1992. Immunological adjuvants. *Adv Exp Med Biol* 327: 133–141.

Allison AG, Gregoriadis G. 1974. Liposomes as immunological adjuvants. *Nature* 252(5480): 252.

Almeida AJ, Alpar HO. 1996. Nasal delivery of vaccines. *J Drug Target* 3: 455–467.

Bozkir A, Hayta G, Saka OM. 2004. Comparison of biodegradable nanoparticles and multiple emulsions (water-in-oil-in-water) containing influenza virus antigen on the in vivo immune response in rats. *Pharmazie* 59(9): 723–725.

Bozkir A, Hayta G. 2004. Preparation and evaluation of multiple emulsions water-in-oil-in-water (W/O/W) as delivery system for influenza virus antigens. *J Drug Target* 12(3): 157–164.

Bozkir A, Saka OM. 2004. Chitosan-DNA nanoparticles: effect on DNA integrity, bacterial transformation and transfection efficiency. *J Drug Target* 12(5): 281–288.

Briones C, Mas A, Perez-Olmeda M, Altisent C, Domingo E, Soriano V. 2001. Prevalence and genetic heterogeneity of the reverse transcriptase T69S-S-X insertion in pretreated HIV-infected patients. *Intervirology* 44(6): 339–343.

Chapel, HM, August PJ. 1976. Report of nine cases of accidental injury to man with Freund's complete adjuvant. *Clin Exp Immunol* 24: 538–541.

Ciardi JE, McGhee JR, Keith JM. 1992. Summary and recommendations for future research. *Adv Exp Med Biol* 327: 310–314.

Cournarie F, Savelli MP, Rosilio V, Bretez F, Vauthier C, Grossiord JL, Seiller M. 2004. Insulin-loaded W/O/W multiple emulsions: Comparison of the performances of systems prepared with medium-chain-triglycerides and fish oil. *Eur J Pharm Biopharm* 58(3): 477–482.

Couvreur P, Blanco-Prieto MJ, Puisieux F, Roques B, Fattal E. 1997. Multiple emulsion technology for the design of microspheres containing peptide and oligopeptides. *Adv Drug Deliv Rev* 28: 85–96.

Elson CO, Tomasi M, Dertzbaugh MT, Thaggard G, Hunter R, Weaver C. 1996. Oral-antigen delivery by way of a multiple emulsion system enhances oral tolerance. *An NY Acad Sci* 778: 156–162.

Eratalay A, Oner F, Ozcengiz E, Alpar R, 2004 "Adjuvant effects of niosome and water/oil/water multiple emulsions carrier systems for recombinant hepatitis B surface antigen" *Hacettepe University, J Faculty Pharm* 24(2): 81–94.

Floyd AG. 1999. Top ten considerations in the development of parenteral emulsions. *Pharm Sci Technol Today* 4(2): 134–143.

Garti N, Aserin A. 1996. Pharmaceutical emulsions, double emulsions and microemulsions. In: *Microencapsulation: Methods and Industrial Applications.* Benita S, ed. New York: Marcel Dekker Inc., pp 411–518.

Gregoriadis G, Tan L, Xiao Q. 1989. The immunoadyuvant actions of liposomes: Role of structural characteristics. In: *Immunological Adjuvants and Vaccines.* New York: Plenum Press, pp. 79–94.

Hearn TL, Olsen M, Hunter RL. 1996. Multiple emulsions oral vaccine vehicles for inducing immunity or tolerance. *An NY Acad Sci* 778: 388–389.

Herbert WJ. 1965. Multiple emulsions: A new form of mineral-oil antigen adjuvant. *Lancet* 2(7416): 771.

Hilgers LA, Lejeune G, Nicolas I, Fochesato M, Boon B. 1999. Sulfolipo-cyclodextrin in squalane-in-water as a novel and safe vaccine adjuvant. *Vaccine* 17(3): 219–228.

Hilgers LA, Platenburg PL, Luitjens A, Groenveld B, Dazelle T, Ferrari-Laloux M, Weststrate MW. 1994a. A novel non-mineral oil-based adjuvant. I. Efficacy of a synthetic sulfolipopolysaccharide in a squalane-in-water emulsion in laboratory animals. *Vaccine* 12(7): 653–660.

Hilgers LA, Platenburg PL, Luitjens A, Groenveld B, Dazelle T, Weststrate MW. 1994b. A novel non-mineral oil-based adjuvant. II. Efficacy of a synthetic sulfolipopolysaccharide in a squalane-in-water emulsion in pigs. *Vaccine* 12(7): 661–665.

Hilleman MR. 1966. Critical appraisal of emulsified oil adjuvants applied to viral vaccines. *Prog Med Virol* 8: 131–182.

Hirabayashi Y, Kurata H, Funato H, Nagamine T, Aizawa C, Tamura S, Shimada K, Kurata T. 1990. Comparison of intranasal inoculation of influenza HA vaccine combined with cholera toxin B subunit with oral or parenteral vaccination. *Vaccine* 8(3): 243–248.

Hjorth RN, Bonde GM, Piner ED, Goldberg KM, Levner MH. 1997. The effect of Syntex adjuvant formulation (SAF-m) on humoral immunity to the influenza virus in the mouse. *Vaccine* 15(5): 541–546.

Jiskoot W, Gideon FA, Kersten E, Beuvery EC. 2002. *Vaccines: Pharmaceutical Biotechnology*, 2 ed. Taylor and Francis Group, pp 259–280.

Keitel W, Couch R, Bond N, Adair S, Van Nest G, Dekker C. 1993. Pilot evaluation of influenza virus vaccine (IVV) combined with adjuvant. *Vaccine* 11(9): 909–913.

Klegerman ME. 1992. Vaccines. In: *Pharmaceutical Biotechnology*. Klegerman ME, Groves MJ, eds. Buffalo Grove, IL: Interpharm Press, pp 64–67.

Lidgate DM, Trattner T, Shultz RM, Maskiewicz R. 1992. Sterile filtration of a parenteral emulsion. *Pharm Res* 9(7): 860–863.

Maloy KJ, Donachie AM, O'Hagan DT, Mowart A. 1994. Induction of mucosal and systemic immune responses by immunization with ovalbumin entrapped in poly(lactide-co-glycolide) microparticles. *Immunology* 81: 661–667.

Michalek SM, Childers NK, Katz J, Dertzbaugh M, Zhang S, Russell MW, Macrina FL, Jackson S, Mestecky J. 1992. Liposomes and conjugate vaccines for antigen delivery and induction of mucosal immune responses. *Adv Exp Med Biol* 327: 191–198.

Minutello M, Senatore F, Cecchinelli G, Bianchi M, Andreani T, Podda A, Crovari P. 1999. Safety and immunogenicity of an inactivated subunit influenza virus vaccine combined with MF59 adjuvant emulsion in elderly subjects, immunized for three consecutive influenza seasons. *Vaccine* 17(2): 99–104.

Mishra B, Pandit JK. 1990. Multiple water-oil-water emulsions as prolonged release formulations of pentazocine. *J Cont Rel* 14: 53–60.

Miyakawa T, Zhang W, Uchida T, Kim NS, Goto S. 1993. In vivo release of water-soluble drugs from stabilized water-in-oil-in-water (W/O/W) type multiple emulsions following intravenous administrations using rats. *Biol Pharm Bull* 16(3): 268–272.

Montagne JR, Rabinovich R. 1992. Challenges and opportunities in vaccine research. *Adv Exp Med Biol* 327: 293–299.

Newman MJ, Todd CW, Balusubramanian M. 1998. Design and development of adjuvant-active nonionic block copolymers. *J Pharm Sci* 87(11): 1357–1362.

O'Hagan DT, MacKichan ML, Singh M. 2001. Recent developments in adjuvants for vaccines against infectious diseases. *Biomol Eng* 18(3): 69–85.

Okochi H, Nakano M. 2000. Preparation and evaluation of W/O/W type emulsions containing vancomycin. *Adv Drug Deliv Rev* 45(1): 5–26.

Omotosho JA, Whateley TL, Florence AT. 1989. Methotrexate transport from the internal phase of multiple W/O/W emulsions. *J Microencapsul* 6(2): 183–192.

Podda A. 2001. The adjuvanted influenza vaccines with novel adjuvants: Experience with the MF59-adjuvanted vaccine. *Vaccine* 19(17–19): 2673–2680.

Poland GA, Couch R. 1999. Intranasal influenza vaccine: adding to the armamentarium for influenza control. *JAMA* 282(2): 182–184.

Rabinovich NR, McInnes P, Klein DL, Hall BF. 1994. Vaccine technologies: View to the future. *Science* 265(5177): 1401–1404.

Saatci F, Bozkir A. 2003. Nasal delivery of vaccines. *Ankara University, J Faculty Pharm* 32(3): 175–192.

Shalaby WS. 1995. Development of oral vaccines to stimulate mucosal and systemic immunity: Barriers and novel strategies. *Clin Immunol Immunopathol* 74(2): 127–134.

Tamura S, Yajima A, Hatori E, Tamura S, Asanuma H, Suzuki Y, Aizawa C, Kurata T. 1997. Effects of frequent intranasal administration of adjuvant-combined influenza vaccine on the protection against virus infection. *Vaccine* 15(16): 1784–1790.

Tedajo GM, Bouttier S, Fourniat J, Grossiord JL, Marty JP, Seiller M. 2005. Release of antiseptics from the aqueous compartments of a W/O/W multiple emulsion. *Int J Pharm* 288(1): 63–72.

Todd CW, Pozzi LA, Guarnaccia JR, Balasubramanian M, Henk WG, Younger LE, Newman MJ. 1997. Development of an adjuvant-active nonionic block copolymer for use in oil-free subunit vaccines formulations. *Vaccine* 15(5): 564–570.

Tomasi M, Dertzbaugh MT, Hearn T, Hunter RL, Elson CO. 1997. Strong mucosal adjuvanticity of cholera toxin within lipid particles of a new multiple emulsion delivery system for oral immunization. *Eur J Immunol* 27(10): 2720–2725.

Vaziri A, Warburton B. 1994. Slow release of chloroquine phosphate from multiple taste-masked W/O/W multiple emulsions. *J Microencapsul* 11(6): 641–648.

Vila A, Sanchez A, Tobio M, Calvo P, Alonso MJ. 2002. Design of biodegradable particles for protein delivery. *J Control Release* 78: 15–24.

Walker RI. 1994. New strategies for using mucosal vaccination to achieve more effective immunization. *Vaccine* 12(5): 387–400.

Yannarell DA, Goldberg KM, Hjorth RN. 2002. Stabilizing cold-adapted influenza virus vaccine under various storage conditions. *J Virol Meth* 102(1–2): 15–25.

Yarkoni E, Rapp HJ. 1978. Toxicity of emulsified trehalose-6,6′-dimycolate (cord factor) in mice depends on size distribution of mineral oil droplets. *Infect Immun* 20(3): 856–860.

INDEX

WILEY SERIES ON SURFACE AND INTERFACIAL CHEMISTRY

Series Editors:

Ponisseril Somasundaran
Columbia University

Nissim Garti
The Hebrew University of Jerusalem

Multiple Emulsions: Technology and Applications. Edited by Abraham Aserin